Astrophysics and Space Science Proceedings

Volume 52

More information about this series at http://www.springer.com/series/7395

Massimiliano Vasile • Edmondo Minisci
Leopold Summerer • Peter McGinty
Editors

Stardust Final Conference

Advances in Asteroids and Space Debris
Engineering and Science

Springer

Editors
Massimiliano Vasile
Aerospace Centre of Excellence
Department of Mechanical and
Aerospace Engineering
University of Strathclyde
Glasgow, UK

Edmondo Minisci
Aerospace Centre of Excellence
Department of Mechanical and
Aerospace Engineering
University of Strathclyde
Glasgow, UK

Leopold Summerer
Advanced Concepts Team
ESA-ESTEC
Noordwijk, The Netherlands

Peter McGinty
Aerospace Centre of Excellence
Department of Mechanical and
Aerospace Engineering
University of Strathclyde
Glasgow, UK

ISSN 1570-6591   ISSN 1570-6605   (electronic)
Astrophysics and Space Science Proceedings
ISBN 978-3-319-69955-4   ISBN 978-3-319-69956-1   (eBook)
https://doi.org/10.1007/978-3-319-69956-1

Library of Congress Control Number: 2017963996

© Springer International Publishing AG 2018
This work is subject to copyright. All rights are reserved by the Publisher, whether the whole or part of the material is concerned, specifically the rights of translation, reprinting, reuse of illustrations, recitation, broadcasting, reproduction on microfilms or in any other physical way, and transmission or information storage and retrieval, electronic adaptation, computer software, or by similar or dissimilar methodology now known or hereafter developed.
The use of general descriptive names, registered names, trademarks, service marks, etc. in this publication does not imply, even in the absence of a specific statement, that such names are exempt from the relevant protective laws and regulations and therefore free for general use.
The publisher, the authors and the editors are safe to assume that the advice and information in this book are believed to be true and accurate at the date of publication. Neither the publisher nor the authors or the editors give a warranty, express or implied, with respect to the material contained herein or for any errors or omissions that may have been made. The publisher remains neutral with regard to jurisdictional claims in published maps and institutional affiliations.

Printed on acid-free paper

This Springer imprint is published by Springer Nature
The registered company is Springer International Publishing AG
The registered company address is: Gewerbestrasse 11, 6330 Cham, Switzerland

Preface

In recent years, it has become clear that the increasing amount of space debris could lead to catastrophic consequences in the near term, and although statistically less likely to occur, an asteroid impact would have devastating consequences for our planet.

In order to address these issues, many countries have created dedicated programmes such as the Space Situational Awareness or the Clean Space programme of the European Space Agency, or they have invested significant funding in long-term research projects. Among the projects supported by the European Commission was Stardust, a 4-year, European Union-wide programme funded through the FP7 Marie Skłodowska Curie Initial Training Networks (ITN). It was a unique training and research network devoted to developing and mastering techniques for asteroid and space debris monitoring, removal/deflection, and exploitation. By pushing the boundaries of space research with innovative ideas and visionary concepts, Stardust worked to turn the threat posed by asteroids and space debris into an opportunity.

Stardust recruited 15 research fellows who developed new ideas and methods and explored advanced concepts and solutions in the three main areas covered by the Stardust research programme: modelling and simulation, orbit and attitude determination and prediction, and active removal and deflection of uncooperative targets. The training and research programme included the organisation of several events through schools, workshops, and conferences, to disseminate the scientific results generated within Stardust and stimulate a discussion between scholars and major players in the fields.

The Final Stardust Conference was the last and largest event organised within the framework of the Stardust research and training programme. The conference was organised around a number of parallel symposia and keynote lectures and was open to everybody who wished to present recent advancements on asteroids and space debris. In particular, the conference focused on:

- Orbital and Attitude Dynamics Modelling
- Long-Term Orbit and Attitude Evolution
- Particle Cloud Modelling and Simulation

- Collision and Impact Modelling and Simulation
- Re-entry Modelling and Simulation
- Asteroid Origins and Characterisation
- Orbit and Attitude Determination
- Impact Prediction and Risk Analysis
- Mission Analysis—Proximity Operations
- Active Removal/Deflection Control under Uncertainty
- Active Removal/Deflection Technologies
- Asteroid Manipulation

The Final Stardust Conference was jointly organised by the University of Strathclyde, ESA-ESTEC, the Universita' di Roma Tor Vergata, the Deutsche Forschungszentrum für Künstliche Intelligenz (DFKI), the Astronomical Observatory of Belgrade, the Universidad Politécnica de Madrid, the Universita' di Pisa, Deimos Space, the Institute of Celestial Mechanics and Calculation of Ephemerides (IMCCE on behalf of the KePASSA committee), the Centro Nazionale delle Ricerche (CNR), Dinamica, and the University of Southampton. The conference took place at ESA-ESTEC in the Netherlands from the 31st of October to the 3rd of November 2016.

Among the 104 contributions to the conference, 18 extended papers were selected for publication in this volume after peer review by the members of the organisation committee. The 18 papers are grouped into the following parts:

- Mission to Asteroids
- Orbit and Uncertainty Propagation
- Space Debris Monitoring, Mitigation, and Removal
- Re-entry Analysis and Design for Demise

The organisers are grateful to all the members of the Stardust network, its attendees, and especially to the keynote speakers.

Moreover, the editors acknowledge Maury Solomon, Hannah Kaufman, and Springer for their interest in publishing the selected contributions of the Final Stardust Conference.

Glasgow, UK	Massimiliano Vasile
Glasgow, UK	Edmondo Minisci
Noordwijk, The Netherlands	Leopold Summer
Glasgow, UK	Peter McGinty

Contents

Part I Mission to Asteroids

Optimization of Asteroid Capture Missions Using Earth Resonant Encounters 3
Rita Neves and Joan Pau Sánchez

Evaluating Proximity Operations Through High-Fidelity Asteroid Deflection Evaluation Software (Hades) 17
Massimo Vetrisano, Juan L. Cano, and Simone Centuori

Prediction of Orbital Parameters for Undiscovered Potentially Hazardous Asteroids Using Machine Learning 45
Vadym Pasko

Part II Orbit and Uncertainty Propagation

Exploring Sensitivity of Orbital Dynamics with Respect to Model Truncation: The Frozen Orbits Approach 69
Martin Lara

A Parametric Study of the Orbital Lifetime of Super GTO and SSTO Orbits Based on Semi-analytical Integration 85
Denis Hautesserres, Juan F. San-Juan, and Martin Lara

On the Use of Positive Polynomials for the Estimation of Upper and Lower Expectations in Orbital Dynamics 99
Massimiliano Vasile and Chiara Tardioli

Part III Space Debris Monitoring, Mitigation, and Removal

Trajectory Generation Method for Robotic Free-Floating Capture of a Non-cooperative, Tumbling Target 111
Marko Jankovic and Frank Kirchner

**Taxonomy of LEO Space Debris Population for ADR Capture
Methods Selection** 129
Marko Jankovic and Frank Kirchner

**Remote Sensing for Planar Electrostatic Characterization Using
the Multi-Sphere Method** 145
Heiko J. A. Engwerda, Joseph Hughes, and Hanspeter Schaub

**Active Debris Removal and Space Debris Mitigation using Hybrid
Propulsion Solutions** 163
Stefania Tonetti, Stefania Cornara, Martina Faenza, Onno Verberne,
Tobias Langener, and Gonzalo Vicario de Miguel

**The Puzzling Case of the Deep-Space Debris WT1190F: A Test Bed
for Advanced SSA Techniques** 181
Alberto Buzzoni, Siwei Fan, Carolin Frueh, Giuseppe Altavilla,
Italo Foppiani, Marco Micheli, Jaime Nomen, and Noelia Sánchez-Ortíz

Development of a Debris Index 191
Francesca Letizia, Camilla Colombo, Hugh G. Lewis, and Holger Krag

Part IV Re-entry Analysis and Design for Demise

A Multi-disciplinary Approach of Demisable Tanks' Re-entry 209
C. Bertorello, C. Finzi, P. Perrot-Minnot, G. Pinaud, J. M. Bouilly,
and L. Chevalier

Design-for-Demise Analysis using the SAM Destructive Re-Entry Model. 233
James C. Beck, Ian E. Holbrough, James A. Merrifield,
and Nicolas Leveque

**Low-Fidelity Modelling for Aerodynamic Characteristics
of Re-Entry Objects** 247
Gianluca Benedetti, Nicole Viola, Edmondo Minisci, Alessandro Falchi,
and Massimiliano Vasile

**Re-entry Predictions of Potentially Dangerous Uncontrolled
Satellites: Challenges and Civil Protection Applications** 265
Carmen Pardini and Luciano Anselmo

**Uncertainty Quantification for Destructive Re-Entry Risk Analysis:
JAXA Perspective** 283
Keiichiro Fujimoto, Hiroumi Tani, Hideyo Negishi, Yasuhiro Saito,
Nobuyuki Iizuka, Koichi Okita, and Akira Kato

**HDMR-Based Sensitivity Analysis and Uncertainty Quantification
of GOCE Aerodynamics Using DSMC** 301
Alessandro Falchi, Edmondo Minisci, Martin Kubicek, Massimiliano
Vasile, and Stijn Lemmens

Part I
Mission to Asteroids

Optimization of Asteroid Capture Missions Using Earth Resonant Encounters

Rita Neves and Joan Pau Sánchez

Abstract This paper describes a robust methodology to design Earth-resonant asteroid capture trajectories leading to Libration Point Orbits (LPOs). These trajectories consider two impulsive manoeuvres; one occurring before the first Earth encounter and a final one that inserts the asteroid into a stable hyperbolic manifold trajectory leading to an LPO of the Sun-Earth system. The first manoeuvre is key to exploit the chaotic perturbative effects of the Earth and obtain important reductions on the cost of inserting the asteroid into a manifold trajectory. The perturbative effects caused by the Earth are here modelled by means of a Keplerian Map approximation, and these are a posteriori compared with the dynamics of the Circular Restricted Three-Body Problem. Savings in the order of 50% of total Δv are computed for four different asteroids.

1 Introduction

Asteroid capture and retrieval missions have been getting the attention of the scientific community for some years. There are thousands of asteroids in orbits relatively close to the Earth and new ones are discovered often; as of January 2017, there are over 15,000 observed near-Earth asteroids (NEA), from which 5% were only reported the year before [5]. The characteristics of most NEA are still unknown, from size to material composition. As such, these bodies are considered very interesting targets for investigation. From data collection, to technology demonstrations or in situ resource utilization, there are many scientific operations that can be undertaken, which presents an opportunity for challenging mission scenarios.

Asteroid capture missions are characterized by the rendezvous of a spacecraft with an asteroid and moving it to an orbit in the vicinity of the Earth. The spacecraft

---

R. Neves · J. P. Sánchez (✉)
Centre for Autonomous and Cyber-Physical Systems, Cranfield University, College Road, Cranfield MK43 0AL, UK
e-mail: r.neves@cranfield.ac.uk; jp.sanchez@cranfield.ac.uk

is utilised to modify the celestial body's trajectory in such a way as to make it enter the target orbit; the utilization of Solar Electric Propulsion (SEP) is one possible technology for this endeavour [9], although some others have been studied.

This work proposes to minimise the total fuel consumption, here regarded as Δv_C, of capturing an asteroid into a Libration Point Orbit (LPO). For this objective to be achieved, a manoeuvre that takes the asteroid from its nominal orbit to the destination has to be performed. This is the case studied by Yárnoz et al. [11], which considers a single Δv change that alters the asteroid's orbit to the one of the invariant manifold leading to the LPO, creating a database for Easily Retrievable Asteroids (EROs) by noting the capture Δv of several bodies.

This work intends to exploit the chaotic nature of our Solar System and its numerous gravitational perturbations to find low-energy trajectories that lead to the capture of NEA into LPO. In this way, a different approach is proposed: the application of an initial manoeuvre Δv_M for an optimal passage near the Earth, which is thereafter referred as an *Earth-resonant encounter*, and the final Δv_I insertion into the LPO. It is proposed that the initial manoeuvre Δv_M can be optimized in such a way that the resonant encounter with the Earth impacts the asteroid's orbital elements optimally, so that the total cost of the trajectory is lower than for a direct capture.

In order to model an asteroid's motion, the Keplerian Map (KM) equations are used. This is a perturbative model that allows for the simulation of Earth's gravitational influence on the body's orbit around the Sun, while being less computationally expensive than higher fidelity models such as the Circular Restricted Three-Body Problem (CR3BP). Given that the number of asteroids to be considered for capture is very high, utilizing this model is a way to decrease computational time. This is essential in space mission design, where several variables may have to be taken into account and, thus, extensive search spaces must be explored.

The present paper is organized as follows. Section 2 makes an overview on manifold theory and LPO, as well as detailing the dynamical models used for the presented asteroid capture trajectories, namely the KM. Section 3 analyses the full trajectory design and explains the procedure to obtain the best solution and its refinement with a higher order model. Section 4 reports the results for four different asteroids, studying the savings in Δv_C and the impact on the capture's time of flight. Finally, Sect. 5 evaluates the implications of these developments and highlights some points that may benefit from further work.

2 Near-Earth Asteroids

NEA have been so far classified according to their orbits and divided into four main categories: Amors, Apollos, Atens and Atiras. Amor asteroids stay always outside Earth's orbit and never cross it; Apollos and Atens cross Earth's orbit, but the former still have a wider orbit than the planet, while the latter are characterized for staying longer inside Earth's orbit and having smaller semi-major axes; Atiras remain confined inside Earth's orbit throughout their motion. These categories are depicted in Fig. 1.

Fig. 1 Orbits of different NEA categories in the Solar System

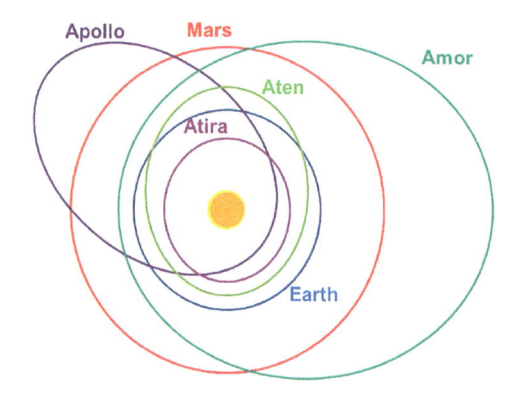

The purpose of this work is to develop low-energy trajectories, using Earth-resonant encounters, that lead a NEA to an LPO. The latter are not the only near-Earth orbits that can be used for asteroid capture, but the asteroid population that can be cheaply moved into such an orbit may not be the same as into others (such as Distant Retrograde Orbits or DRO), as reported by Sánchez and Yárnoz [6], making these interesting targets for investigation.

2.1 LPO and Invariant Manifolds

The Libration or Lagrangian points are positions in space where an object of negligible mass, affected by the gravitational interactions between two larger bodies (the primary and the secondary), can maintain a stationary position. These points are generally represented in a synodic reference frame, in which the primaries appear to be static while the third body rotates around them. They are very attractive for a great number of missions, namely to hold telescopes or other observation-type spacecraft, since the fuel consumption required to perform station-keeping is very low.

Several types of periodic orbits can be found around these points, from which we highlight three: Horizontal Lyapunov orbits, which are in the ecliptic plane, Vertical Lyapunov orbits, that are horizontally symmetric and shaped like a figure-eight, and Halo orbits, which bifurcate from the Horizontal Lyapunov orbit family; these can be seen on Fig. 2. An infinite number of quasi-periodic orbits can also be found, divided into two families: Lissajous around the Vertical Lyapunov orbits, and the Quasi-Halos around the Halo orbits [1].

Hyperbolic invariant manifolds, dynamical structures composed of countless orbits, are connected to the LPOs [2]. Mathematically, these are defined as sets of points in the system's phase space that tend toward a given limit as time tends to plus or minus infinity; they exist for a range of energies and form a series of 'tubes' connecting different regions around the primaries. These invariant manifold tubes can be used to explore new spacecraft trajectories with interesting characteristics: by moving one body to an invariant manifold orbit connected to an LPO, it will arrive there without any further manoeuvring.

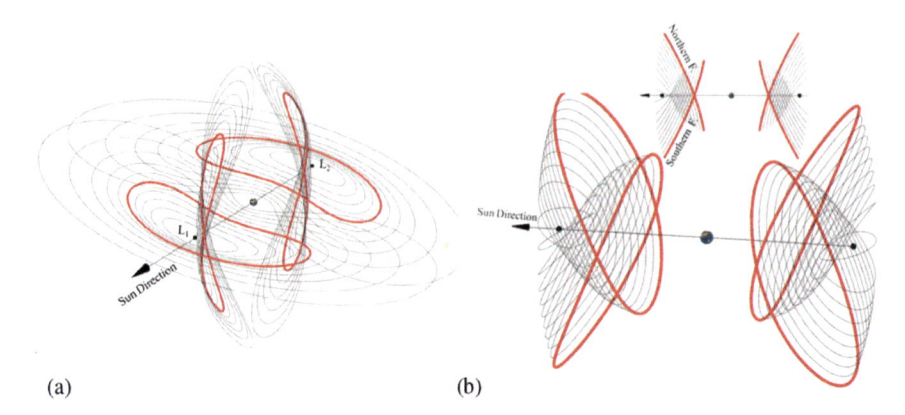

Fig. 2 Libration point orbits associated with the Sun-Earth L_1 and L_2 points [6]. (**a**) Planar and vertical Lyapunov orbits. (**b**) Northern and southern families of halo orbits

2.2 The Keplerian Map

Considering that the asteroid encounters the Earth at some point along its trajectory, it is necessary to take into account its perturbative influence on the mission design. Since the considered asteroids move outside Hill sphere, it is infeasible to use a patched-conics method; therefore, we resort to the KM, a perturbation model for the motion of an object orbiting a central body.

The KM influence is factored in using a first-order approximation of Picard's iteration on Lagrange's planetary equations. Its equations can be used to calculate the changes in orbital elements caused by the perturbing object, which are computed at each periapsis passage of the body, and then added to the previously known orbital elements [4]. In this way, the action of the KM can be represented by the mapping $\mathscr{K}$:

$$\mathscr{K} : \{a, e, i, \omega | \alpha\} \mapsto \{\Delta a, \Delta e, \Delta i, \Delta \omega | \alpha\} \tag{1}$$

The parameter α accounts for the phasing of the perturbed body with the one provoking the disturbance: in a synodic reference frame with the Sun (central body) and the Earth (perturbing body) as primaries, it is the angle in between the Sun-Earth line and the projection of the Sun-asteroid line in the ecliptic plane. Since the KM is only computed at α values in which a periapsis passage occurs, these are uniquely named α_P. Considering the asteroid's movement, the value of $\Delta \alpha_P$ has also to be updated to represent the following periapsis passage, using this equation:

$$\alpha_{Pn+1} = \alpha_{Pn} + 2\pi \left| \sqrt{\frac{a_{n+1}^3}{1-\mu}} - 1 \right| \tag{2}$$

Fig. 3 Phasing α_P for an Earth-resonant encounter for asteroid 2016RD34

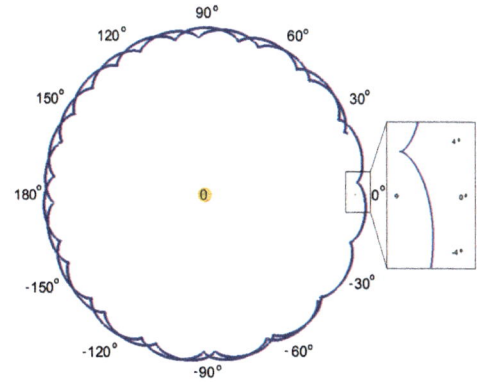

Fig. 4 Kick-map: change in Δa with α for a resonant encounter of asteroid 2016RD34

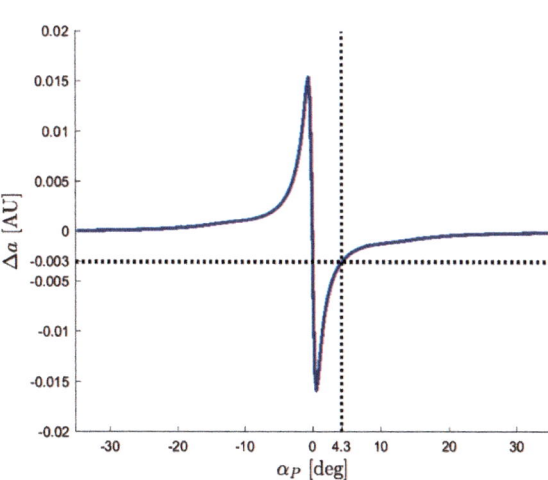

in which n indicates in which time step the computation is being made, a represents the asteroid's semi-major axis and μ is the normalized gravitational parameter of the system.

On Fig. 3, we can observe the movement of asteroid 2016RD34 in the synodic reference frame; polar axes were juxtaposed to these, showing the range of α. One of the periapsis passages happening during the Earth-resonant encounter is highlighted, revealing $\alpha_P = 4.3°$.

An interesting application of the KM is the kick-map, a visual representation of the orbital elements changes as a function of the object's phasing with the perturbing body. As an example, Fig. 4 shows the semi-major axis change undergone by asteroid 2016RD34 depending on the angle α_P; this is, then, the kick-map that matches the movement shown in Fig. 3. In this way, the value of Δa corresponding to $\alpha_P = 4.3°$ can be simply taken from this plot—on Fig. 4, it is represented by the crossing of dotted lines.

3 Trajectory Design

The proposed trajectory consists on manoeuvring the asteroid for an optimal resonant encounter with the Earth and posterior capture into an LPO. Therefore, it can be divided into three distinct sections, highlighted in Fig. 5. The first section, Phase A, starts when the asteroid is at the periapsis, right outside the perturbative region of influence of the Earth; at this point, the asteroid's velocity is changed by Δv_M, altering its path. The second section, Phase B, corresponds to the resonant encounter with the Earth, in which the asteroid is affected by its perturbation. This region was defined by $|\alpha| = \frac{\pi}{8} + \frac{\Delta\alpha}{2}$, which delimits a sufficiently large zone to encompass all α_P in which the object's motion is noticeably perturbed. The third section, Phase C, ends at the insertion of the asteroid into an invariant manifold connected to an LPO by performing a manoeuvre of cost Δv_I. The final capture Δv_C is the added total of the two different manoeuvres, Δv_M and Δv_I.

The asteroid's motion during Phases A and C is Keplerian around the Sun; its path is only altered by Δv_M on the former case. However, due to the close proximity with the Earth, Phase B has the object in a three-body configuration, where its movement is modelled with the KM. One simplification must be mentioned: although Δv_M will cause a change in the orbital elements before the Earth-resonant encounter, it is so small that makes no difference in the application of the KM model. In this way, the mapping shown on Eq. (1) was performed with the original set of orbital parameters, regardless of the application of Δv_M.

Fig. 5 Phases of the capture trajectory with an Earth-resonant encounter

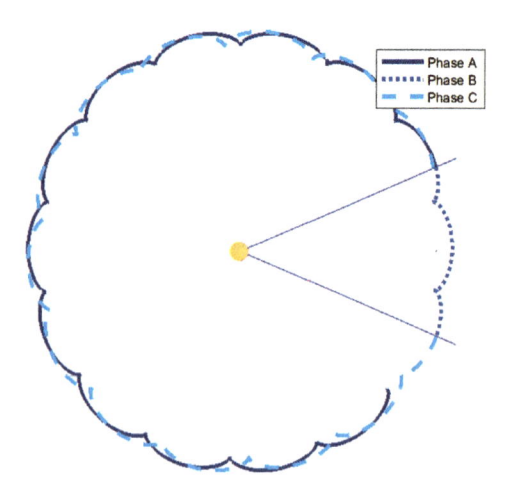

3.1 Initial Filtering

In order to assess which asteroids show reductions in capture cost by the implementation of this trajectory, a list of all discovered NEA orbital elements was collected from the Minor Planet Center's database. Considering over 15,000 candidates, pruning had to be performed for this study to be feasible.

For this purpose, a filter that computes an estimate of the insertion cost Δv_I in the LPO, based on a direct capture using a bi-impulsive manoeuvre, was designed. This was first described by Sánchez et al. [7] and later expanded [6]. It has proved to be a good lower threshold of the real capture manoeuvre and, as such, is used in the current paper to estimate Δv_I for the computed trajectories. For completeness, this section includes a brief summary of how the filter works.

The bi-impulsive manoeuvre considers one burn on the perihelion and one on the aphelion, in which only one of the two is responsible for an inclination correction, and both include a semi-major axis change. This is described by Eq. (3):

$$\Delta v_I = \sqrt{\Delta v_{a1}^2 + \Delta v_{i1}^2} + \sqrt{\Delta v_{a2}^2 + \Delta v_{i2}^2} \qquad (3)$$

in which Δv_a is the classical change in semi-major axis manoeuvre, whereas Δv_i is the inclination change.

Thus, there are four computed values for Δv_I, depending on whether the perihelion or aphelion burn is the first and which of them will include the inclination correction; the lowest value out of these will be the filter output.

The filter application allowed, using an established ceiling of 1.2 km/s, to restrict our search to 61 asteroids. Posteriorly, the ephemerides from this reduced list were taken from the Horizons JPL database [3]; the time period for data collection was from 2020 to 2100. One synodic period, out of all comprised in this time span, was chosen: the one with highest optimization sensitivity, corresponding to the one where the orbital elements suffer the greatest change as caused by the Earth.

3.2 Grid Search

In Fig. 6, we observe the application of Δv_M changes the asteroid's movement and encounter with the Earth, as opposed to its original path. It is important to denote that, for all the figures in this paper, the Earth is not in scale, but was plotted instead with the radius of Hill sphere. We can discern how a small change of the asteroids' orbital elements leads to a different encounter; therefore, our purpose is to develop a trajectory that leads to the cheapest capture possible by exploiting this effect. In order to achieve this, a grid search was performed for the Δv_M corresponding to the lowest Δv_C.

As mentioned in Sect. 3, the KM is computed only for the initial asteroid's elements; as such, the orbital element changes will repeat themselves after the

Fig. 6 Comparison of asteroid 2016RD34 trajectory with different initial values of Δv_M

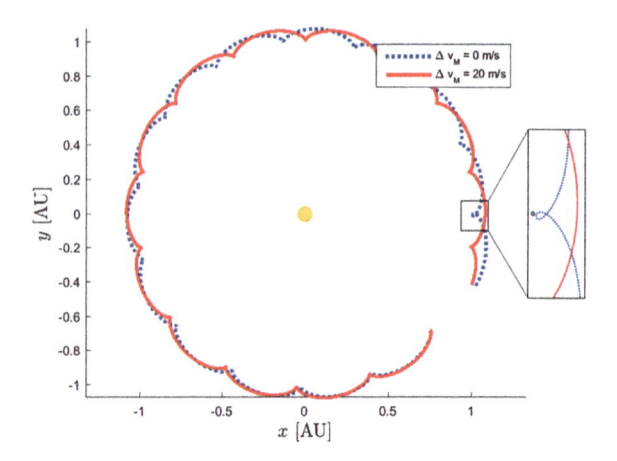

asteroid's orbit is moved forwards or backwards one epicycle, since α_P values will be the same. Thus, the analysis is restricted to Δv_M inside limits that correspond to the asteroid moving backwards or forwards one epicycle. These are easily computed resorting to Gauss' form of the variational equations [7], as following:

$$\Delta v_M = \frac{\mu_{Sun} \Delta a}{2a^2 v_P} \tag{4}$$

where μ_{Sun} is the Sun's gravitational constant, v_P is the velocity at the periapsis and Δa represents the variation in initial semi-major axis corresponding to the addition of an extra epicycle to the asteroid's motion. The latter is computed using the equations:

$$\Delta a = \frac{\sqrt{1-\mu}\,\Delta\alpha}{3\pi n_P \sqrt{a}} \tag{5}$$

where μ is the normalized gravitational constant of the system, n_P is the number of periapsis passages occurring from the manoeuvre to the target point and $\Delta\alpha$ is the angular span of one epicycle:

$$\Delta\alpha = 2\pi \left(\sqrt{\frac{a^3}{1-\mu}} - 1 \right) \tag{6}$$

Once the limits are obtained, all the values of Δv_C are computed for Δv_M inside the established limits, with a step change of 0.2 m/s.

3.3 Refinement

After obtaining a solution using the method detailed in Sect. 3.2, the trajectory was refined with a higher fidelity model: namely, the CR3BP. This dynamical model is very well established for orbital motion, having been used countless times in mission analysis [10].

For the purpose of obtaining a more refined solution, the CR3BP simulation has to be matched with the KM motion. Due to the higher sensitivity of the former, this may not be achieved by propagating the asteroid's motion using the orbital elements obtained after the Δv_M manoeuvre.

In general terms, the closer the asteroid is to the perturbing body, the greater its influence on it; as such, the object may undergo several periapsis passages in the region of the Earth's perturbation, but the one that will exert the most significant impact on its motion will be the one in which α_P is the closest to zero. Following this logic, in order to get a similar orbital change for the CR3BP as from the KM, the closest of its periapsis passages, $\alpha_{closest}$, should be the same as in the KM.

In order to target $\alpha_{closest}$, we employ a bisection method: this is implemented by defining upper and lower bounds to a trial Δv_M and propagating the motion in the CR3BP using their mean value. Depending on whether the asteroid has surpassed or fallen behind $\alpha_{closest}$, the limits to the manoeuvre are changed and the consequent trial Δv_M is altered accordingly.

On Fig. 7, we can observe the evolution of the semi-major axis throughout time, for the entire trajectory, as depicted in Fig. 5. Three plots can be distinguished: the propagation using the KM with the grid search solution ($\Delta v_M = -7.4$ m/s), with the CR3BP using the same manoeuvre and, finally, using the Δv_M provided by the targeting method ($\Delta v_M = -6.6$ m/s). The latter option matches the first one much more accurately, reinforcing the choice to apply the targeting procedure.

4 Results and Discussion

From the list of 61 asteroids that were filtered with a threshold of $\Delta v_I = 1.2$ km/s, the ones presented in this paper are the four that benefit the most from the Earth-resonant encounter trajectory. In order to find these, three capture scenarios were compared and shown in Fig. 8. Case 1 considers the cost of capturing an asteroid without any resonant encounter with the Earth; Case 2 examines a capture after the Earth-resonant encounter, but with no interference on the asteroid's path. Both these cases are used as benchmarks for comparison with our studied trajectory—the resonant capture with optimal manoeuvring Δv_M, depicted by Case 3.

The asteroids selected were the ones presenting the highest savings in fuel consumption as compared to Cases 1 and 2. These results are presented on Table 1:

Two fuel reduction computations were obtained from comparing Case 3 to Cases 1 and 2; the value shown in Table 1 is the lowest of them, in order to highlight the asteroids in which Case 3 is clearly the most cost efficient. There are,

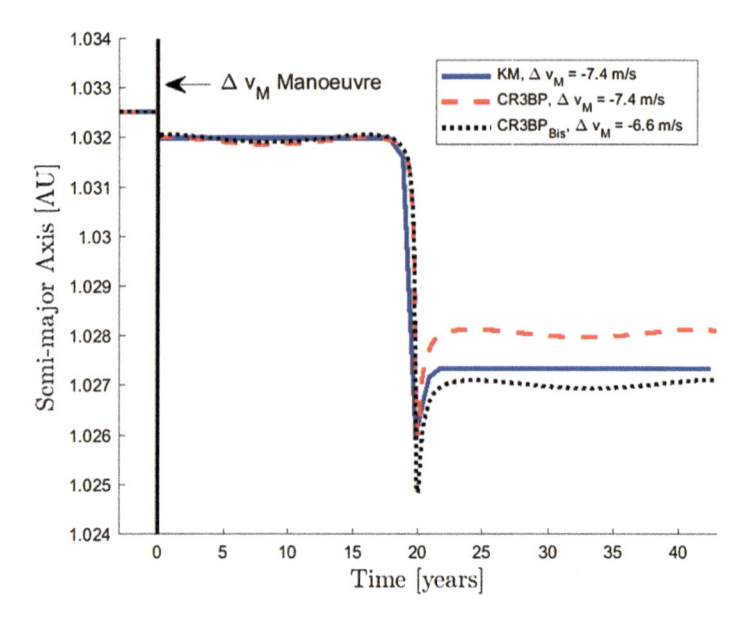

Fig. 7 Evolution of the semi-major axis of asteroid 2011BL45 throughout time

however, asteroids which benefit greatly from a resonant trajectory, but in which the manoeuvre is not essential (i.e. asteroid 2011MD), which means that the Δv of Cases 2 and 3 is similar—these were, consequently, left out of this discussion.

From Table 1, we observe that the obtained fuel reduction is very high, ranging from 38.5% to 60.6%. The Δv obtained for Cases 1 and 2 are very different; they correspond, in fact, to capture on two distinct synodic periods where substantial perturbation by the Earth was occurring regardless of any manoeuvre. However, in the depicted cases, a very small Δv_M corresponded to a significant decrease in Δv for Case 3, as can be observed in Fig. 9.

In these figures, we can distinguish two coloured areas, *Flag1* and *Flag2*. *Flag1* is raised when the asteroid is moving inside the Hill sphere, whereas *Flag2* appears when the perturbation is so strong that a transition occurs, meaning the asteroid's semi-major axis decreases from greater to smaller than one, or vice-versa; in both these cases, the KM should not be used to compute the body's motion, as previously mentioned in Sect. 2.2. It is possible to see how chaotic the plot behaviour becomes in these areas, corroborating this decision.

The refinement of Δv_{Case3} values with the CR3BP is shown on Table 1, represented by the parameter Δv_{CR3BP}. The targeting of $\alpha_{closest}$ using the bisection method converges quickly with great results: the error for the targeted angle reduces to less that 10^{-8} rad in a very short computational time—this corresponds to a distance of about 80 km in the orbital motion.

For the first three cases, the values of Δv_{CR3BP} appear very similar to Δv_{Case3}; however, for asteroid 2016RD34, the difference is considerable. We conjecture that this has to do with the close proximity to the Hill radius of this solution, as we can see on Fig. 9b. As such, for this specific case, we have allowed for a relaxation of

Optimization of Asteroid Capture Missions Using Earth Resonant Encounters

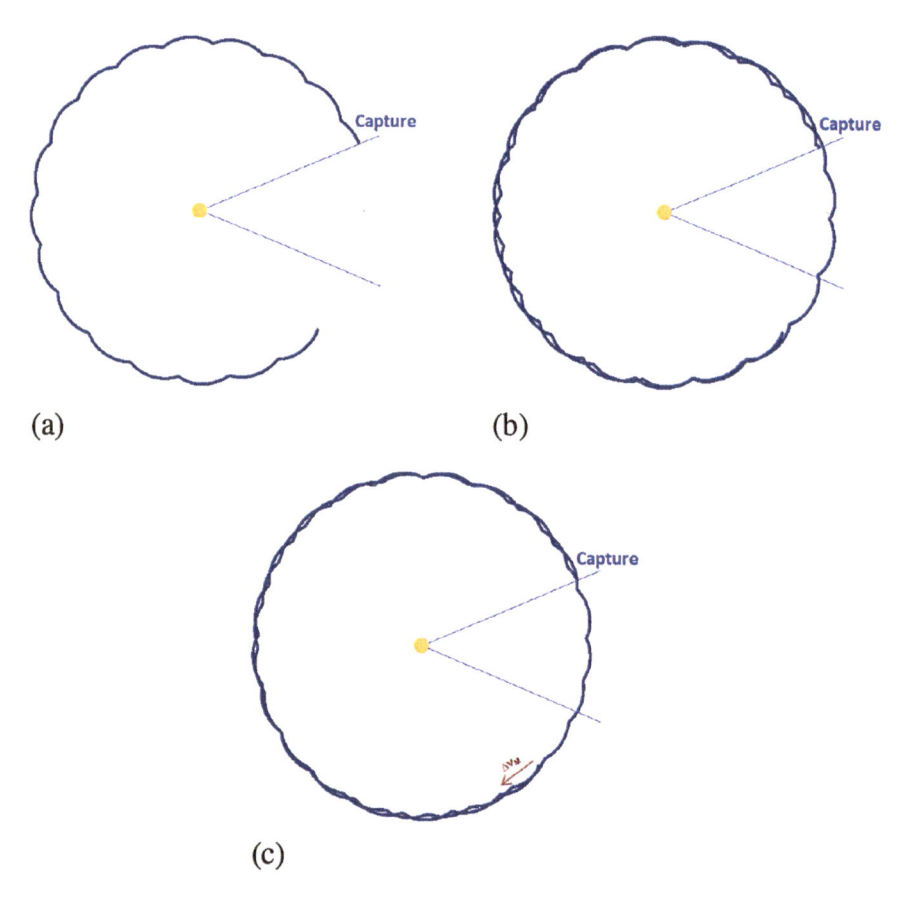

Fig. 8 Capture cases for asteroid 2011BL45. (**a**) Case 1: direct capture on first synodic period. (**b**) Case 2: capture with zero Δv_M, after earth resonance. (**c**) Case 3: capture with optimal Δv_M, after earth resonance

Table 1 Asteroid capture costs for each Case, in [m/s]

Asteroid	Δv_M	Δv_{Case1}	Δv_{Case2}	Δv_{Case3}	Δv_{CR3BP}	Fuel reduction
2011BL45	−7.4	88.0	116.1	48.4	45.1	45.0%
2010VQ98	−2.6	255.3	470.4	112.2	120.1	56.1%
2008UA202	−13.6	307.3	307.6	189.0	214.5	38.5%
2016RD34	−2.8	317.6	508.0	125.3	281.2	60.6%

the targeting error up to 10^{-3} rad (about four times the previous distance) in order to check similar solutions. The obtained Δv_{CR3BP} becomes 90 m/s, a fact that reiterates the increased sensitivity of the motion around Hill sphere and the need for cohesive establishment of the limits in which adequate solutions can be found, something that has been brought to attention by Sánchez et al. [8].

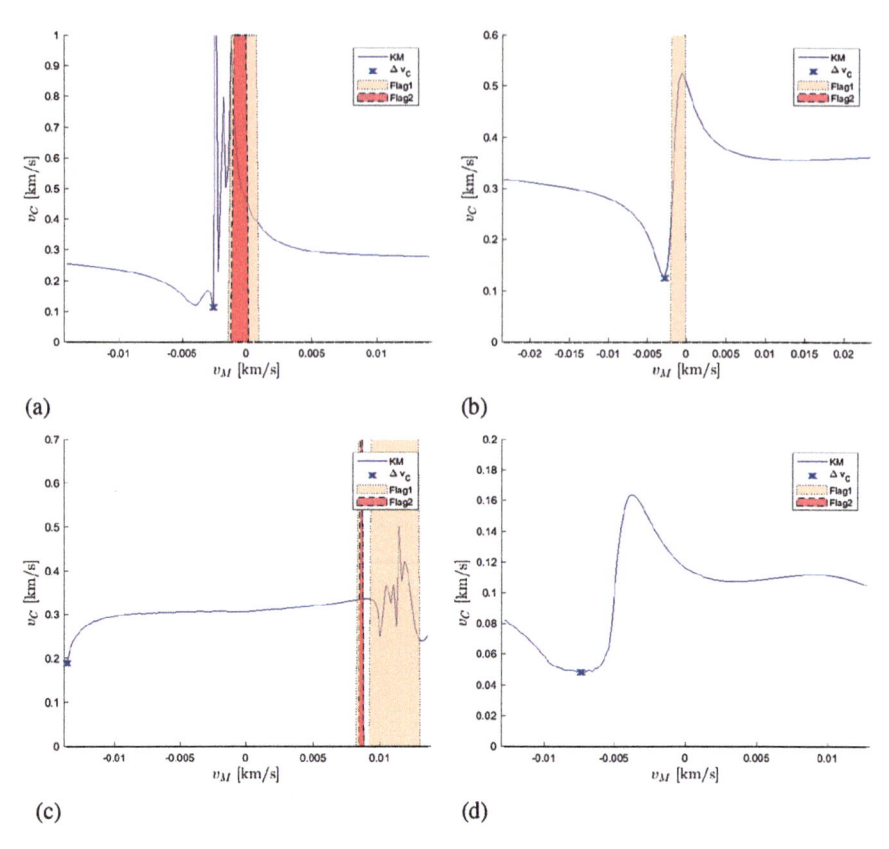

Fig. 9 Asteroid capture Δv_C as a function of the initial manoeuvring Δv_M for the bodies: (**a**) 2010VQ98. (**b**) 2016RD34. (**c**) 2008UA202. (**d**) 2011BL45

Table 2 Asteroid data and times of flight for capture manoeuvre, in years

Asteroid	NEA category	Capture LPO	Starting date	TOF_{Case1}	TOF_{Case3}
2011BL45	Amor	VL$_2$[a]	19/08/2073	17.6	42.4
2010VQ98	Apollo	VL$_2$[a]	10/11/2063	16.6	46.3
2008UA202	Apollo	PL$_2$[b]	29/02/2020	29.4	41.3
2016RD34	Amor	VL$_2$[a]	18/10/2033	12.9	35.4

[a] Vertical Lyapunov in L_2
[b] Planar Lyapunov in L_2

On Table 2, it is possible to see the categories these asteroids fall into, the target LPO and the times of flight of the capture trajectory with and without the resonant encounter. It is obvious that the savings in fuel cost are contrasted by the increased time of flight taken by the capture; since we are considering one extra synodic period, the trajectory takes roughly twice as long.

5 Conclusions and Future Work

The trajectories analysed show great promise in reducing Δv costs of capturing asteroids into LPO. This is relevant in terms of mission design, since fuel consumption is one of the primary impactors of the cost of a space mission, and as such one of the main constraints limiting their boldness. However, the savings in fuel cost have to be weighted against the increased time of flight spent in the entire capture. This is a matter of trade-off analysis that can be performed for a specific mission design. Furthermore, a careful selection of the synodic period of the capture, by itself, is also valuable for cost reduction.

In summary, we have presented a new tool, supported by a dynamical model of motion of low computational cost, that proves to be very efficient in the design of optimal resonant encounters. This tool can be applied to the study of several interesting cost-saving trajectories in other planetary configurations and missions, such as Jovian moon tours.

The KM is shown to behave very similarly to the CR3BP and to adequately model the complexity of low energy resonant motion in asteroids where there is no transition and its distance to the Earth is always greater than Hill radius. Further work should consider the analysis of the boundaries of the KM, meaning the exact limits where it stops being a good approximation to a higher fidelity model.

In regards to resonant capture trajectories, further work will focus on the entire optimization of the manoeuvres, without resorting to the filter described in Sect. 3.1; instead of admitting a bi-impulsive manoeuvre estimate, an optimized Lambert arc will be considered. Furthermore, the use of low-thrust systems in the computation of these trajectories will also be studied and compared to current chemical thrust solutions.

References

1. Gómez, G., Koon, W.S., Lo, M.W., Marsden, J.E., Masdemont, J., Ross, S.D.: Connecting orbits and invariant manifolds in the spatial restricted three-body problem. Nonlinearity **17**(5), 1571–1606 (2004)
2. Koon, W.S., Lo, M.W., Marsden, J.E., Ross, S.D.: Heteroclinic connections between periodic orbits and resonance transitions in celestial mechanics. Chaos Interdiscip. J. Nonlinear Sci. **10**, 427–469 (2000)
3. Park, P., Chamberlin, A.: Horizons JPL. Asteroid database. http://ssd.jpl.nasa.gov/horizons.cgi#top (2017). Accessed 2017-01-10
4. Ross, S.D., Scheeres, D.J.: Multiple gravity assists, capture, and escape in the restricted three-body problem. SIAM J. Appl. Dyn. Syst. **6**, 576–596 (2007)
5. Rudenko, M.: Minor planet center. Asteroid database. http://www.minorplanetcenter.net/iau/TheIndex.html (2017). Accessed 2017-01-10
6. Sánchez, J.P., Yárnoz, D.G.: Asteroid retrieval missions enabled by invariant manifold dynamics. Acta Astronaut. **127**, 667–677 (2016)
7. Sánchez, J.P., Alessi, E.M., Yárnoz, D.G., McInnes, C.: Earth resonant gravity assists for asteroid retrieval missions. In: 64th International Astronautical Congress 2013, p. 13 (2013)

8. Sánchez, J.P., Colombo, C., Alessi, E.M.: Semi-analytical perturbative approaches to third body resonant trajectories. In: Proceedings of the International Astronautical Congress, IAC, vol. 7, pp. 5504–5515 (2015)
9. Strange, N., Landau, D., Lantoine, G., Lam, T., McGuire, M., Burke, L., Martini, M., Dankanich, J.: Overview of mission design for NASA asteroid redirect robotic mission concept. In: 33rd International Electric Propulsion Conference (2013)
10. Szebehely, V.: Theory of Orbit. Academic Press, London (1969)
11. Yárnoz, D.G., Sánchez, J.P., McInnes, C.R.: Easily retrievable objects among the NEO population. Celest. Mech. Dyn. Astron. **116**, 367–388 (2013)

Evaluating Proximity Operations Through High-Fidelity Asteroid Deflection Evaluation Software (Hades)

Massimo Vetrisano, Juan L. Cano, and Simone Centuori

Abstract The High-fidelity Asteroid Deflection Evaluation Software (HADES) deals with the high-fidelity modelling of spacecraft operations at irregular shape asteroids. The software can handle any operational orbit, with particular care paid to fixed hovering configurations. Different control techniques based on both continuous and discrete methods have been considered and implemented. The manoeuvre execution itself can be affected by errors in magnitude and direction.

The spacecraft orbit determination can be performed either through a performance model or by on-board measurements, a navigation camera and a LIDAR, which are processed by an Unscented H-infinity Filter (UHF). HADES can employ different levels of accuracy between the assumed environment knowledge and the real world. The aim of this paper is to discuss in details the models that can be used to describe the dynamics and the estimation of a spacecraft hovering at an irregular object. It will show how the various modelling assumptions can affect the results regarding the control budget and on-board estimation in the highly perturbed environment of the comet 67P/Churyumov–Gerasimenko.

1 Introduction

In the last 20 years, there has been considerable progress in the exploration of the minor celestial objects by spacecraft. Recently the most remarkable mission has been Rosetta, which arrived at Comet 67P/Churyumov–Gerasimenko on 6 August 2014. The lander Philae achieved the first-ever soft landing on the surface of the comet on 12 November 2014. As shown by the difficulties on the identification of the final landing spot of the probe, the environment near minor bodies is pretty

---

M. Vetrisano (✉) · J. L. Cano · S. Centuori
Deimos Space S.L.U, Tres Cantos, Madrid, Spain
e-mail: massimo.vetrisano@deimos-space.com; juan-luis.cano@deimos-space.com; simone.centuori@deimos-space.com

complex because of the lack of precise data where simplification regarding the shape and composition of the asteroid can drive to a completely incorrect picture of the dynamics.

The navigation in close proximity of asteroids can be complicated due to the fact that the environment is uncertain, especially if the asteroid presents an irregular shape and rotation state. The motion of the spacecraft around the asteroid is, thus, highly nonlinear. Generally, the gravitational harmonics of the celestial minor bodies are estimated from on-board data collected during a close fly-by [1], during approach phases [2] or by ground-based radar imaging data [3]. Thus, it is necessary to evaluate possible different navigation strategies to increase the mission reliability and the possibility to cope with both unknown environment and system performance uncertainties.

One important aspect when designing proximity operations is to evaluate how the different control techniques and on-board instruments affect the performance of the system. The manoeuvre execution itself can be affected by errors in magnitude and direction.

This kind of missions typically requires the spacecraft to fly in a tight formation relatively close to the asteroid, so on-board estimation capabilities are desirable and indeed required when the delay time between ground and the spacecraft is too high to ensure the safety of operations.

Methods based on optical navigation camera and laser light radar (LIDAR) or laser range finder (LRF) integrated measurements have been proved to be a feasible option for a single spacecraft to approach or land on an asteroid [4, 5].

The idea beneath this paper is to describe the operating environment a spacecraft will face at the asteroid, to show how the different assumption can affect the outcomes of the simulation. This can advise the reader on how to handle with the results and margins when it comes to the control budget for instance.

HADES is a high-fidelity simulation tool to assess GNC close proximity operations. Detailed models about the close proximity environment about Near Earth asteroids (NEA) and the involved operations are required during preliminary assessment of mission requirements especially under the presence of uncertainties. The implemented spacecraft dynamics considers the most relevant perturbations, i.e. third body effect from the Sun, solar radiation pressure (SRP) and irregular gravity field of the rotating asteroid. The software uses both spherical harmonics and actual asteroid's shape. In the first case the coefficients can be given from actual data or they are calculated on a user-defined ellipsoid; in the second case the gravity field is reconstructed from the asteroid tetrahedral mesh. The software can handle any operational orbit, with particular care paid to inertial and body fixed hovering. Different control techniques based on both continuous and discrete methods have been considered and implemented. HADES has a detailed model of camera and LIDAR, where the actual illumination and visibility conditions are modelled using real asteroid shape data. At initial stage one can also assume a performance model, but we will see how this can produce misleading results in reality.

We want to underline that we devote detailed explanations of the models in order for the interested readers to use them and recreate the results of the simulations, rather than retrieving all the information from diverse sources. Thus this paper is organised as follows. Section 2 explains the different dynamic models and main modelling. In Sect. 3, the control technique used to maintain the spacecraft on its reference trajectory is briefly explained. Section 4 shows the estimation process through the Unscented H-Infinity filter and the assumed measurement models. Finally, Sect. 5 shows some obtained results. In particular, all the analyses for the GNC case are applied to the scenario of comet 67P/Churyumov–Gerasimenko whose shape model is well known after the visit of the mission Rosetta in 2015. The shape of the comet magnifies the possible source of mis-modelling which can affect the overall navigation. We considered inertial hovering configuration where the spacecraft maintains its location fixed with respect to the object. In particular we place the probe at 2.7 km, which is the periapsis distance of Rosetta's orbit.

2 Dynamic Models

In this section we want to give an accurate description of the models used to describe the dynamics of the spacecraft.

2.1 Hill's Reference Frame

In this section, we introduce the motion dynamics of spacecraft and asteroid in the non-inertial Hill's reference Frame (see Fig. 1). It is assumed that the asteroid body frame (later described) is coincident with this frame at the beginning of the simulations.

The spacecraft is subjected to the force due to solar gravity, solar radiation pressure and the asteroid's irregular gravity. The nonlinear relative equations of motion are given by [6]:

Fig. 1 Hill reference frames

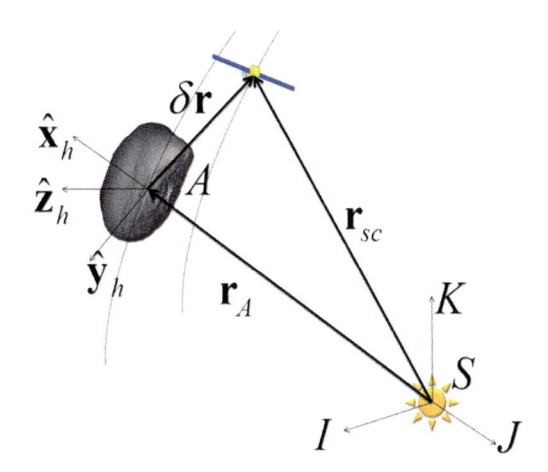

$$\ddot{\mathbf{r}} + 2\dot{\theta}\mathbf{x}\dot{\mathbf{r}} + \dot{\theta}\mathbf{x}\left(\dot{\theta}\mathbf{x}\mathbf{r}\right) = -\frac{\mu_a}{r^3}\mathbf{r} + \mu_{Sun}\left(\frac{\mathbf{r}_a}{r_a^3} - \frac{\mathbf{r}_{sc}}{r_{sc}^3}\right) + \frac{\partial U}{\partial (\mathbf{r})} + SRP(\mathbf{r}_{sc}) + \mathbf{u} \quad (1)$$

where μ_{Sun} is the Sun gravity constant, μ_a is the gravity constant of the asteroid, $\mathbf{r}_a$ and $\mathbf{r}_{sc}$ are the positions of the asteroid and spacecraft with respect to the Sun. r is the relative distance between spacecraft and asteroid. $\dot{\theta}$ represents the instantaneous angular velocity with which the asteroid (i.e. the reference frame) rotates around the Sun. $SRP(\mathbf{r}_{sc})$ is the solar radiation pressure; $\mathbf{u} = [u_x, u_y, u_z]$ is a control input for continuous control. In the case of impulsive control this term is null and impulsive variation of velocity is applied at the time of the manoeuvre. U is the higher order potential of the asteroid.

2.2 Asteroid Motion Around the Sun

The motion of the asteroid with respect to the Sun is given as:

$$\ddot{\mathbf{r}}_a = -\frac{\mu_{Sun}}{r_a^3}\mathbf{r}_a \quad (2)$$

Note that no perturbations acting on the asteroid are considered (i.e. Keplerian motion is assumed).

2.3 Solar Radiation Pressure

The SRP depends on the distance from the Sun as the spacecraft exposed area, the reflectivity coefficient and the mass:

$$SRP(\mathbf{r}_{sc}) = C_r S_{srp} \left(\frac{r_{1AU}}{r_{sc}}\right)^2 \frac{\mathbf{r}_{sc}}{r_{sc}} \frac{A}{m_{sc}} \quad (3)$$

A and m_{sc} are the spacecraft cross section area and mass, respectively, CR is the reflectivity coefficient, S_{srp} is the solar radiation pressure at 1 AU and r_{1AU} is equivalent to the astronomical unit in km.

2.4 Gravity Field Through Spherical Harmonics

The asphericity of these bodies gives raise to perturbations that affect all orbital elements, especially at low altitude. The model that has been considered to describe these effects is based on the standard Legendre polynomials of the gravity field potential as defined by Cunningham [7]. The model works nicely when outside the object circumscribing sphere while it is completely unreliable inside.

The use of Legendre polynomials allows an efficient computation of the potential and resulting perturbation as a function of the Cartesian coordinates in the body fixed reference frame. The gravity potential U can be written as:

$$U = \frac{\mu_a}{R_a} \sum_{n=0}^{\infty} \sum_{m=0}^{\infty} C_{nm} V_{nm} + S_{nm} W_{nm} \qquad (4)$$

where R_a is the body's mean radius respectively, C_{nm}, S_{nm} are the potential coefficients, also known as Stokes' coefficients, that describe the distribution of the mass within the body. V_{nm} and W_{nm} satisfy the recurrence relations:

$$\begin{aligned} V_{mm} &= (2m-1)\tfrac{R_a}{r^2}\left(xV_{m-1,m-1} - yW_{m-1,m-1}\right) \\ W_{mm} &= (2m-1)\tfrac{R_a}{r^2}\left(xW_{m-1,m-1} - yV_{m-1,m-1}\right) \\ V_{nm} &= \tfrac{R_a}{r^2}\left(\tfrac{2n-1}{n-m} z V_{n-1,m} - \tfrac{n+m-1}{n-m} R_a V_{n-2,m}\right) \\ W_{nm} &= \tfrac{R_a}{r^2}\left(\tfrac{2n-1}{n-m} z W_{n-2,m} - \tfrac{n+m-1}{n-m} R_a W_{n-2,m}\right) \end{aligned} \qquad (5)$$

where r is the distance of the spacecraft form the centre of mass of the body. The set of equations hold for $n = m+1$ if $V_{n-1,m}$ and $W_{n-1,m}$ are set to zero. Furthermore, the initial value $V_{00} = \frac{R_A}{r}$, $W_{00} = 0$ are known. The recursions used here are stable, which means that small numerical errors in the computation of the low-order terms do not lead to affect results for high orders. The overall acceleration $\ddot{r}$ is equal to the gradient U and can be directly calculated from V_{nm} and W_{nm} as

$$\ddot{x} = \sum_{nm} \ddot{x}_{nm}, \quad \ddot{y} = \sum_{nm} \ddot{y}_{nm}, \quad \ddot{z} = \sum_{nm} \ddot{z}_{nm},$$

with partial accelerations

$$m = 0: \begin{cases} \ddot{x}_{n0} = -\frac{\mu_a}{R_a^2} C_{n0} V_{n+1,1} \\ \ddot{y}_{n0} = -\frac{\mu_a}{R_a^2} C_{n0} W_{n+1,1} \end{cases}$$

$$m > 0: \begin{cases} \ddot{x}_{nm} = \tfrac{1}{2}\tfrac{\mu_a}{R_a^2} \begin{bmatrix} -C_{nm} V_{n+1,m+1} - S_{nm} W_{n+1,m+1} + \\ \tfrac{(n-m+2)!}{(n-m)!}\left(C_{nm} V_{n+1,m-1} - S_{nm} W_{n+1,m-1}\right) \end{bmatrix} \\ \ddot{y}_{nm} = \tfrac{1}{2}\tfrac{\mu_a}{R_a^2} \begin{bmatrix} -C_{nm} V_{n+1,m+1} + S_{nm} W_{n+1,m+1} + \\ \tfrac{(n-m+2)!}{(n-m)!}\left(S_{nm} V_{n+1,m-1} - C_{nm} W_{n+1,m-1}\right) \end{bmatrix} \end{cases}$$

$$\ddot{z}_{nm} = \frac{1}{2}\frac{\mu_a}{R_a^2}\left[(n-m+1)\left(-C_{nm} V_{n+1,m} + S_{nm} W_{n+1,m}\right)\right] \qquad (6)$$

being the acceleration defined in the planet fixed reference frame a rotation to the inertial frame is necessary.

In the case of an ellipsoidal shape an analytical formula for calculating the even terms of the matrix C was obtained from [8]:

$$C_{2l,2m} = \frac{3}{R^{2l}} \frac{l!\,(2l-2m)}{2^{2m}\,(2l+3)\,(2l+1)!}\,(2-\delta_{0m})$$

$$\times \sum_{i=0}^{\mathrm{int}\left(\frac{l-m}{2}\right)} \frac{\left(c_1^2 - c_2^2\right)^{m+2i}\left[c_3^2 - \frac{1}{2\left(c_1^2+c_2^2\right)}\right]^{l-m-2i}}{16^i\,(l-m-2i)!\,(m+i)!\,i!} \qquad (7)$$

c_1, c_2 and c_3 are the semi-axes of a triaxial ellipsoid.

2.5 Gravity Field Through Shape Model

The gravity model works for an arbitrary shape and was implemented from the equations used in [9]. This model assumes a uniform asteroid density and allows expressing the local acceleration in an arbitrary location in space with respect to the asteroid's centre of mass. It is especially suited for proximity operations, where the harmonic techniques fail to provide an accurate representation of the gravity field.

With reference to Fig. 2, the local acceleration is given by the gradient of the potential field U by

$$\nabla U = -G\rho \sum_{e \in edges} L_e \tilde{E}_e r_e + G\rho \sum_{f \in faces} \omega_f \tilde{F}_f r_f \qquad (8)$$

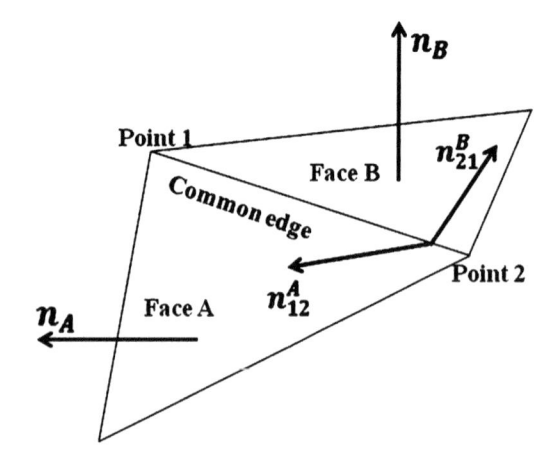

Fig. 2 Reference directions and notations for the shape model equations

where $\vec{r}_e$ is the distance of a generic point from the edge, and $\vec{r}_f$ is the distance from the centre of the facet. $\tilde{E}_e$ and $\tilde{F}_f$, respectively the edge and the facet dyads in this expression, are computed on each triangular face with reference to Fig. 2 as follows:

$$\tilde{E}_e = n_A n_{21}^A + n_b n_{12}^B$$
$$\tilde{F}_f = n_f n_f \tag{9}$$

The edge dyads use the normals to faces and edges, $\vec{n}_f$ and $\vec{n}_{ij}^f$ of two adjacent faces. The normals to faces and edges are calculated from the coordinates of the vertices as:

$$n_f = (r_2 - r_1) \times (r_3 - r_2)$$
$$n_{ij}^f = (r_j - r_i) \times n_f \tag{10}$$

Dimensionless factors L_e and ω_f are then given by

$$L_e = \ln \frac{r_i + r_j + e_{ij}}{r_i + r_j - e_{ij}}$$
$$\omega_f = 2\tan^{-1} \frac{r_i \cdot r_j \times r_k}{r_i r_j r_k + r_i(r_j r_k) + r_j(r_i r_k) + r_k(r_j r_i)} \tag{11}$$

where e_{ij} is the length of the edge.

As an example Fig. 3 shows the error between the harmonics gravity field and the shape based model on the circumscribing sphere around an ellipsoid of semi-axes [3, 2, 0.5] km. A 4th order gravity field produces a relative error up to 40% along c_1,

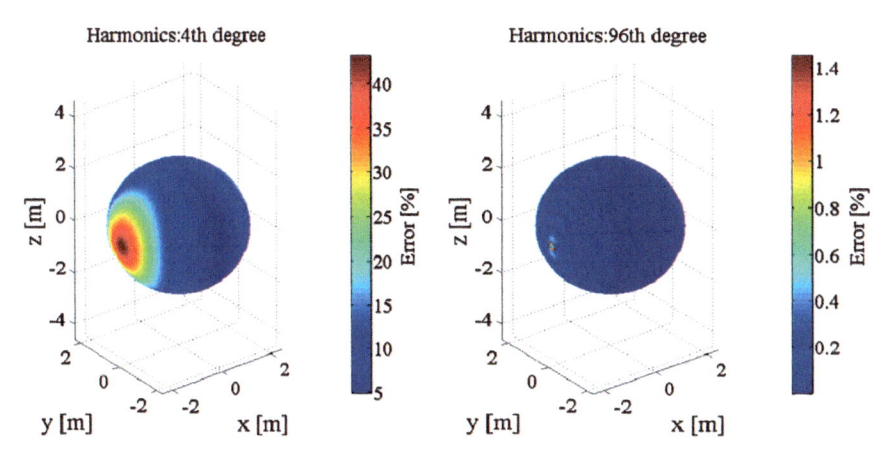

Fig. 3 Error between harmonics and shape model gravity field on the asteroid circumscribing sphere

while a ninty-sixth is pretty much coincident with the shape model with higher error localized along a small area in $c1$. Note that as we move further from the surface this difference cancels out and can be assimilated to a system noise.

As a concluding remark, the shape model can be used to calculate the celestial body inertia, which we exploited to simulate the rigid body rotation over the simulation period.

3 Controller

The controller implemented in this paper is a discrete LQR with integrative contribution, which is described in the following section along with the manoeuvre error modelisation.

3.1 Discrete Lyapunov Controller

We want to calculate the optimal gain matrix $\mathbf{K}$ such that the state-feedback law $\mathbf{u_k} = -\mathbf{K}\delta\mathbf{x_k}$ (where k is the discrete step) minimizes the quadratic cost function

$$J(\mathbf{u}) = \sum_{k=1}^{\infty} x_k^T Q x_k + u_k^T R u_k$$

for the discretized state-space model of Eq. (6). Also in this case we neglected the contribution of the Coriolis force. For convenience we report the results:

$$\dot{\delta x} = \begin{bmatrix} 0_{3x3} & I_{3x3} \\ -\frac{\mu_a}{\delta r_0^3}[I - 3\hat{r}_0\hat{r}_0] & 0_{3x3} \end{bmatrix} \delta x + \begin{bmatrix} 0_{3x3} \\ I_{3x3} \end{bmatrix} u = \begin{bmatrix} 0_{3x3} & I_{3x3} \\ \beta & 0_{3x3} \end{bmatrix} \delta x + \begin{bmatrix} 0_{3x3} \\ I_{3x3} \end{bmatrix} u \quad (12)$$

$\beta = \text{diag}([\beta_1, \beta_2, \beta_3])$. Integrating the equations of motion for a time step Δt using explicit Euler, one obtains that

$$\delta x_{k+1} = \begin{bmatrix} I + \beta\frac{\Delta t^2}{2} & \Delta t I_{3x3} \\ \beta \Delta t & I_{3x3} \end{bmatrix} \delta x_k + \begin{bmatrix} \Delta t I_{3x3} \\ I_{3x3} \end{bmatrix} u_k = A_k \delta x_k + B_k u_k \quad (13)$$

$\mathbf{A_k}$ and $\mathbf{B_k}$ are constant discrete state matrix in this case.

Then the state-feedback gain $\mathbf{K}$ results:

$$K = \left(B_k^T S B_k + R\right)^{-1} B_k^T S \quad (14)$$

Where **S** is given by the solution of the discrete-time Riccati equation:

$$A_k^T S^T A_k - S - A_k^T S A_k (B_k^T S B_k)^{-1} B_k^T S A_k + Q = 0 \tag{15}$$

Although the controller will work using only proportional correction manoeuvre, we decided to add the integrative contribution due to the action of the gravity field during the interval between corrections. The integrative contribution improves the accuracy because otherwise the spacecraft will tend to move towards an artificial equilibrium point where

$$\frac{c_p}{c_d} = -\frac{\delta v_{est} - \delta v_{ref}}{\delta r_{est} - \delta r_{ref}} \tag{16}$$

We assume that the overall effect from the other forces is negligible and the LQR is able to cope with those perturbations. The integrative contribution is calculated assuming a constant acceleration:

$$u_k = K \delta x_k - \bar{a}_p \Delta \widehat{tr}_0 \tag{17}$$

Where $\bar{a}_p$ is the mean value of acceleration as measured at centre, superior and inferior edge of the control box defined by a characteristic length, b, used to calculate the gain. The integrative contribution was added only when contribution of the perturbations does not work to reduce the position error. This is done simply to include the fact that the gravity acts favourably by attracting the spacecraft towards the reference position when the spacecraft is above the nominal altitude. We required the controller to perform the maximum delta-v manoeuvre equal to $2v_{acc}$ when the error in position (on components base) is equal to b and the velocity error is v_{acc} (where v_{acc} is the velocity acquired by constant gravitational acceleration during a free fall along b) This means that:

$$\begin{aligned} Q &= \text{diag}\left(\begin{bmatrix} b & b & b & v_{acc} & v_{acc} & v_{acc} \end{bmatrix}^{-2}\right) \\ R &= \text{diag}\left(\begin{bmatrix} 2v_{acc} & 2v_{acc} & 2v_{acc} \end{bmatrix}^{-2}\right) \end{aligned} \tag{18}$$

In practice, if the spacecraft started moving from the nominal state, a reflection manoeuvre would be performed at the edge of the control box.

3.2 Manoeuvre Errors

Error in the manoeuvre execution has been modelled in terms of magnitude and direction as:

$$\Delta v = R(\theta, \varphi) \Delta v_{nom} (1 + r_{ex}) \tag{19}$$

where Δv_{nom} is the nominal manoeuvre, and r_{ex} is the execution error and θ, φ are the error angles on two directions. The errors are generated randomly consistently with the assumed execution error statistics.

4 Navigation Models

The Navigation Module conceptually contains two trajectory estimation routines

1. A performance model based on typical knowledge of the spacecraft trajectory.
2. A real-time on board filter based on the Unscented H-infinity Filter (UHF) which uses LIDAR and camera measurements.

4.1 Performance Model

There is a simple performance model for the orbit determination which consists of pseudo state vector measurements z simply given as.

$$\mathbf{z} = [\text{r v}] + \mathbf{x}_{\text{meas}} \qquad (20)$$

x_{meas} is the measurement error, randomly generated at each time consistently with the assumed noise error. Note that the measurement error is given along track and cross track, thus a transformation from the local rotating frame to the Cartesian one is performed.

4.2 Unscented H-Infinity Filter

The hypothesis underneath the generic Kalman filter is that the noise in measurements, dynamic model and priors is Gaussian in nature. This might not be the case in general and even though the UKF has proven to work reasonably well when the Kalman filter hypotheses are not satisfied, a better alternative would be to use a H_∞ filter, also called minmax filter. The H_∞ filter does not require prior assumptions on the nature of the noise, and minimizes the worst-case estimation error. The choice of the H_∞ filter is preferable when the Gaussian hypothesis cannot be fully guaranteed, for example when biases in the instruments are not detected [10, 11]. In our case, besides biases affecting all the instruments, the LIDAR measurements are affected by the camera process and errors. Therefore, the noise introduced by the LIDAR cannot be modelled as an uncorrelated white noise.

In order to deal with nonlinearities, one can use an extension to the H_∞ filter, the Extended H_∞ Filter (EHF), analogous to the extended Kalman filter. In this case, however, some hypotheses need to be introduced on the smoothness and regularity of the process and measurements. An alternative is to introduce the unscented transformation in the H_∞ filter to avoid the approximation of the Jacobian matrices [11] and build an Unscented H_∞ Filter.

As well as the UKF, the UHF works on the premises that one can well approximate the posteriori covariance by propagating a limited set of optimally chosen samples. The UHF is hereafter briefly described. Using the estimation theory formalism, the nonlinear process estimation process composed of dynamics and measurements equations can be discretized in time and written as:

$$\begin{aligned} x_{k+1} &= f(x_k, u_k, w_k) \\ y_k &= h(x_k, v_k) \end{aligned} \quad (21)$$

Where w_k is the process noise and v_k is the measurement noise. The process noise could belong to a generic distribution but in the following we restrict the analyses to the case in which $w_k \sim N(0, Q_k)$, with Q_k the process noise covariance at time step k. The quantity u_k represents the control input required to counteract the perturbations acting on the spacecraft.

The estimated motion between t_k and t_{k+1} in which the measurements are received and processed, is simply given by the integration of Eq. (21) without the contribution of w_k. The initial conditions are the estimated position and velocity from the filter at time t_k. Similarly to the UKF, the UHF relies on the unscented transformation to propagate a set of suitable sigma points, drawn from the apriori covariance matrix.

The set of sigma points $\mathcal{X}_i$ are given as:

$$\mathcal{X}_i = \begin{cases} \tilde{x}_k & i = 0 \\ \tilde{x}_k^- + \left(\sqrt{(n + k_{ukf}) P_k + Q_k}\right) & i = 1, 2, ..n \\ \tilde{x}_k^- - \left(\sqrt{(n + k_{ukf}) P_k + Q_k}\right) & i = n+1, ..2n \end{cases} \quad (22)$$

where $\mathcal{X}_i$ is a matrix consisting of $(2n + 1)$ vectors, with $k_{ukf} = \alpha_{ukf}^2 (n + \lambda_{ukf}) - n$, k_{ukf} is a scaling parameter, constant α_{ukf} determines the extension of these vectors around $\tilde{x}_k$. We set α_{ukf} equal to 10^{-3} and λ_{ukf} is set equal to 3-n.

The sigma points are transformed or propagated through the nonlinear function, the so-called unscented transformation, to give:

$$\begin{aligned} \mathcal{X}_{k+1} &= f(\mathcal{X}_k, u_k) \\ \mathcal{Y}_k &= h(\mathcal{X}_k) \end{aligned} \quad i = 0, \ldots, 2n \quad (23)$$

Then the mean value and covariance of $\mathcal{Y}$ are approximated using the weighted mean and covariance of the transformed vectors:

$$\begin{aligned} \bar{y} &= \sum_{i=0}^{2n} W_i^{(m)} \mathcal{Y}_i \\ P_k &= \sum_{i=0}^{2n} W_i^{(c)} (\mathcal{Y}_i - \bar{y})(\mathcal{Y}_i - \bar{y})^T \end{aligned} \quad (24)$$

where $W_i^{(m)}$ and $W_i^{(c)}$ are the weighted sample mean and covariance given by:

$$\begin{aligned} W_i^{(m)} &= \frac{k_{ukf}}{n+k_{ukf}} \\ W_i^{(c)} &= \frac{k_{ukf}}{n+k_{ukf}} + \left(1 - \alpha_{ukf}^2 + \beta_{ukf}\right) \\ W_i^{(m)} &= W_i^{(c)} = \frac{k_{ukf}}{2(n+k_{ukf})} \quad , i = 1, 2, \ldots, 2n \end{aligned} \qquad (25)$$

and β_{ukf} is used to incorporate prior knowledge of the distribution with $\beta_{ukf} = 2$. The predicted mean of the state vector $\tilde{x}_k^-$, the covariance matrix P_k^- and the mean observation $\tilde{y}_k^-$ can be approximated using the weighted mean and covariance of the transformed vectors:

$$\begin{aligned} \mathcal{X}_{k|k+1}^i &= f\left(\mathcal{X}_{k+1}^i, u_k\right) \\ \tilde{x}_k^- &= \sum_{i=0}^{2n} W_i^{(m)} \mathcal{X}_{k|k+1}^i \\ P_k^- &= \sum_{i=0}^{2n} W_i^{(c)} \left(\mathcal{X}_{k|k+1}^i - \tilde{x}_k^-\right) \left(\mathcal{X}_{k|k+1}^i - \tilde{x}_k^-\right)^T + Q_k \\ \mathcal{Y}_{k|k-1}^i &= h\left(\mathcal{X}_{k|k+1}^i\right) \\ \tilde{y}_k^- &= \sum_{i=0}^{2n} W_i^{(m)} \mathcal{Y}_{k|k-1}^i \end{aligned} \qquad (26)$$

The updated covariance $P_{y,k}$ and the cross correlation matrix $P_{xy,k}$ are:

$$\begin{aligned} \mathbf{P}_{y,k} &= \sum_{i=0}^{2n} W_i^{(c)} \left(\mathcal{Y}_{k|k-1}^i - \tilde{y}_k^-\right) \left(\mathcal{Y}_{k|k-1}^i - \tilde{y}_k^-\right)^T \\ \mathbf{P}_{xy,k} &= \sum_{i=0}^{2n} W_i^{(c)} \left(\mathcal{X}_{k|k+1}^i - \tilde{x}_k^-\right) \left(\mathcal{Y}_{k|k-1}^i - \tilde{y}_k^-\right)^T \end{aligned} \qquad (27)$$

Finally, the filter state vector $\tilde{x}_k$, and covariance updated matrix P_k are represented as follows

$$\begin{aligned} \tilde{x}_k &= \tilde{x}_k^- + \mathbf{K}\left(y_k - \tilde{y}_k^-\right) \\ (P_k)^{-1} &= (P_k^-) + (P_k^-)^{-1} P_{xy,k} R_k^{1-} \left[(P_k^-)^{-1} P_{xy,k}\right]^T - \theta_k I_d \\ \mathbf{K} &= \mathbf{P}_{xy,k} \mathbf{P}_{y,k}^{-1} \end{aligned} \qquad (28)$$

where $\mathbf{K}$ is the Kalman gain matrix and θ_k is the performance bound of the H_∞ filter. $\mathbf{R}_k$ is a suitable matrix which in the case of normal distribution coincides with the measurement noise covariance matrix at time step k. In order to assure that the covariance matrix is positive definite, this value is calculated at each iteration as:

$$\theta_k^{-1} = \xi \max\left(\text{eig}\left((P_k^-)^{-1}(P_k^-)^{-1} P_{xy,k} R_k^{1-}\left[(P_k^-)^{-1} P_{xy,k}\right]^T\right)^{-1}\right) \qquad (29)$$

with ξ a scaling parameter. For small values of θ_k the terms $\theta_k \mathbf{I}_d$ tends to 0, and the covariance update in Eq. (24) is equivalent to the one in the UKF. As one can see from the set of equations the performance bound has not a direct effect on the calculation of the gain and on the update step for the estimated state. Nonetheless θ_k modifies the shape of covariance matrix update in Eq. (28), which, in turn, generates a different distribution of the sigma points. In this way the propagation and the update step at the following time step will be directly influenced by the value of the performance bound.

4.3 Instruments Model

4.3.1 Camera Model

In order to develop the measurement model of the camera, two intermediate reference frames are required as shown in Fig. 4:

1. Spacecraft coordinate system $SC\{x_{SC}, y_{SC}, z_{SC}\}$: the origin of this frame lies in the spacecraft's mass centre, three body axes of symmetry are defined as three coordinate axes.
2. Camera coordinate system $C\{\widehat{x}_c, \widehat{y}_c, \widehat{y}_c\}$: the centre C is the perspective projection of the camera. $\widehat{x}$-axis is parallel to the optical axis of the camera and directed to the centre of the asteroid. Image plane is defined on the $\widehat{y}_c - \widehat{z}_c$ plane. To simplify mathematics, it is assumed that the spacecraft and camera coordinate systems are coincident.

Fig. 4 Pin-Hole camera model

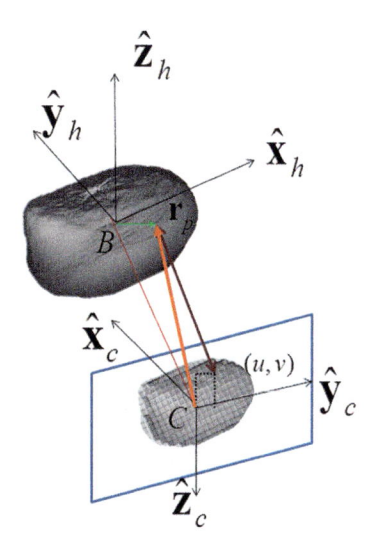

It is assumed that the attitude of each spacecraft is known with a level of precision corresponding to the one of the star tracker on two axes.

Having identified the two reference frames, one can define the geometric relationship in the asteroid Hill rotating reference frame. The position vector of the *i*-th feature is $x^j_{surface}$, which is selected randomly on the asteroid surface. The spacecraft position vector with respect to the asteroid is defined as δr_{SC}, while $x^j_{Surf-SC}$ refers to the position vector from the estimated spacecraft to the feature.

With reference to Fig. 4, assuming a pinhole model for the camera [12], a point on the surface of the asteroid, with position $r_p = [x_c, y_c, z_c]$ in the reference frame of the camera, has coordinates on the image plane given by:

$$\begin{bmatrix} u \\ v \end{bmatrix} = \frac{f}{x_c} \begin{bmatrix} y_c \\ z_c \end{bmatrix} \qquad (30)$$

where x_c is the distance of the point from the image plane along the boresight direction and f is the focal length of the camera.

Without taking into account the effect of the attitude error, the position in the camera reference frame would be given by:

$$r_p = R_{HC} x_{Surf-SC} \qquad (31)$$

where R_{HC} is the rotation matrix from the Hill's reference frame to the camera frame and:

$$x_{Surf-SC} = x_{Surface} - \delta r_{SC} \qquad (32)$$

with $x_{Surface}$ the vector position of the points with respect to the centre of the Hill's reference frame.

The coordinates of the point on the image plane measured in pixels are given by:

$$\begin{aligned} x_{screen} &= u/p_{width} \\ y_{screen} &= v/p_{width} \end{aligned} \qquad (33)$$

with p_{width} the pixel width. If one considers the effect of the attitude errors $\lambda_{1,2}$ around $\hat{z}_c$ and $\hat{y}_c$, the vector r_p of each feature will be subject to a random rotation in the reference frame of the camera given by:

$$r_p^a = R_{attitude} r_p = \begin{bmatrix} \cos\lambda_1\cos\lambda_2 & \sin\lambda_1 & \cos\lambda_1 \\ -\sin\lambda_1\cos\lambda_2 & \cos\lambda_1 & \sin\lambda_2 \\ -\sin\lambda_1 & 0 & \cos\lambda_2 \end{bmatrix} \begin{bmatrix} x_c \\ y_c \\ z_c \end{bmatrix} \qquad (34)$$

In the case of small error in the pointing angles, substituting r_p^a into Eq. (30) gives:

$$\begin{bmatrix} u \\ v \end{bmatrix} = \frac{f}{x_c + \lambda_1 y_c + \lambda_2 z_c} \begin{bmatrix} y_c - \lambda_1 x_c \\ z_c - \lambda_2 x_c \end{bmatrix} \tag{35}$$

where the attitude error contribution increases with the distance. If one neglects the terms multiplying angles in the denominator, Eq. (35) leads to:

$$\begin{bmatrix} u \\ v \end{bmatrix} = \frac{f}{x_c} \begin{bmatrix} y_c - \lambda_1 x_c \\ z_c - \lambda_2 x_c \end{bmatrix} + f \begin{bmatrix} -\lambda_1 x_c \\ -\lambda_2 x_c \end{bmatrix} \tag{36}$$

The mean position of all the points on the image plane of the camera defines the coordinates of the centroid of the asteroid $[x_c^c, y_c^c] = [u_c^c, v_c^c] p_{width}$. If we consider only the visible points we will obtain the centre of brightness. Then by measuring the angular position of the centroid one can estimate the angular position of the centre of mass in the reference frame of the camera. The azimuth and elevation angles of the centroid are given by:

$$\begin{aligned} \phi &= \tan^{-1} \frac{x_c^c}{f} \\ \psi &= \tan^{-1} \frac{y_c^c}{\sqrt{(x_c^c)^2 + f^2}} \end{aligned} \tag{37}$$

It is assumed that the centroid of the asteroid identifies the position of the centre of mass with some uncertainty. The measurement from the camera is affected by the spacecraft attitude pointing, the pixelization and the centroiding errors (where the last one is the mismatch between centroid and centre of mass). The pixelization error is due to the fact that the image of the asteroid is formed by a discrete number of pixels.

Note that here the illumination conditions are not considered, so it is assumed that each spacecraft sees the whole visible surface from its position. This is sensible if one assumes that a complementary map could be built while starting the orbit acquisition, combining the pictures from the whole formation. In the absence of a map that relates the centroid to the centre of mass the navigation system would rely on the centre of brightness which, depending on shape and solar aspect angle, could introduce a bias in the determination of the relative position.

4.3.2 LIDAR Model

In general, the LIDAR provides range from the spacecraft to a point on the surface of target object and works at a range from 50 m to 50 km. It is assumed that the LIDAR illuminates the point on the surface that corresponds to the centroid derived from the elaboration of the images acquired by the camera [13]. This distance is simply given by:

$$l = \| \delta r_{SC} - x_{surface}^c \| \tag{38}$$

where $x^c_{surface}$ is the position of a point on the asteroid's surface along the centroid direction. The observation equation of the LIDAR including the measurement noise reads:

$$y_l = h_l(\delta r_{SC}) + \varsigma_l \qquad (39)$$

with ς_l the measurement noise. The accuracy of this measurement depends on the characteristics error of the sensor, along with a bias defined by the mounting error of the instrument. If the range l is pre-processed in combination with the angular measurements from Eq. (37), a relative position vector from the spacecraft to the point on the surface can be constructed as

$$z \begin{bmatrix} l \\ \phi \\ \psi \end{bmatrix} = h(\delta r_{SC}) + \varsigma \qquad (40)$$

where z is the measurement vector obtained from the combination of camera and LIDAR, h(x) is the vector containing the measurement model and ς is the total measurement noise vector. It is important to remark that in the simulations, the errors in the two angles are derived from Eqs. (33)–(36), while the observation equations used in the filter are Eq. (37).

The actual illumination and visibility condition are considered such that the image on the screen of the camera will be as shown in an example of Fig. 5, where the centre of brightness has been represented along with the footprint of the LIDAR on the surface as taken around it.

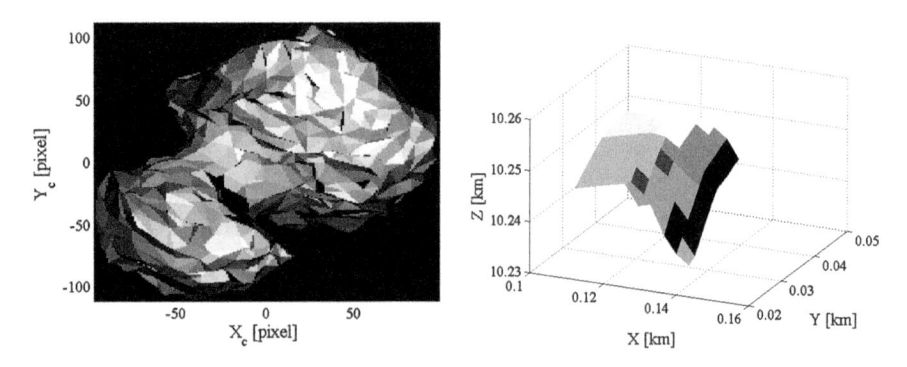

Fig. 5 Example of image as seen on the screen of the camera generated using the Comet 67P/Churyumov–Gerasimenko (left), footprint of the LIDAR around the corresponding centre of brightness

5 Case Studies

In the following, we will move from simpler to more complex models, showing how the different assumptions can lead to dissimilar results, especially for what concerns the control budget. We considered a number of four different cases:

1. The comet shape is assumed to be an ellipsoid, thus gravity field is described as an 8th order ellipsoidal field and a performance model is used for the estimated trajectory.
2. The gravity field of the comet is given by the actual spherical harmonics, and a performance model is employed for navigation purposes.
3. Same as above, but the estimated trajectory is obtained through filtering where the gravity order is reduced with respect to the real one, and the shape of the comet is an ellipsoid whose mean radius for the measurements model differs by 1% error from the actual radius.
4. In this case the dynamics and the measurements are given by the actual shape of body, while the filter relies on the harmonics, and the mean radius for its measurement model differs by 1% error from the actual one. The gravity field in the real world is generated using the actual shape, while the filter relies only on a third order gravity field.

In the following, we first introduce the environment the spacecraft is operating in Sect. 5.1 and its operative conditions in Sect. 5.2. We then analysis the characteristic trend for the single simulation in Sect. 5.3 and eventually we discuss the results with the aid of a Monte Carlo simulation to draw some statistical conclusions in Sect. 5.3.5.

5.1 Comet Comet 67P/Churyumov–Gerasimenko

In the followings, the analysed methods are tested. Besides the calculation of mere control figures as the navigation budget, the comparison is based also on the capability to control the spacecraft with a limited number of actuations. The minor body selected was the Comet 67P/Churyumov–Gerasimenko, whose Keplerian elements are reported in see Table 1. The motion of the asteroid around the Sun was considered purely Keplerian without any perturbation and simulations start from perigee. Moreover there is no effect of the coma included in the simulation. The asteroid was assumed to be shaped as an ellipsoid of semi-axes

$$[c_1 \ c_2 \ c_3] = [2.530 \ 1.857 \ 1.656] \text{ km}$$

Table 1 Comet 67P/Churyumov–Gerasimenko from JPL database

a [AU]	3.464805313920435
e	0.6414365761974745
i [deg]	7.04529818125678
ω [deg]	50.08466699140272
Ω [deg]	12.84210194638212
C20	-7.93×10^{-2}
C22	2.71×10^{-2}
C30	-1.36×10^{-2}
C31	10^{-2}
C32, C33, S31, S32, S33	10^{-3}
C22	10^{-4}
C11, S21	10^{-13}
C10, S11, S21, S22	10^{-14}

Fig. 6 Shape of Comet 67P/Churyumov–Gerasimenko

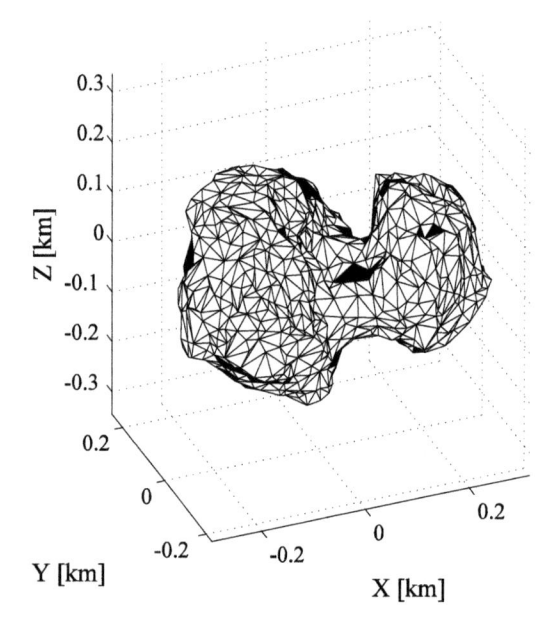

For reference to the next analyses, where the shape of the comet will be used in the dynamics as well as in the measurements generation, Fig. 6 shows the 3D mesh of the well-known duck-shape comet.

Assuming such a shape allows calculating the gravitational harmonics analytically using the results of [8]. When we will consider the actual shape, we used the denormalized Stokes coefficients for the third order degree from [14], reported in Table 1. The gravity constant from the asteroid is thus 6.67259×10^{-7} km/s^3. The asteroid rotates around c_3 axis every 12.4043 h with the equatorial plane coincident with the asteroid orbital plane at the beginning of the simulation (c_1 and c_2 aligned with x–y of the Hill reference frame).

5.2 Spacecraft Characteristics

The initial nominal condition of the spacecraft was randomly generated around the nominal operational trajectory in the local Hill's frame (radial, tangential and out of plane components – position in km, velocity in km/s):

$$[x\ y\ z\ \ v_x\ v_y\ v_z] = [-5.2\ 0\ 0\ 0\ 0\ 0]$$

We arbitrarily decided to place the spacecraft at about 2.7 km from the surface of the comet (considering the maximum semi-axis c_1) as it was the periapsis distance of Rosetta. Table 2 reports the characteristics of the sensors assembly. We assumed a 40° wide angle camera; otherwise the asteroid would not be contained in the camera screen.

The spacecraft is assumed to have a ballistic coefficient of 0.0393 m²/kg and an equivalent reflectivity coefficient (given by reflection and diffusion) of 1.3. We considered an actuation error of 3% (3σ) on magnitude and 2° on angles (3σ).

For the navigation we used the performance model where the pseudo state vector was known with 20 m along track, 10 m cross track in position and 2 mm/s along track and 1 mm/s cross track in velocity (all the quantities are 1σ). We used an error of 20 m just to have an along track error comparable to the one obtained using extensive radiometric measurements.

For what concerns the controller, the gain b was set to 30 m and manoeuvres below 1 cm/s were not executed. The actuation time was coincident with the measurements frequency of 10 min.

5.3 Assessing the Performance

In this section, we want to see how different dynamics and measurements model affect the performance of the state estimate along with the navigation budget. In the following we will show the trend for the controlled and estimated trajectory when a filter is employed. In any case the random number generator was set to the same value at the beginning of simulation in order to present consistent results. The

Table 2 Measurements assembly characteristics

Lidar mounting error	0.001 deg
Lidar range error	20 m
Lidar range bias	1 m
Number of pixels per side	2048
Camera FoV	40 deg
Camera side	10 cm
Attitude error	0.0057 deg
Attitude bias	0.0006 deg

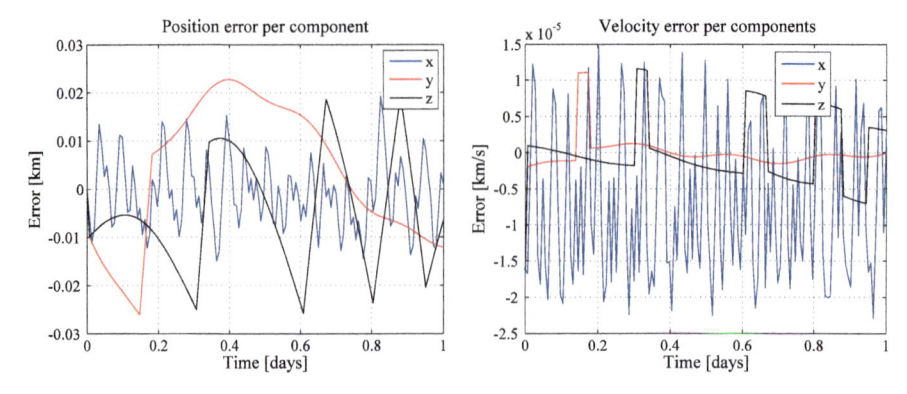

Fig. 7 Case 1: controlled position (left) and velocity (right)

spacecraft is placed on the nominal trajectory, while in Sect. 5.3.5 we will report the results a statistical control budget obtained by modifying the initial conditions, accordingly with the initial filtering guess of 50 m on position and 2 cm/s in velocity for each component.

5.3.1 Case 1

The first analysed case can be considered a first guess, where the control performance can be preliminary tested, under a typical trajectory knowledge from ground station. The fact comet shape is assumed to be an ellipsoid and the gravity field is described as an eighth order ellipsoidal field simplifies considerably the dynamics.

Figure 7 shows the trend for the controlled position and velocity. As one can see, the trajectory error is confined between 30 m boundaries and the maximum velocity error is in the range of 2 cm/s, basically due to a combination of control capabilities and actuation error. The peak along y direction is due to the rotation of the comet with subsequent strong variation of the gravity field after 12 h when the distance of the spacecraft with respect to the surface is the minimum.

5.3.2 Case 2

Although the spacecraft motion results confined as in Case 1 the spacecraft is not confined, the variation of the gravity field of due to the actual spherical harmonics is stronger driving the spacecraft outside the 30 m boundaries, see Fig. 8. This case also shows the limitation of using the linearized approach of Sect. 3.1 to calculate the controller gains. Time varying gains accordingly to the relative configuration of the spacecraft could improve the overall controller performance and could be analysed in future works.

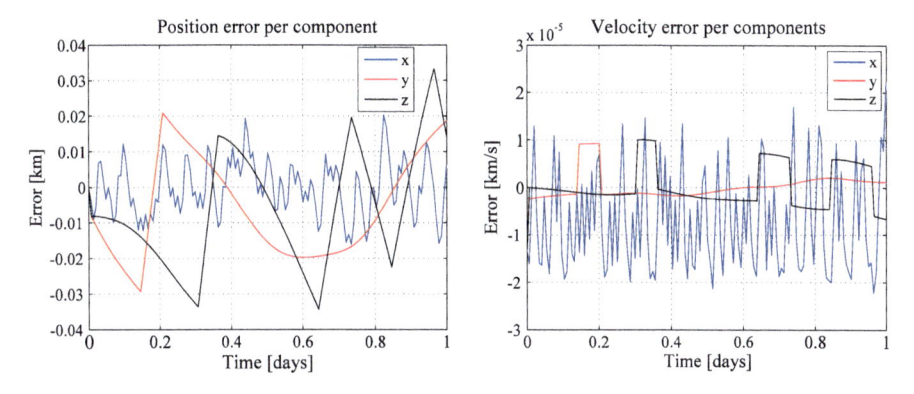

Fig. 8 Case 2: controlled position (left) and velocity (right)

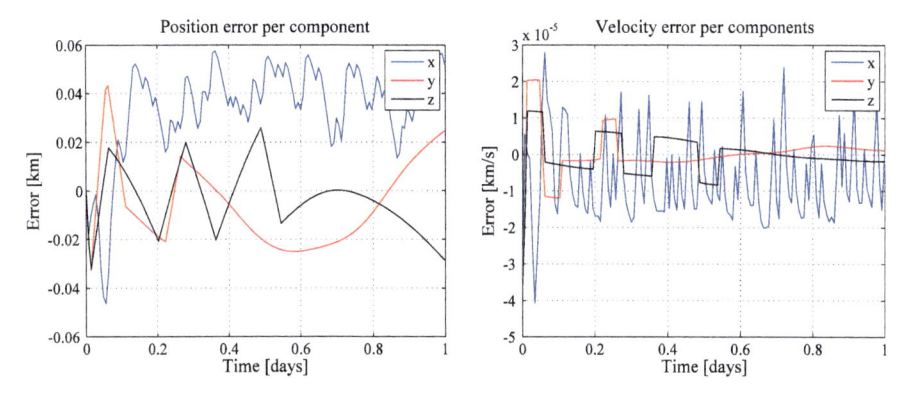

Fig. 9 Case 3: controlled position (left) and velocity (right)

5.3.3 Case 3

In this case, we employ the on-board system to estimate the trajectory of the spacecraft. The gravity field in the filter is known to second degree harmonics, while shape of the comet is assumed to be an ellipsoid whose mean radius for the measurements model differs by 1% error from the actual radius. Figure 9 shows the controlled trajectory where one can see that the controlled position error is biased along the x direction where the combination of the gravity field and the lack of shape knowledge produce the maximum effect. In fact from Fig. 10, we see that the estimation error itself is biased along the x-axis by a value that in mean terms is about 25 m that is consistent with 1% size error. Note that the lack of harmonics leads the filter to place the spacecraft closer to the comet (negative sign along x error). In Fig. 9 right, we see that as a consequence of the estimation error also the controlled velocity error is in general increased with respect to Case 2. If we compare its trend with the one of Fig. 8 we see that the difference is mainly due to the estimated velocity (Fig. 10 right).

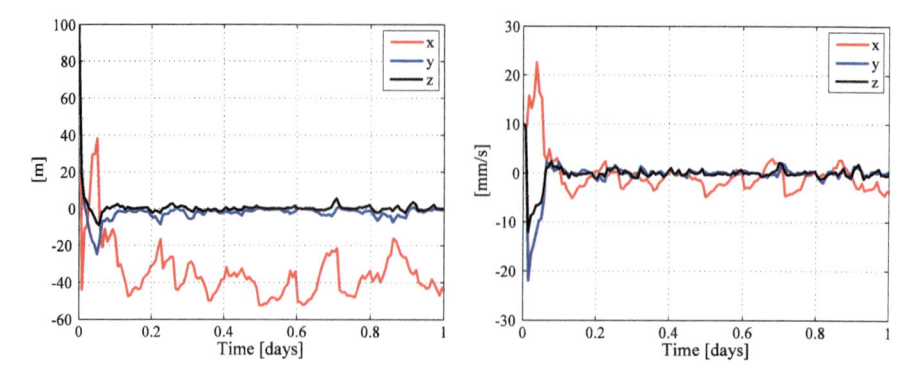

Fig. 10 Case 3: estimated position (left) and velocity (right)

5.3.4 Case 4

In this case the dynamics and the measurements are given by the actual shape of body, while the filter relies the third order gravity field of Table 1, and the mean radius for its measurement model differs by 1% error from the actual one. This case is the most interesting because it allows us to

1. show the effect of the size of the asteroid in terms of accuracy of the mesh; for computational convenience we use a baseline coarse mesh in the filter of 100 facets and we compare the results with respect to a 1000 facets shape.
2. evaluate the impact of the on-board processing on the performance; spacecraft computer have limited computational capabilities, so fixed-time step integrator are used to reduce the computation burden. We used a fixed step Runge-Kutta of order five for all the time-integrations required in the filter.

Figures 11 and 12 show the controlled and estimated trajectory trends in the case the shape model used for both the real world camera and the filter is a 1000 facets polyhedron. On the contrary of Case 3, the filter positions the spacecraft farther from surface (Fig. 12 left) and the error is in the range of 25, which is 1% of the major semi-axis. This produces also a velocity error which is biased along x. The higher control errors are as a consequence along x (see Fig. 11). Nonetheless the overall system manages to contain the spacecraft within 50 m error from the reference position.

Figures 13 and 14 shows that similar trend are obtained when the on-board system relies on simplified surface models with 100 facets polyhedron. The main difference can be seen on the controlled and estimated trajectory peaks which are slightly higher than the ones in the 1000 facets case.

Eventually, if the computational capabilities are limited to a fixed-step time integrator, we obtain that the overall errors are further amplified. With reference to Figs. 15 and 16, we see that the controller and the filter errors are above the 50 m we have seen in the previous case. Also the velocity periodically exceeds 1 cm/s error.

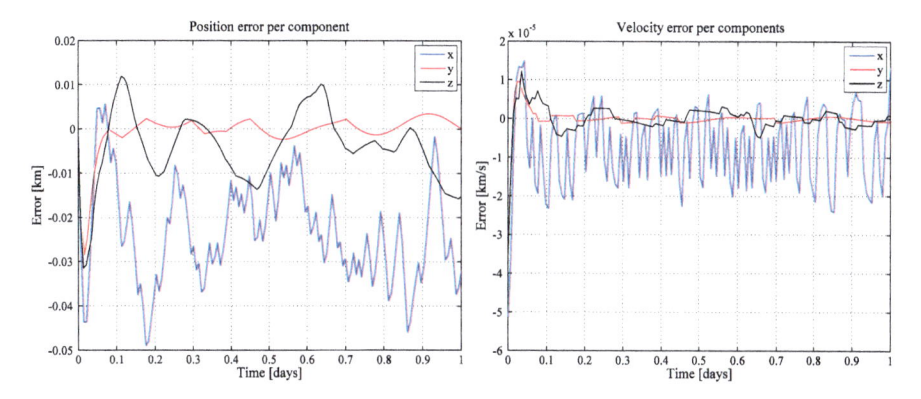

Evaluating Proximity Operations Through High-Fidelity Asteroid Deflection... 39

Fig. 11 Case 4 using a 1000 faces mesh for the asteroid surface: controlled position (left) and velocity (right)

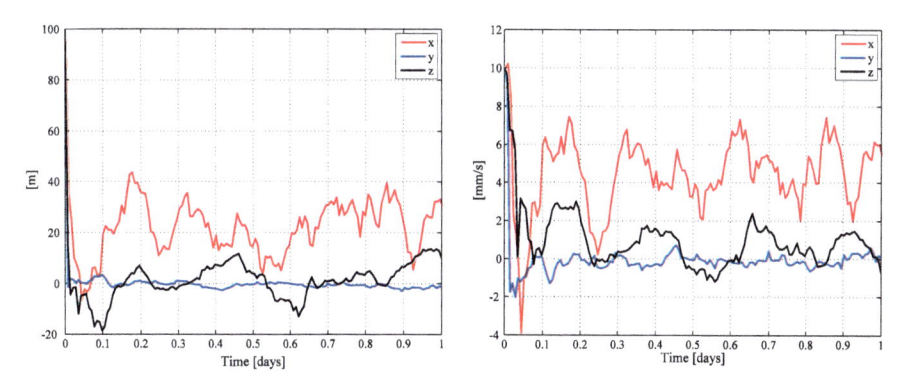

Fig. 12 Case 4 using a 1000 faces mesh for the asteroid surface: estimated position (left) and velocity (right)

This is due to the discrepancy between the measurement model in the filter and the actual model, as well to the strong sensitivity of the dynamics to the variation of the conditions and numerical errors which mislead the filter to wrong estimate.

5.3.5 Summary

An important factor is represented by the impact of the environment as well as the system assumptions on the control budget. Of course the number of uncertain parameters which can affect this figure tends to diminish thanks to extensive ground support prior the start of and during operations. Nonetheless good models can help mission designer and system engineers to use more refined values and predict which will be the behavior of the system.

For the above cases, we performed a Monte Carlo analysis consisting in 200 independent realizations where the initial trajectory and initial filter guess was

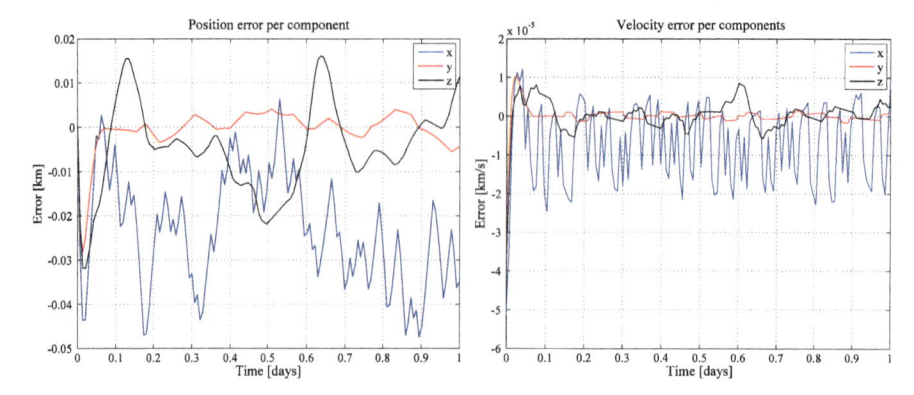

Fig. 13 Case 4 using a 100 faces mesh for the asteroid surface: controlled position (left) and velocity (right)

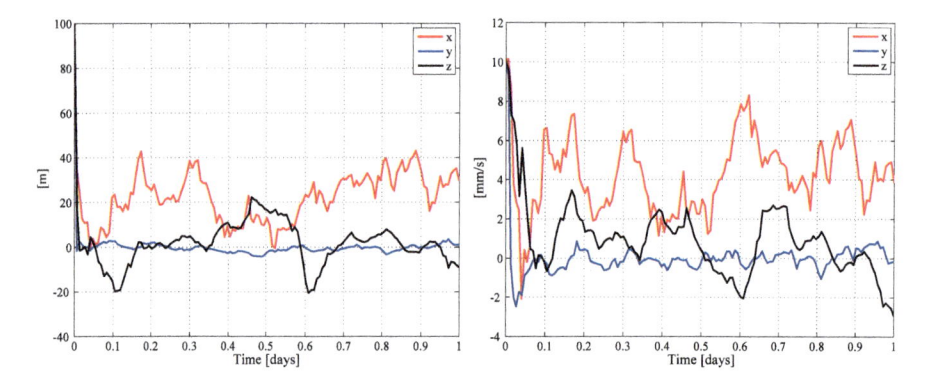

Fig. 14 Case 4 using a 100 faces mesh for the asteroid surface: estimated position (left) and velocity (right)

drawn from a normal distribution with a dispersion of 50 m and 2 cm/s in position and velocity, respectively. We did not consider the 1000 facets polyhedron case because the difference with respect to the 100 facets was marginal, as well for the high computational cost. The effect on the Δv in terms of mean and standard deviation is reported in Table 3. We see that as we move from simpler models to more complex ones, the control budget tends to increase monotonically. Also the dispersion appears to increase, although Case 2 displays a slightly higher dispersion with respect to Case 3, which is caused by the fact that in this configuration the spacecraft is systematically place closer to the surface, thus the deviation are magnified.

If we consider a level of confidence of 99.7 (corresponding to 3-sigma), we see that the maximum difference in control budget is about 5%.

Evaluating Proximity Operations Through High-Fidelity Asteroid Deflection... 41

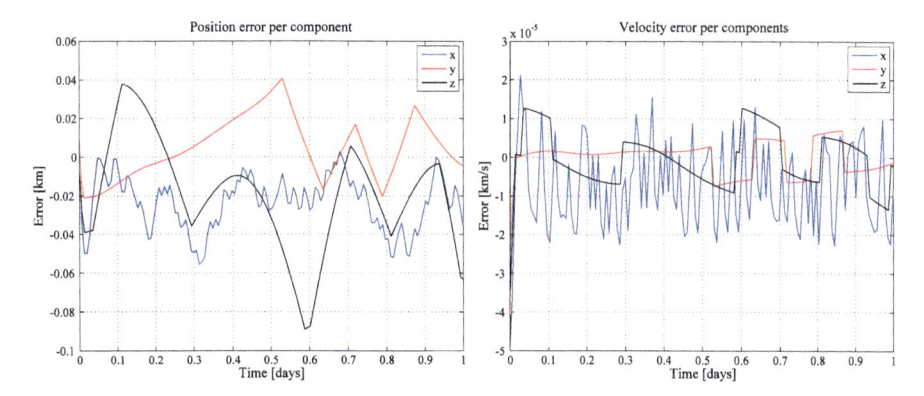

Fig. 15 Case 4 using a 100 faces mesh for the asteroid surface and on-board fixed step integrator: controlled position (left) and velocity (right)

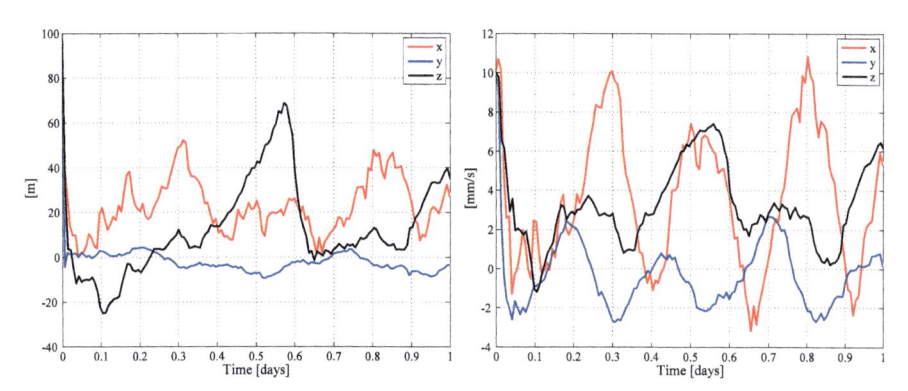

Fig. 16 Case 4 using a 100 faces mesh for the asteroid surface and on-board fixed step integrator: estimated position (left) and velocity (right)

Table 3 Impact of different modelling assumptions of the performance of the control budget

Scenario	Mean Δv [m/s]	1-sigma Δv [m/s]
Case 1	2.1816	0.023907
Case 2	2.1922	0.024455
Case 3	2.2094	0.019932
Case 4	2.2431	0.019559
Case 4 OB	2.2748	0.025643

6 Conclusions

This paper presented a comparative assessment on how the different modelling assumptions can affect the overall control and navigation performance. For this purpose we used the main features of the High-fidelity Asteroid Deflection Evaluation Software developed at Deimos Space S.L.U. for close proximity operations.

In order to stress the effects from the uncertain environment we considered the duck-shaped 67P/Churyumov–Gerasimenko, whose gravity field cannot easily model. In this way we could assess how the knowledge of the environment affects the navigation and the control budget. We showed and compare several cases and we focused on the navigation performance for different level of environment knowledge, assuming shape and harmonics models for the gravity field and the measurements generation. In general if we use a performance model and the environments is quite predictable or well known the performance of the controller does not differ much from the one obtained with relatively more accurate models. When the environment is pretty unknown the difference could differ significantly, although where we place the spacecraft in principle affects the subsequent results. From the analysed case we have seen that the control budget difference can be as high as 5%. Although this level is often absorbed by safety margin which can be 20% for known environment or as high as 100% for highly perturbed and uncertain minor objects, this work and HADES can be applied to several other analysis. For instance if the size of the asteroid is roughly known as well as it composition and rotational motion, we could perform extensive analysis based on different shapes, masses and angular velocities to draw accurate predictions of the control and navigation performance.

Acknowlegements The presented work has been developed within the European Commission's Framework Programme 7 funded Stardust project, through the Stardust Marie Curie Initial Training Network, FP7-PEOPLE-2012-ITN, Grant Agreement 317185.

References

1. Morley, T., Budnik, F.: Rosetta navigation for the fly-by of the asteroid 2867 steins. In: Proceedings of the 21st International Symposium on Space Flight Dynamics. Toulouse, France, 2009
2. Scheeres, D.J., Gaskell, R., Abe, S., Barnouin-Jha, O., Hashimoto, T., Kawaguchi, J., Kubota, T., Saito, J., Yoshikawa, M., Hirata, N., Mukaik, T., Ishiguro, M., Kominato, T., Shirakawa, K., Uo, M.: The actual dynamical environment about Itokawa. In: AIAA/AAS Astrodynamics Specialist Conference and Exhibit, Keystone, Colorado, 21–24 August 2006
3. Scheeres, D.J., Broschart, S., Ostro, S.J., Benner, L.A.: The dynamical environment about asteroid 25143 Itokawa: target of the Hayabusa mission. In: AIAA/AAS Astrodynamics specialist conference and exhibit. Providence, Rhode Island, 16–19 August 2004
4. Kubota, T., Hashimotoa, T., Sawai, S., Kawaguchib, J., Ninomiyaa, K., Uoc, M., Babac, K.: An autonomous navigation and guidance system for MUSES-C asteroid landing. Acta Astronaut. **52**, 125–131 (2003)
5. Li, S., Cui, P., Cui, H.: Autonomous navigation and guidance for landing on asteroids. Aerosp. Sci. Technol. **10**(3), 239–247 (2006)
6. Scheeres, D.J.: Orbital Motion in Strongly Perturbed Environments. Applications to Asteroid, Comet and Planetary Orbiters, 1st edn. Springer-Praxis, Chichester (2011)
7. Montenbruck, O., Gill, E.: Satellite Orbits: Models, Methods and Applications. Springer, Heidelberg (2005)

8. Boyce, W.: Comment on a formula for the gravitational harmonic coefficients of a triaxial ellipsoid. Celest. Mech. Dyn. Astron. **67**(2), 107–110 (1997)
9. Winkler, T., Kaplinger, B., Wie, B.: Optical navigation and fuel-efficient orbit control around an irregular-shaped asteroid. In: AIAA Guidance, Navigation, and Control (GNC) Conference, Boston, USA, 08/2013
10. Simons, D.: Optimal State Estimation, Kalman, and Non-linear Approaches. Wiley, Hoboken (2006)
11. Li, W., Jia, Y.: H-infinity filtering for a class of nonlinear discrete-time systems based on unscented transform. Signal Process. **90**, 3301–3307 (2010)
12. Oh, S.M., Johnson, E.N.: Relative motion estimation for vision-based formation flight using unscented Kalman filter. In: AIAA Guidance, Navigation and Control Conference and Exhibit, Hilton Head, South Carolina, 2007
13. Dionne, K.: Improving autonomous optical navigation for small body exploration using range measurements. In: AIAA 2009-6106. AIAA Guidance, Navigation, and Control Conference, Chicago, Illinois, 10–13 August 2009
14. Lhotka, C., Reimond, S., Souchay, J., Baur, O.: Gravity field and solar component of the precession rate and nutation coefficients of Comet 67P/Churyumov–Gerasimenko. Mon. Not. R. Astron. Soc. **455**(4), 3588–3596 (2015)

Prediction of Orbital Parameters for Undiscovered Potentially Hazardous Asteroids Using Machine Learning

Vadym Pasko

Abstract The purpose of this study is to make a prediction of combinations of orbital parameters for yet undiscovered potentially hazardous asteroids (PHAs) with the use of machine learning algorithms. The proposed approach aims at outlining subgroups of all major groups of near-Earth asteroids (NEAs) with high concentration of PHAs in them. The approach is designed to obtain meaningful results and easy-understandable boundaries of the PHA subgroups in 2- and 3-dimensional subspaces of orbital parameters. Boundaries of these PHA subgroups were found mainly by the use of Support Vector Machines algorithm with RBF kernel. Additional datasets of virtual asteroids were generated to handle sufficient amount of training and test data, as well as to emulate undiscovered asteroids. This synthetic data helped in revealing 'XX'-shaped region with high concentration of PHAs in the (ω, q) plane. Boundaries of this region were used to split all NEAs into several domains. For each domain the subgroups of PHAs were outlined in different subspaces of orbital parameters. Extracted subgroups have high PHA purity ($\sim$90%) and contain $\sim$90% of all real and virtual PHAs. Obtained results can be useful for planning future PHA discovery surveys or asteroid-hunting space missions.

1 Introduction

The increasing rate of asteroid discovery has a limit connected with constraints of using ground-based telescopes for near-Earth asteroid (NEA) observations [1, 2]. Discovery of asteroids with space-based telescopes [3–5], being a promising solution, is currently limited by the low number of hardware launched to date. Development of new asteroid-hunting tools, both ground-based and space-based, requires sophisticated analysis of asteroid trajectories, which can be used for

V. Pasko (✉)
Yuzhnoye State Design Office, Dnipro, Ukraine
e-mail: mail@vadym-pasko.com; keenon3d@gmail.com

efficient survey planning, softening requirements to space-based hardware, and thus reducing the total cost of future asteroid-hunting space missions.

A list of works aimed at revealing true size frequency and orbital distributions of all existing NEAs ([6–10] and others) is growing with an increasing pace. The techniques typically adopted to meet these goals include the characterization of the detection efficiency of a reference survey and subsequent simulated detection of a synthetic population, or the statistical tracking of NEAs from their source regions in the main belt to the inner solar system and subsequent comparison to the detections by a reference survey, or the combination of these two approaches [10].

While each new approach of finding best debiased estimate for the whole NEA population is essential for better understanding the amount of yet undiscovered NEAs of particular sizes [21], successful planning of asteroid surveys requires more thorough analyses, capable of predicting orbital distributions of the NEA subpopulations. These subpopulations or groups[1], namely Atiras, Atens, Apollos and Amors contain asteroids that cross Earth orbit at small distances (less than 0.05 AU) which makes them objects of the top interest, since they may evolve into potential impactors within the foreseeable future. Such objects are known as potentially hazardous asteroids (PHAs) and are defined as asteroids with an Earth Minimum Orbit Intersection Distance (MOID) of 0.05 AU or less and an absolute magnitude H = 22.0 ($\sim$140 m in diameter) or less. The estimation of PHA distributions and prediction of orbital parameters for undiscovered PHAs is placed in the center of the current research.

An effort of analyzing orbital distributions of PHAs has already been made by Mainzer et al. [9], where authors estimated the entire population of PHAs larger than 100 m. In this regard they extended the definition of PHA to include objects with diameters down to 100 m. In the current study the limit on asteroid size is omitted from the PHA definition. This is done with consideration that even small objects (30–50 m or even smaller) can cause major regional damage in the event of an Earth impact [5]. And since the current research is focused on the introduction of a different method of the PHA orbital distribution analysis, the issue of the survey biases that put the limit of 100 m in the work of Mainzer et al. [9] has been left behind the scene, but will be incorporated in the future work. Thus, here and after, any close-approaching asteroid with MOID <0.05 AU is referred as PHA.

While Mainzer et al. [9] treated PHAs as a separate subpopulation of NEAs, in the current research PHAs are examined with respect to each NEA group listed above. The approach of extracting and analyzing smaller subpopulations of NEAs allows to reveal hidden peculiarities in their size and orbital distributions specific to these particular subpopulations, and thus obtain new insights that can be used to build more accurate models. The examples of such analyses includes works of Granvik et al. [11] and Fedorets et al. [12], where authors calculated the population characteristics of the Earth's irregular natural satellites (NES) that are temporarily captured from the NEA population.

---

[1]http://neo.jpl.nasa.gov/neo/groups.html

Being devoted to the analysis of orbital distributions of PHAs inside groups of NEAs, the approach presented in the current paper differs from the common statistical treatment of NEAs and relies mainly on the application of machine learning techniques. It is aimed at revealing correlations between orbital parameters of PHAs that help to outline subgroups inside each group of NEAs with high concentration of PHAs in them. In order to obtain better insight into these correlations the approach incorporates generation of virtual asteroids using simplified models of orbital distributions of all known NEAs.

The majority of related works provide estimations of NEA orbital distributions with regard to reduced subsets of orbital parameters, considering uniform distribution of NEAs by the argument of perihelion and longitude of the ascending node. In the current work all 5 parameters that define heliocentric orbit were taken into consideration, which has been justified be revealing dependency of the PHA orbital distribution on the argument of perihelion and longitude of the ascending node for the Amors and non-uniform distribution of all observed NEAs by the last parameter. Thus, the correlation analyses were performed for pairs of 5 orbital parameters, namely: semi-major axis (a), perihelion distance (q), inclination (i), argument of perihelion (ω) and longitude of the ascending node (Ω).

Considering strict dependence of MOID on these 5 orbital parameters, we can assume that there exists a boundary in the space of these parameters that divides PHAs from non-hazardous asteroids (NHAs). This boundary is essentially a surface of the hypersolid that encompasses all possible combinations of orbital parameters that define PHA in our formulation. In order to obtain meaningful results the process of outlining populations of PHAs, presented here, is based on the application of several consequent cuts of this hypersolid by finding boundaries in low-dimensional projections of NEA orbital distributions (2D and 3D). This approach allows to obtain regions (or subgroups) of high PHA purity (~90%) that together contain ~90% of all existing and hypothetical PHAs. The ensemble of these regions provides a unique insight into the possible residences of yet undiscovered PHAs, which, as believed by the author, can facilitate future discoveries of PHAs.

The structure of the paper is as follows. Section 2 is devoted to a brief overview of the machine learning and its applications in solving modern problems of astronomy. This section also contains brief descriptions of two machine learning algorithms used in the current work. In Sect. 3 we'll dive into the analysis of NEAs' orbital parameters and their correlations for PHAs. Here we will generate virtual asteroids and observe an interesting structure of PHAs in the (ω, q) projection. In the Sect. 4 we will split all NEAs into 4 domains. Section 5 is devoted to the divisions of PHAs from NHAs inside each domain and outlining PHA subgroups. At the end of this section we'll make a summary of the divisions' qualities and purities of the PHA subgroups.

2 Machine Learning in Astronomy

Since the advent of astrophotography and spectroscopy over a century ago, astronomers have faced the challenge of characterizing and understanding vast numbers of asteroids, stars, galaxies and other cosmic populations [13].

Various mathematical methods were invented and applied for interpreting astronomical data. A long way from the least-squares and maximum likelihood to the inverse probability and Bayesian methods have led to a rapid expansion in the diversity of numerical methods that are used nowadays by astronomers for data analysis.

Since the middle of the last century statistical modeling has become a crucial component in building an inference from incomplete observational data. But the continuously growing size, sources and diversity in spectrum of the astronomical data has led to the need of applying different techniques. Over the past two decades a significant progress in data analysis and, particularly in image processing, has been achieved with the use of machine learning, which is essentially the study of software that learns from experience. The machine learning approach is rather new for astronomy but has a great potential in bringing brand new inference to old problems as well as in discovering new dependencies in structure and behavior of celestial objects.

Machine learning is a method of data analysis, aimed at finding hidden insights by the means of using algorithms that iteratively learn from data rather than being explicitly programmed where to look. Machine learning can appear in many guises. But two types of problems in machine learning—classification and clustering refer to the most common problems in astronomy.

2.1 Classification and Support Vector Machines

Classification (or supervised learning) is a classic problem that goes back a few decades. In supervised learning a training set of examples with the correct responses (targets) are provided and, based on this training set, the algorithm generalizes to respond correctly to all possible inputs. A wide range of supervised learning algorithms includes but not limited to k-Nearest Neighbors, Decision Trees, Random Forests and Support Vector Machines (SVM). A tangible difference between all of them is the shape of the decision boundary that the algorithm can learn (to split data of different classes). In a classification problem with two classes, a decision boundary or decision surface is a hypersurface that partitions the underlying vector space (feature space) into two sets, one for each class (Fig. 1).

In high-dimensional spaces, data can more easily be separated linearly and the simplicity of classifiers such as linear SVMs might lead to better generalization than is achieved by other classification algorithms [14]. On the other hand, in low-dimensional spaces smooth nonlinear decision boundaries can provide not only

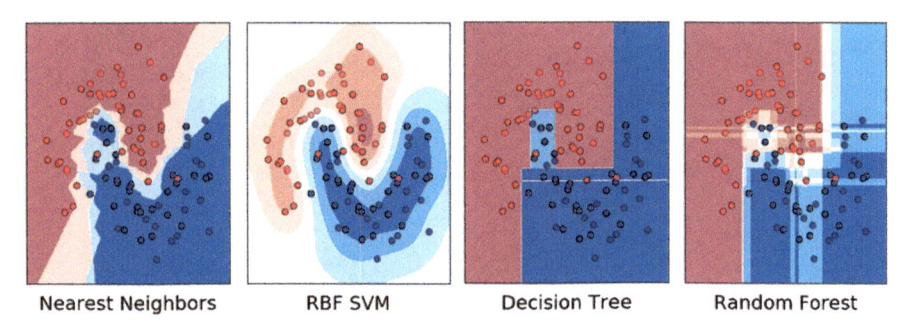

Prediction of Orbital Parameters for Undiscovered PHAs... 49

Fig. 1 Decision boundaries produced by different classification algorithms [14]

Fig. 2 Maximum margin separation of two linearly-separable classes

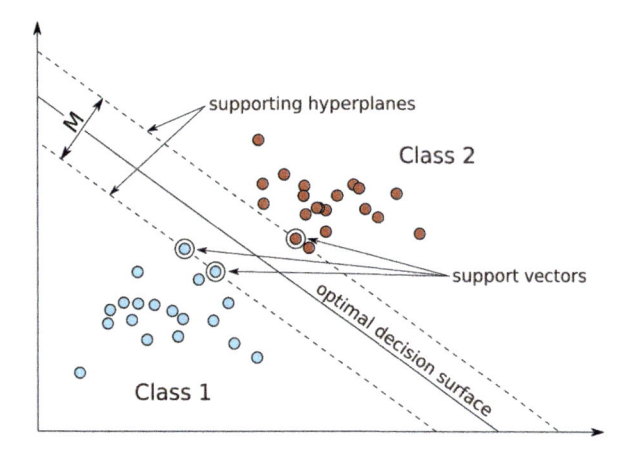

better classification accuracy, but a simpler insight. Nonlinear decision boundaries with controllable smoothness can be provided by Support Vector Machines algorithms that are usually applied in cases of overlapping classes. The core operational principle of all SVM algorithms is to find a hyperplane (or hypersurface in non-linear case) that separates the feature space with the maximum margin (Fig. 2).

The data samples closest to the decision hyperplane are known as support vectors [15]. In the two-class problems the support vectors define two parallel supporting hyperplanes equidistant from the decision hyperplane. Distance between these planes is essentially the margin M that has to be minimized.

The SVMs with non-linear kernel functions produce nonlinear boundaries by constructing a linear boundary in a large, transformed version of the feature space. One of the widely-used nonlinear kernel functions is the Gaussian radial basis function (RBF). Particularly the SVM with RBF kernel is the main tool in the current study.

Selection of the appropriate kernel is not the only way of getting desired behavior of the decision boundary. In some cases there might be a need to push the decision boundary towards one of the data classes. Such a need may arise when one of the

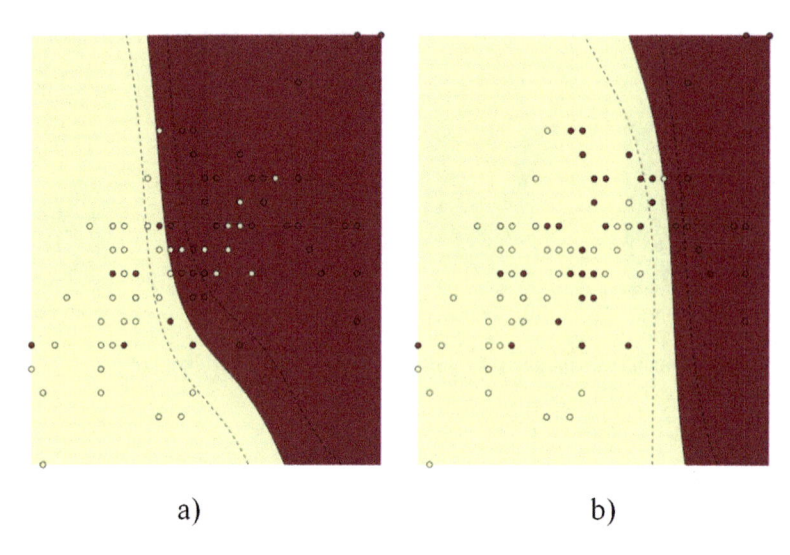

Fig. 3 Decision boundary produced by the RBF SVM for the same training data and different class weights: (**a**)—no class weights assigned, (**b**)—class weight of the yellow (light) samples is higher

classes is more important than the other, or if we want to increase the purity of one class sacrificing the purity of another. This can be achieved by manipulating the class weights (Fig. 3). We will use the trick with assigning different class weights several times in the current work in order to achieve desired classification purity.

Still there are much more options to tune SVM. That is the reason why SVM is a popular choice for classification.

The computational complexity of the SVM algorithms consists of the kernel complexity, which is $O(m^2 n)$ for the RBF kernel, and the factorization complexity, which is $O(m^3)$ in general [16]. Here m is the number of data points (samples) and n is the dimensionality, or in other words—number of features. This is why the SVM is very expensive to use for large datasets. And this is exactly what we will do in the current work. Thus, some classification operations described below require significant computation time, and, depending on the processor and operating system may take up to 1 h or even more.

2.2 Clustering and DBSCAN Algorithm

When working with large datasets it is in most scenarios useful to be able to break data into several groups (clusters) and eventually, to do class identification. This objective can be efficiently achieved with the use of clustering techniques, which are applied to search for groupings of multivariate data points by proximity of objects or other criteria. This task is also known in the world of machine learning as unsupervised learning.

In the unsupervised learning correct responses are not provided, instead the algorithm tries to identify similarities between the inputs so that inputs that have something in common are categorized together in clusters [16].

The wide range of clustering algorithms includes k-means, k-medoids, SNN, MCLUST and many others [17]. In the current work, in order to perform efficient density-based clustering, we will use DBSCAN (Density Based Spatial Clustering of Applications with Noise) algorithm.

The DBSCAN algorithm can identify clusters by looking at the local density of data points [18]. DBSCAN can find clusters of arbitrary shape. However, clusters that lie close to each other tend to belong to the same class.

The central component of the DBSCAN is the concept of core samples, which are samples that are in areas of high density. A cluster is therefore a set of core samples, each close to each other (measured by some distance measure) and a set of non-core samples that are close to a core sample (but are not themselves core samples). There are two parameters to the algorithm, μ—minimal number of samples to form core and ε—radius of the core sample neighborhood. Higher μ or lower ε indicate higher density necessary to form a cluster [14].

More formally, core sample is a sample in the dataset such that there exist μ other samples within a distance of ε, which are defined as neighbors of the core sample. This ensures that the core sample is in a dense area of the vector space. A cluster is a set of core samples that can be built by recursively taking a core sample, finding all of its neighbors that are core samples, finding all of their neighbors that are core samples, and so on. A cluster also has a set of non-core samples, which are samples that are neighbors of a core sample in the cluster but are not themselves core samples. Intuitively, these samples are on the fringes of a cluster.

Any core sample is part of a cluster, by definition. Any sample that is not a core sample, and is distant from any core sample at least at ε, is considered an outlier by the algorithm (Fig. 4).

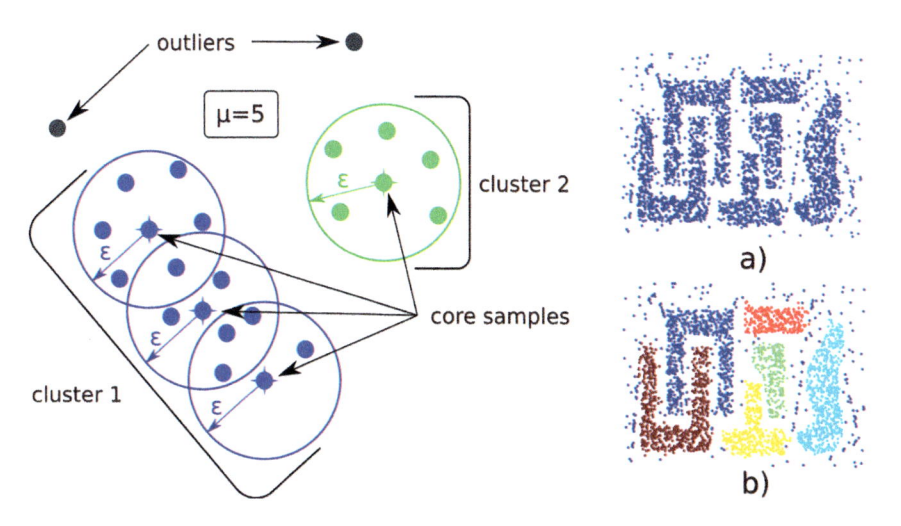

Fig. 4 Formation of clusters with DBSCAN. On the right: (**a**)—original points, (**b**)—clusters

3 Analysis of NEAs' Orbital Distributions

The typical application of machine learning algorithms in astronomy lies in the area of image processing. Automated classification tools have become increasingly useful with the growth of megadatasets in astronomy, often from wide-field surveys of the optical sky. Just as an example, the most elementary need is to discriminate galaxies, which are typically resolved blurry objects, from stars, which are unresolved [13].

Nevertheless, it is far not the only possible application of machine learning techniques in astronomy. Particularly, the interdisciplinary problem of detection, classification and orbit determination of near-Earth asteroids can be solved more efficiently using a unique inference that can be provided by the deep analysis of the existing asteroid database.

Asteroid database that we use in the current work counts over 600,000 asteroids including 14,858 NEAs, and was compiled from the Jet Propulsion Laboratory's Small Body Database and Minor Planet Center by Ian Webster—developer of the Asterank[2] (a web service for ranking asteroids by mining profit). All NEAs present in the database were split into two subsets—PHAs and NHAs by the threshold value of MOID (0.05 AU). Distributions of asteroids were analyzed for all possible combinations of two orbital parameters separately for PHAs and NHAs.

Particularly, orbital distribution of NEAs in the (ω, q) plane reveals correlation for PHAs, that gather into the M-shaped structure (Fig. 5). A similar structure has

Fig. 5 Correlation between two orbital parameters for PHAs. On the figure orange (light) dots—PHAs, blue (dark) dots—NHAs

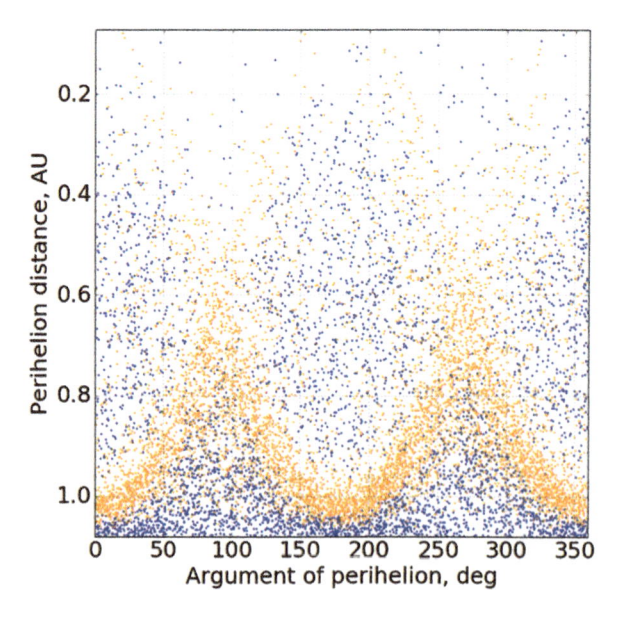

---

[2]http://www.asterank.com/

Table 1 Groups of NEAs

Group name	Definition	Population	Relative population (%)
Atiras	a < 1.0 AU, Q < 0.983 AU	16	0.1
Atens	a < 1.0 AU, Q > 0.983 AU	1087	7
Apollos	a > 1.0 AU, q < 1.017 AU	7968	54
Amors	a > 1.0 AU, 1.017 < q < 1.3 AU	5774	38.9

Q stands for aphelion distance

already been shown in the work of Gronchi and Valsecchi [19], where authors provided a nice explanation of the M-shaped structure for the faint asteroids and its dependency on the orbit distance.

Other pairs of orbital parameters don't provide more distinctive separation of PHAs from NHAs, so this pattern will serve us as a starting point for further divisions.

3.1 Virtual Asteroids

All NEAs, which are essentially asteroids with perihelion distance q < 1.3 AU, are represented by 4 groups[3] with different populations (Table 1). The quality of the analysis with machine learning depends on the number of samples and for small groups like Atiras and Atens it is highly desirable to get more data. More of that, even in the case of more numerous groups, additional data can reveal some yet unseen patterns.

Luckily our response parameter (0 for NHAs and 1 for PHAs) explicitly depends on the input parameters so, we can synthetically increase amount of data by generating virtual asteroids and computing MOID for them. In order to get better insight we will generate two additional datasets of asteroids with different distributions of orbital parameters: one with uniform distribution and another with distributions that approximate distributions of real NEAs. These additional datasets will be referred as uniform and non-uniform respectively.

Despite the constraints on the possible combinations of semi-major axis and perihelion distance for elliptical orbits, we will make an assumption that they are independent and will fix failed generated orbits (with negative eccentricity) by regenerating them. This simple iterative approach will help us preserve physical sense of generated orbits without increasing complexity of the process.

In the case of the non-uniform dataset, first we need to find continuous distributions that approximate distributions of real NEAs. Then virtual asteroids can be generated using these approximations.

---

[3]http://neo.jpl.nasa.gov/neo/groups.html

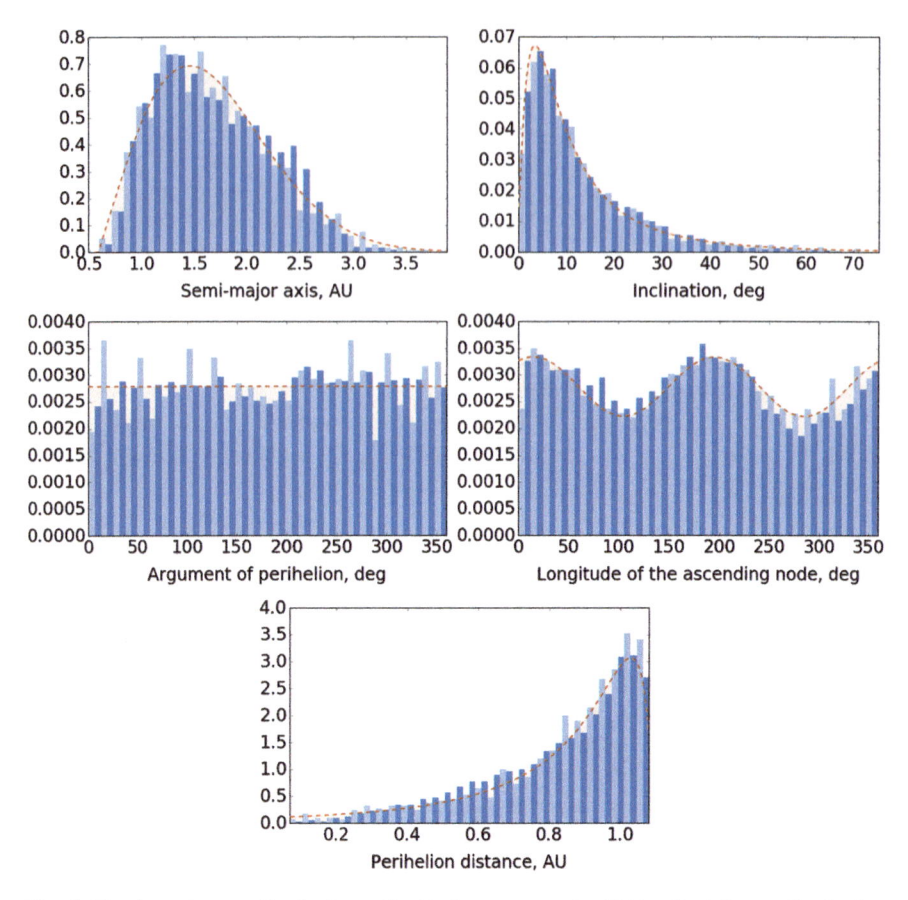

Fig. 6 Fitted continuous distributions of orbital parameters for NEAs. Dark bars—distribution density of real asteroids, light bars—same for virtual asteroids generated using continuous distributions (red dashed lines). Next models were used for approximations: semi-major axis—Rayleigh distribution; inclination—log-normal distribution; argument of perihelion—uniform distribution; longitude of the ascending node—harmonic distribution; perihelion distance—Johnson's SU-distribution

A rich set of probability distributions (81 continuous distributions) embedded in the SciPy library [20] along with the tools for curve-fitting enables a quick and efficient selection of the best models (Fig. 6).

So, by these means additional datasets of 30,000 uniform and 200,000 non-uniform virtual NEAs were generated. The amounts of virtual asteroids were selected from next considerations: to obtain uniformly filled space of orbital parameters without significant increase in density; and to obtain distributions similar to real asteroids but with more than 10 times higher density.

In order to find out PHAs amongst generated asteroids we need to calculate MOID for them. This can be achieved by the use of numerical optimization algorithms. We will use four initial guess points in the minimization problem:

Fig. 7 Positions of the initial guess points on the Earth and asteroid orbits for calculating minimal distances

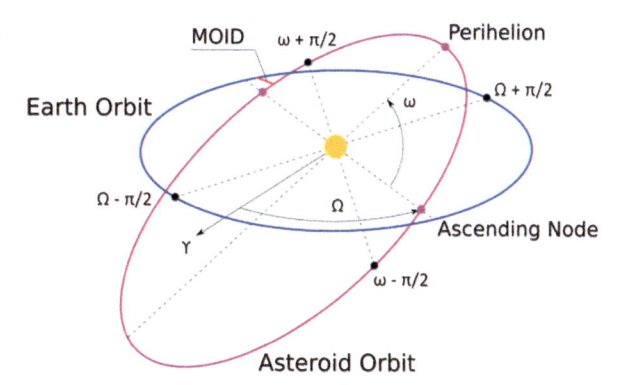

two opposite points on the Earth ellipse, shifted by $\pm\pi/2$ from the direction to the ascending node of the asteroid orbit; and two opposite points on the asteroid ellipse, shifted by $\pm\pi/2$ from the perihelion (Fig. 7). Therefore, we obtain 4 possible combinations for the pairs of initial points, and by iteratively altering their positions on the ellipses will find 4 minimal distances between them. Finally, by definition, MOID is a minimal of four obtained distances.

We use a downhill simplex optimization algorithm to find minimal distances between orbits. This method is commonly applied to find the minimum or maximum of an objective function in a multidimensional space and is embedded in the SciPy library [20] that we use. In our case it's a two-dimensional problem. Computation of MOID has taken 200 s for the uniform dataset and 1447 s for the non-uniform dataset, while running in parallel in 3 threads (CPU Intel Core i7-4510U 2.00GHz×2) on a 64-bit OS Linux Mint 17.2.

After calculating MOID for all virtual asteroids we can split them into the PHA and NHA datasets and take a better look on the distributions of asteroids in the (ω, q) plane. Distributions are shown on the Fig. 7 separately for PHAs (right) and NHAs (left) because of the high distribution density.

Now we are able to see new features of the M-shaped structure of PHAs in a (ω, q) plane. Particularly "upper branches" become visible for both datasets and 'M' morphs to a 'XX' shape, which is better distinguished for the uniform dataset. Areas of extreme PHA purity emerged in the neighborhood of $q = 1\ AU$.

The most obvious inference we get from these pictures is that a large part of PHAs is located in the thin 'XX'-shaped belt. Others are mixed with NHAs and, probably can be separated in other dimensions. According to this hypothesis we will try to split all NEAs into several domains neighboring to the 'XX' structure:

1st domain—what lies under the 'XX' structure;
2nd domain—what is on the left, right and in-between the 'XX' structure;
3rd domain—what is above the 'XX' structure;
4th domain—the 'XX' structure by itself.

4 Finding Boundaries of the NEA Domains

In order to preserve obvious symmetry of the asteroid distributions along the ω axis we will extend generated datasets with symmetric mirrors over vertical planes that cross $\omega = 90°$ and $\omega = 180°$. This operation will increase density of data points and, thus, significantly increase computations time.

As the shape of the 'XX' structure is more accurate for the uniform dataset (Fig. 8), we will use it as the basis for extraction of clusters that represent 4 defined domains. The 1st, 2nd and 3rd domains can be extracted from the dataset of the uniform NHAs by using DBSCAN clustering algorithm. The 4th domain can be

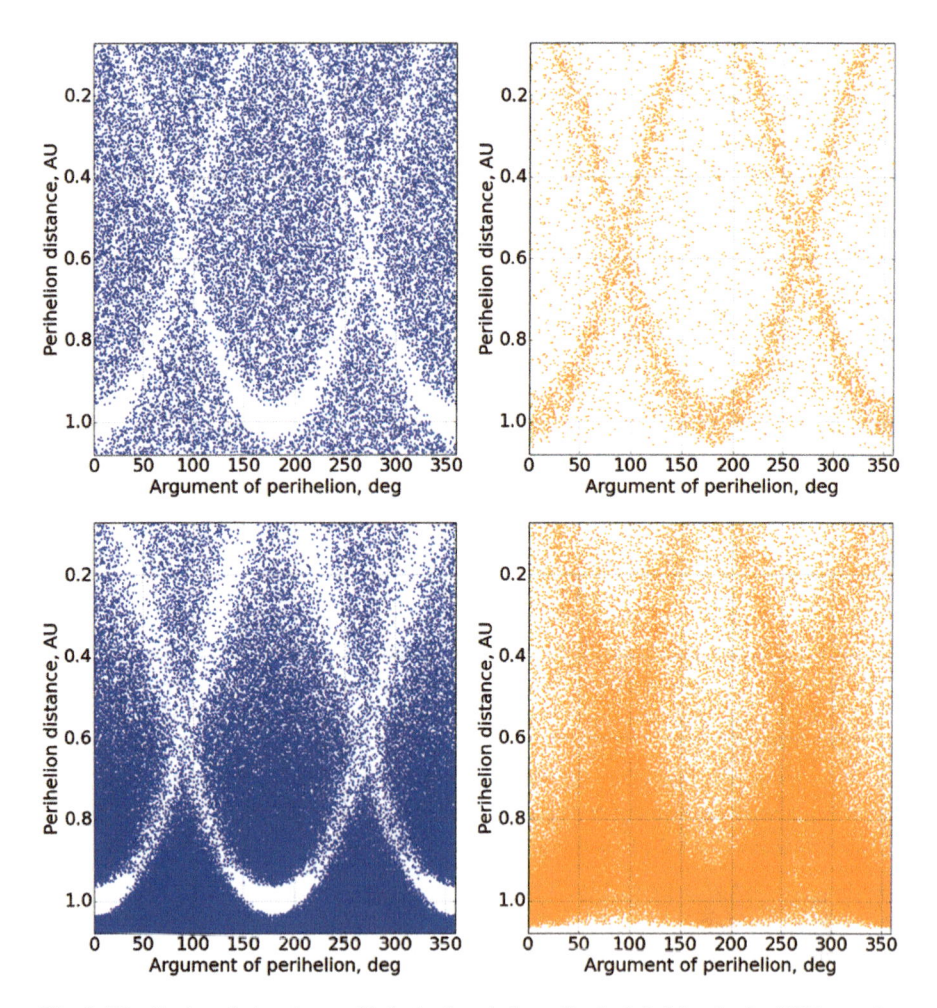

Fig. 8 Distribution of virtual asteroids in the (ω, q) plane. On the left (blue dots)—NHAs, on the right (orange dots)—PHAs. On the top—uniform virtual asteroids, on the bottom—non-uniform virtual asteroids

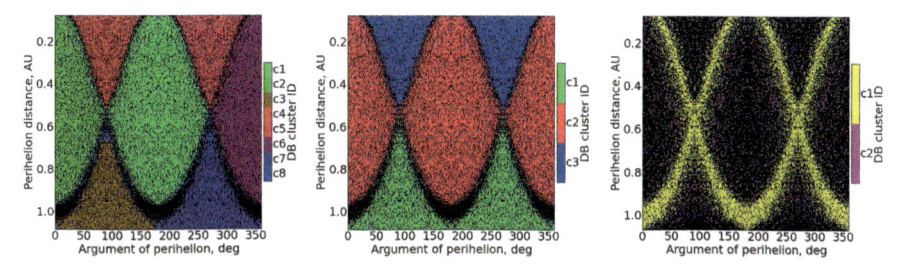

Prediction of Orbital Parameters for Undiscovered PHAs... 57

Fig. 9 Clusters found with DBSCAN. On the left—original 8 clusters (8th is a collection of outliers) found by DBSCAN in the uniform virtual dataset of NHAs; in the center—manually rearranged and merged clusters; on the right—original 'XX' cluster found by DBSCAN in the uniform virtual dataset of PHAs (second cluster is a collection of outliers)

easily extracted by the same means from the dataset of uniform PHAs. As we are using DBSCAN we can't explicitly control the number of generated clusters, what can be done with other clustering algorithms like k-means. So, selection of different values for μ and ε can lead to different number of clusters found.

The selected compromise values of $\mu = 105$ and $\varepsilon = 0.022$ provide us with 8 clusters of NHAs instead of desired 3 (Fig. 9 left). But this can be easily fixed by splitting and merging clusters (Fig. 9 center). What is more important is that we obtain clusters of the desired shape.

Three clusters from the NHA dataset overlap with the 'XX' cluster from the PHA dataset (Fig. 9 right). In order to find a smooth boundary between them we will apply SVM algorithm with RBF kernel. The SVM algorithm implemented in the scikit-learn package can handle multiclass classifications, so we will use cluster IDs as a class reference to find boundaries between 4 desired domains.

The implementation of RBF SVM in the scikit-learn uses two input parameters to control the decision surface shape: C and γ. Intuitively, the γ parameter defines how far the influence of a single training example reaches, with low values meaning 'far' and high values meaning 'close'. The γ parameters can be seen as the inverse of the radius of influence of samples selected by the model as support vectors. The C parameter trades off misclassification of training examples against simplicity of the decision surface. A low C makes the decision surface smooth, while a high C aims at classifying all training examples correctly by giving the model freedom to select more samples as support vectors [14].

So after a series of trials the SVM with values of $C = 10,000$ and $\gamma = 6$ has been selected. It produces smooth boundaries between domains (Fig. 10 right).

Now we can use trained SVM to 'predict' domain membership of any NEA. By passing dataset of real NEAs we can estimate PHA purities of the obtained domains (Fig. 10 right).

It is quite remarkable, that the estimated purity of the 4th domain for real NEAs reaches 0.93. Purities of other domains are significantly lower. In the next sections we will try to separate PHAs from NHAs in these domains in different dimensions and will use the non-uniform dataset of virtual asteroids as a training data for SVMs.

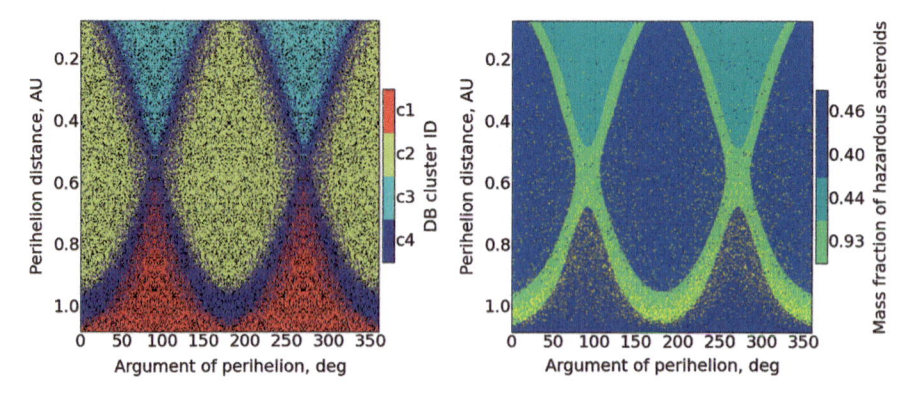

Fig. 10 Classification of NEAs by cluster IDs. On the left—clusters extracted with DBSCAN; on the right—domains outlined by RBF SVM and distribution of real NEAs plotted over (yellow dots—PHAs, blue dots—NHAs)

5 Extraction of PHA Subgroups From NEA Domains

In each domain we will find representatives of the NEA groups (Table 1) and work with them separately (except Atiras and Atens). This approach was proven to be the most successful by numerous failures in trying different strategies. Thus, we will extract subgroups with high PHA purity for each group of virtual non-uniform NEAs.

The representatives of the Amor group are present only in the 1st of 3 'XX'-neighboring domains. Asteroids of other groups are present in all 3 domains.

In some cases we will split PHAs from NHAs linearly, but in most cases we will use RBF SVM to find decision surfaces in 2- and 3-dimensional projections of asteroids' orbital distributions. The summary on the qualities of the divisions made in each domain is represented at the end of the section.

5.1 Domain #1

The Atiras & Atens in the 1st domain can be easily separated by the SVM with a linear kernel in the (a, i) plane (Fig. 11).

Two divisions were applied for the Amors. First—in the (ω, i) plane to separate most part of NHAs. Second decision surface was learned by the RBF SVM in the (ω, Ω, q) space for those classified as PHAs in the first division (Fig. 12). It separates only a half of PHAs with sufficient purity. The dependence of the PHA distribution on the longitude of the ascending node can be explained by the eccentricity of the Earth's orbit and its influence on the values of MOID for asteroids with outer orbits.

Two consequent divisions by RBF SVMs were applied for the Apollos: first in the (ω, q, i) space and the second in the (ω, q, a) space (Fig. 13). Second surface covers most part of PHAs left above the first surface. SVM parameters are presented in the Table 2.

Prediction of Orbital Parameters for Undiscovered PHAs... 59

Fig. 11 Division of the Atiras & Atens into the PHA and NHA regions with linear SVM. Green (light) area—PHA region, blue (dark) area—NHA region; yellow dots—PHAs, blue dots—NHAs

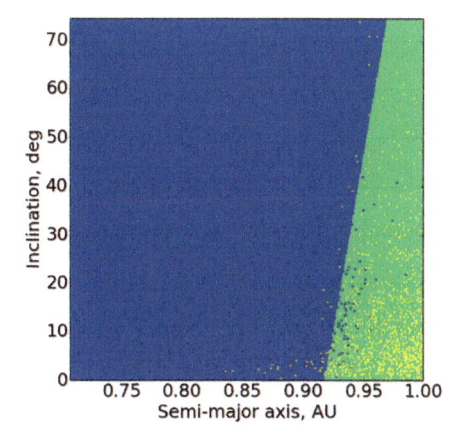

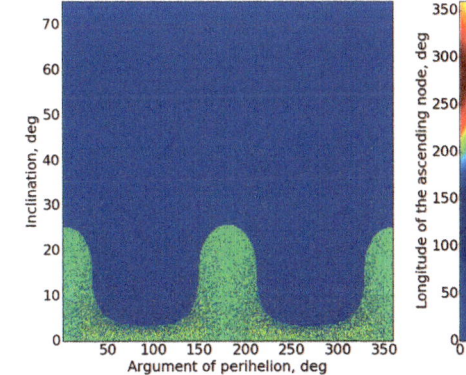

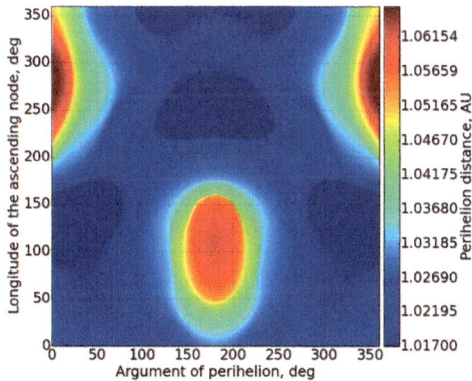

Fig. 12 Decision surfaces between PHAs and NHAs for the Amor group. The surface on the right covers approximately a half of PHAs from the PHA region (green) on the left

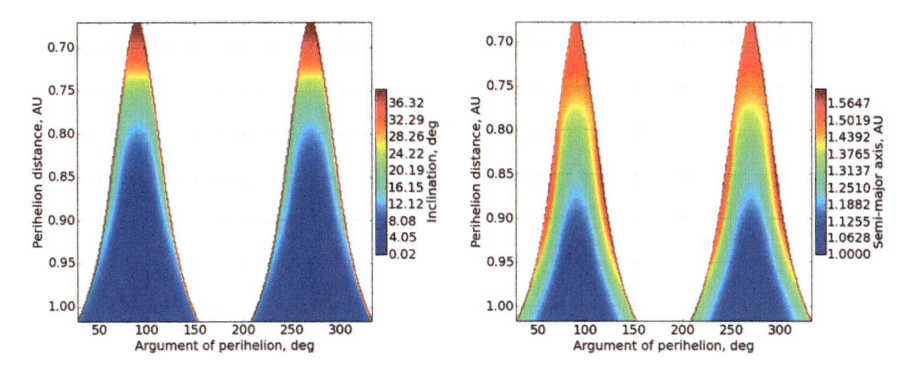

Fig. 13 Two decision surfaces for Apollos. PHA regions are below the surfaces. The second surface (on the right) covers PHAs that are above the first surface (on the left)

Table 2 SVM parameters for divisions in the 1st domain

NEA group	Division	Space	Kernel	γ	C	Class weight
Atiras & Atens	1	(a, i)	Linear	—	1	Equal
Amors	1	(ω, Ω, q)	RBF	20	8	NHA: 2.4
Apollos	1	(ω, q, i)	RBF	40	0.05	NHA: 1.2
	2	(ω, q, a)	RBF	40	0.1	NHA: 1.5

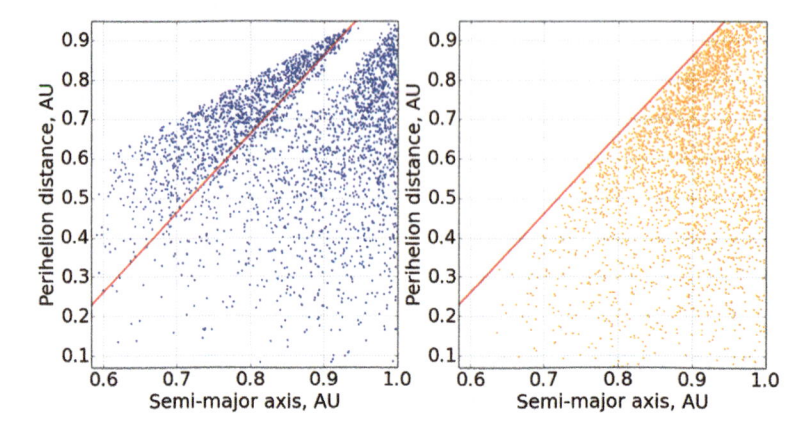

Fig. 14 Atiras & Atens of the 2nd domain. Red line—NHA division plane

5.2 Domain #2

A piece of NHAs of Atiras & Atens in the 2nd domain can be linearly separated from other asteroids. Despite the simplicity of such operation, it has turned out to be a challenging task for the linear SVM. So we make the section manually.

PHAs and NHAs below the red line on the Fig. 14 can be divided by a complex surface in the (ω, a, i) space, produced by the RBF SVM (Fig. 15 left). A small portion of asteroids misclassified by this division gather into a strap in the (a, q) plane. This strap can be outlined by another RBF SVM (Fig. 15 right).

The Apollo asteroids in the 2nd domain can be efficiently divided into PHA and NHA regions by two consequent splits with RBF SVMs—the first in the (ω, i) plane and the second for those left after the first split—in the (ω, q, i) space (Fig. 16). SVM parameters are presented in the Table 3.

5.3 Domain #3

The 3rd domain includes Atiras, Atens and Apollos. Atiras & Atens can be easily divided into PHAs and NHAs by applying single split in the (ω, i) plane with RBF SVM (Fig. 17 left). The boundary between PHAs and NHAs for the Apollo

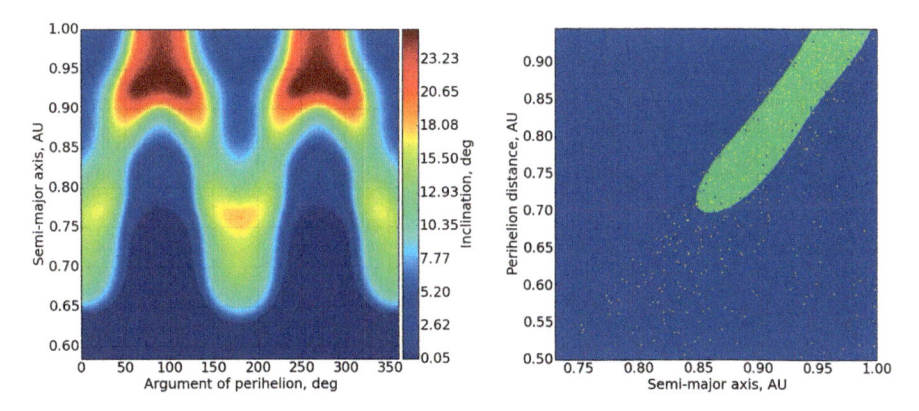

Prediction of Orbital Parameters for Undiscovered PHAs... 61

Fig. 15 PHA regions found by the RBF SVMs for Atiras & Atens

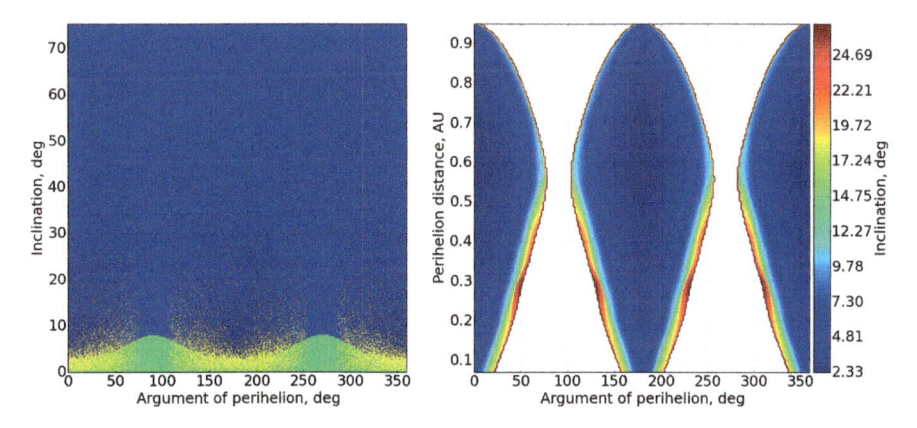

Fig. 16 Two divisions of Apollos. On the right—PHA region is filled with green (light), yellow dots are PHAs and blue dots are NHAs. The surfaces on the left cover PHAs left in the NHA region from the first split (blue area on the left)

Table 3 SVM parameters for divisions in the 2nd domain

NEA group	Division	Space	Kernel	γ	C	Class weight
Atiras & Atens	1	(ω, a, i)	RBF	80	0.1	NHA: 1.5
	2	(a, q)	RBF	8	1000	NHA: 1.5
Apollos	1	(ω, i)	RBF	30	0.1	NHA: 10
	2	(ω, q, i)	RBF	100	2	NHA: 1.5

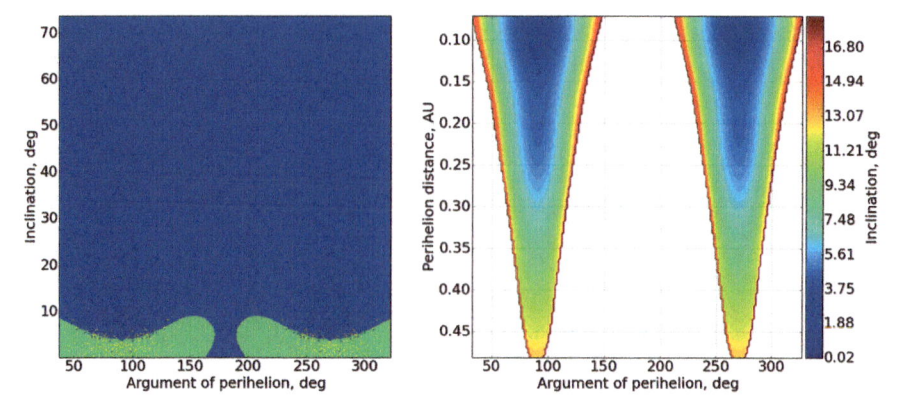

Fig. 17 PHA regions found by RBF SVMs for Atiras & Atens (green area on the left) and Apollos (space under the surface on the right)

Table 4 SVM parameters for divisions in the 3rd domain

NEA group	Division	Space	Kernel	γ	C	Class weight
Atiras & Atens	1	(ω, i)	RBF	80	0.4	NHA: 1.1
Apollos	1	(ω, q, i)	RBF	20	0.5	Equal

asteroids resides in the (ω, q, i) space (Fig. 17 right). SVM parameters are presented in the Table 4.

5.4 Assessment of the Divisions' Qualities

The worst-separable population of asteroids is the group of Amors in the 1st domain. Only 42% of virtual PHAs were separated by the surface in the (ω, Ω, q) space. Other hard-separable populations of asteroids belong to the 2nd domain. In other cases the fraction of correctly classified PHAs is close to 90%.

Most part of divisions was made by training SVM with RBF kernel, while in some cases a linear kernel was used, and once a manual linear separation was made. The non-uniform dataset of virtual asteroids was used for training SVMs. This dataset along with the dataset of real asteroids were used to estimate qualities of the divisions and purities of the PHA subgroups. The summary is depicted in the Table 5.

6 Conclusions

Generation of virtual asteroids and analysis of their orbital distributions revealed a new shape of the known 'M' structure of PHAs in the (ω, q) plane, which morphs into the 'XX' structure with the increase of samples. This 'XX' structure contains

Table 5 Divisions summary

Domain	Groups of NEAs	GW	N	PHA fraction	NHA fraction	PHA purity
1	Atiras & Atens	0.02 (0.01)	1	0.96 (0.97)	0.15 (0)	0.92 (1)
	Apollos	0.63 (0.69)	2	0.93 (0.97)	0.08 (0.12)	0.91 (0.93)
	Amors	0.35 (0.3)	1	0.42 (0.48)	0.01	0.88 (0.91)
	All NEAs	1	4	**0.86 (0.9)**	**0.05 (0.06)**	**0.91 (0.93)**
2	Atiras & Atens	0.09 (0.3)	3	0.86 (0.78)	0.1	0.9 (0.93)
	Apollos	0.91 (0.7)	2	0.88 (0.83)	0.02	0.95 (0.96)
	All NEAs	1	5	**0.88 (0.81)**	**0.03**	**0.94 (0.95)**
3	Atiras & Atens	0.11 (0.22)	1	0.87 (1)	0.03 (0)	0.9 (1)
	Apollos	0.89 (0.78)	1	0.92	0.05 (0.03)	0.94 (0.97)
	All NEAs	1	2	**0.92**	**0.05 (0.02)**	**0.94 (0.97)**
4	**All NEAs**	1	0	1	1	**0.9 (0.93)**
Total			11	0.93	**0.09 (0.1)**	**0.91 (0.93)**

GW stands for a group weight and N for the number of divisions. The fraction of correctly classified PHAs (PHA fraction) was calculated with regard to the total number of PHAs in a group or domain. Same is applied to the fraction of misclassified NHAs (NHA fraction). PHA purity represents a cumulative purity of all PHA regions outlined for a group or domain. Values without parentheses correspond to the virtual non-uniform NEAs and values in parentheses correspond to the real NEAs. If a value for real NEAs is not provided it is the same as for virtual NEAs

approximately a half of all real and virtual PHAs and its purity is around 0.9 (0.93 for real NEAs). The boundaries of the 'XX' structure, found with the application of the Support Vector Machines algorithm, divide all NEAs into 3 domains, while the 4-th domain is the region inside the 'XX'.

Other domains were analyzed in details and the effort of outlining 2- or 3-dimensional regions with high PHA concentration in them was more or less successful. The representatives of the main NEA groups were analyzed separately in each of 3 domains. In most cases separations of PHAs from NHAs were effective, producing PHA subgroups with high purity.

The analysis of PHAs in the Amor group revealed the dependency of PHA distribution density on the argument of perihelion and longitude of the ascending node. This may be explained by the eccentricity of the Earth's orbit and will be covered in more details in the future work.

The summary of all divisions (Table 5) shows that the proposed approach allows to group over 90% of all real and virtual PHAs into regions with ~90% purity. This essentially means that dominant part of all yet undiscovered PHAs resides in these regions. While the knowledge of the shapes of PHA regions can be useful for planning future PHA discovery surveys and future asteroid-hunting space missions, yet some work has to be done to verify obtained results. Particularly, the original dataset of NEAs contains survey biases that may influence the shape of obtained PHA regions. This issue will be addressed in the future work and obtained results will be tested against debiased model of NEA orbital distribution ([7] or other). In the case of significant influence of the survey biases the similar method will be applied to the debiased data to correct the shapes of the PHA regions, preserving their purity.

Tools Used and the Source Code All computations carried out in the frame of the current work were made with the use of Python programming language and open tools for numeric computations (NumPy, SciPy), data analysis (Pandas, SciPy), machine learning (Scikit-learn) and data visualization (Matplotlib).

The code is organized as a collection of Python modules and Jupyter Notebooks as a separate open-source project named Asterion. The project was initiated at NASA Space Apps Challenge global hackathon in April 2016 and became a global finalist in the nomination "Best use of data". The code can be accessed at GitHub[4].

Acknowledgments The author is grateful to: the organizers of the global hackathon NASA Space Apps Challenge 2016 for offering awesome challenges, one of which inspired him for the current research; the officials of Kirovograd Flight Academy of the National Aviation University, who organized and hosted first-ever Space Apps Challenge in Ukraine, particularly—Alexey Izvalov and Sergey Nedelko; the members of the team Asterion—CYA, particularly—Eugene Scherbina and Andrij Blakitnij, who's hard labor and enthusiasm pushed the boundaries of impossible; Ian Webster—developer of Asterank for collecting and sharing the asteroid database and for collaborating with the author on including PHA ranking features to the service, Giovanni F. Gronchi from the University of Pisa for sharing his paper, Carrie R. Nugent from IPAC/Caltech for referring to important papers and pointing the need of testing obtained results on the debiased data, which will be the subject of the future work. The author would also like to show his gratitude to all contributors to the Scikit-learn project—open-source software for machine learning that provides easy access to sophisticated math and encourages experimenting with different algorithms.

References

1. Beeson, C.L., Elvis, M., Galache, J.L.: Scaling near Earth asteroid (NEA) characterization rates. Harv. Undergrad. Res. J. Astrophys. **6**(1), (2013). http://thurj.org/research/2013/05/4458/
2. Galache, J.L., Beeson, C.L., McLeod, K.K., Elvis, M.: The need for speed in near-Earth asteroid characterization. Planet. Space Sci. **111**, 155–166 (2015.) https://arxiv.org/pdf/1504.00712.pdf
3. Mainzer, A., Grav, T., Bauer, J., Conrow, T., Cutri, R.M., Dailey, J., Fowler, J., Giorgini, J., Jarrett, T., Masiero, J., Spahr, T., Statler, T., Wright, E.L.: Survey simulations of a new near-Earth asteroid detection system. Astron. J. **149**(5), 172 (2015.) 17pp
4. Myhrvold, N.: Comparing NEO search telescopes. Publ. Astron. Soc. Pac. **128**(962), 045004 (2016.) http://iopscience.iop.org/article/10.1088/1538-3873/128/962/045004/pdf
5. Shao, M., Turyshev, S.G., Spangelo, S., Werne, T., Zhai, C.: A constellation of SmallSats with synthetic tracking cameras to search for 90% of potentially hazardous near-Earth objects. Astron. Astrophys. **603**, A126 (2017)
6. Bottke, W.F., Morbidelli, A., Jedicke, R., Petit, J.M., Levison, H.F., Michel, P., Metcalfe, T.S.: Debiased orbital and absolute magnitude distribution of the near-Earth objects. Icarus. **156**(2), 399–433 (2002)
7. Granvik, M., Morbidelli, A., Jedicke, R., Bolin, B., Bottke, W.F., Beshore, E., Vokrouhlický, D., Delbò, M., Michel, P.: Super-catastrophic disruption of asteroids at small perihelion distances. Nature. **530**, 303–306 (2016)

---

[4]https://github.com/nomad-vagabond/asterion

8. Jedicke, R., Bolin, B., Granvik, M., Beshore, E.: A fast method for quantifying observational selection effects in asteroid surveys. Icarus. **266**, 173–188 (2016)
9. Mainzer, A., Grav, T., Masiero, J., Bauer, J., McMillan, R.S., Giorgini, J., Spahr, T., Cutri, R.M., Tholen, D.J., Jedicke, R., Walker, R., Wright, E., Nugent, C.R.: Characterizing subpopulations within the near-Earth objects with NEOWISE: preliminary results. Astrophys. J. **752**(2), 110 (2012.), 16pp
10. Tricarico, P.: The near-Earth asteroid population from two decades of observations. Icarus. **284**, 416–423 (2017)
11. Granvik, M., Vaubaillon, J., Jedicke, R.: The population of natural Earth satellites. Icarus. **218**, 262–277 (2012)
12. Fedorets, G., Granvik, M., Jedicke, R.: Orbit and size distributions for asteroids temporarily captured by the Earth-Moon system. Icarus. **285**, 83–94 (2017)
13. Feigelson, E.D., Babu, G.J.: Modern Statistical Methods for Astronomy: With R Applications. Cambridge University Press, New York (2012)
14. Scikit-learn official website. http://scikit-learn.org/stable/
15. Harrington, P.: Machine Learning in Action. Manning, Greenwich (2012)
16. Marsland, S.: Machine learning: an algorithmic perspective Machine Learning & Pattern Recognition, 2nd edn. Chapman & Hall/CRC, Boca Raton (2014)
17. Madhulatha, T.S.: An overview on clustering methods. IOSR J Eng. **2**(4), 719–725 (2012.) https://arxiv.org/ftp/arxiv/papers/1205/1205.1117.pdf
18. Ester, M., Kriegel, H.-P., Sander, J., Xu, X.: A density-based algorithm for discovering clusters in large spatial databases with noise. In: Proceedings of the 2nd ACM International Conference on Knowledge Discovery and Data Mining (KDD), Portland, OR, pp. 226–231. (1996). http://www.dbs.ifi.lmu.de/Publikationen/Papers/KDD-96.final.frame.pdf
19. Gronchi, G.F., Valsecchi, G.B.: On the possible values of the orbit distance between a near-Earth asteroid and the Earth. Mon. Not. R. Astron. Soc. **429**, 2687–2699 (2013)
20. Blanco-Silva, F.J.: Learning SciPy for Numerical and Scientific Computing. Packt Publishing. www.packtpub.com (2013)
21. Schunová-Lilly, E., Jedicke, R., Vereš, P., Denneau, L., Wainscoat, R.J.: The size-frequency distribution of H > 13 NEOs and ARM target candidates detected by Pan-STARRS1. Icarus. **284**, 114–125 (2017)

Part II
Orbit and Uncertainty Propagation

Exploring Sensitivity of Orbital Dynamics with Respect to Model Truncation: The Frozen Orbits Approach

Martin Lara

Abstract The mathematical model used in orbit determination problems must be as close to the actual dynamics as possible. On the contrary, accuracy constraints can be notably relaxed for orbit prediction purposes. For the latter, it is important to determine which is the simplified dynamical model that, while retaining the bulk of the dynamics, allow for faster predictions. Methods for doing that are commonly heuristic. We focus on perturbed Keplerian motion and explore how to ascertain the correct truncation of the dynamical model required in orbit prediction problems by investigating relevant particular solutions of the orbital motion: the so-called frozen orbits.

1 Introduction

The efficiency of an orbit propagator program depends on a variety of facets, like the method used in the orbit modeling (Cowell, Encke, variation of parameters, ...) or the numerical method used in the integration of the differential equations of the flow (multi-step, Runge-Kutta, ...). But it obviously depends also on the dynamical model used for the propagations. Setting a propagation model as much complete as possible could seem the ideal situation. However, depending on orbit geometry, using a full dynamical model may slow down computations by carrying out unnecessary mathematical operations in the evaluation of negligible effects.

The sensitivity of an orbit propagation program with respect to different truncations of the model is customarily assessed by a preliminary inspection of the magnitudes of the different model parameters followed by a variety of propagation trials that may confirm the adequacy of the truncation, in which the addition of as much zonal terms as possible is encouraged [1]. In the case of the Geopotential, this procedure normally offers good results for orbits with low eccentricities,

---

M. Lara (✉)
GRUCACI, University of La Rioja, C/Madre de Dios, 26006 Logroño, Spain

Space Dynamics Group, UPM, Pza. del Cardenal Cisneros, 28040 Madrid, Spain
e-mail: mlara0@gmail.com

but it may be harder to evaluate the influence of the different harmonics on highly eccentric orbits. In reference to perturbed Keplerian motion, we propose to replace the numerical integration tests by the simple evaluation of certain analytical expressions. Indeed, we focus on the zonal model of the Geopotential and show how the simple drawing of inclination-eccentricity diagrams of frozen orbits, which are effortlessly evaluated from standard analytical expressions, can be used to disclose the sensitivity of orbits with different characteristics with respect to the model truncation.

The need of including the effects of the zonal harmonics of higher degree in the frozen orbits problem was pointed out by Rosborough and Ocampo [2] (see also the interesting discussion by Cook [3]). Based on Kaula functions for expressing the gravitational potential in orbital elements [4], these authors derived a general constraint equation for the frozen orbit geometry including any number of zonal harmonics. However, their constraint equation is restricted to the case of small eccentricities. Later, new relations for determining frozen orbit configurations including the effects of higher degree harmonics were derived without limiting to the low eccentricity case [5]. These relations showed useful in the systematic construction of inclination-eccentricity diagrams of frozen orbits, a procedure that revealed definitive in assessing the degree of the Selenopotential truncation required for a realistic modeling of high inclination, low altitude lunar orbits [6] (see also [7]).

As an alternative to equations based on Kaula-type functions, the present computational power and software makes the brut force approach feasible when dealing with higher degrees of the gravitational potential [8, 9]. Still, none of the mentioned studies take second order effects of J_2 into account, an approach that may be correct when dealing with lunar orbits, but cannot be accepted in the case of the Geopotential, where second order effects of J_2 may be crucial to the elliptic frozen orbits behavior [10, 11].

The current investigation revisits the frozen orbits problem with the aim of using inclination-eccentricity diagrams of frozen orbits as a general tool for exploring the sensitivity of the orbital dynamics with respect to Geopotential truncation. The constraint equation on the frozen orbit geometry derived from Kaula functions is complemented with the literal expressions that encapsulate the second order effects of J_2, which, besides, include the necessary long-period effects that make the mean elements to be "centered" [12]. With the new formulation, the instant rendering of inclination-eccentricity diagrams of earth's frozen orbits provides a fast an efficient way of determining the zonal model truncation that may be required for a given orbital configuration.

2 Physical Model

For the sake of alleviating notation, the following convention is used for a generic summation index i and integer m

$$i^\star = i \bmod 2, \qquad i_\pi = \frac{\pi}{2} i^\star, \qquad i_m = \left\lfloor \frac{i-m}{2} \right\rfloor, \qquad i_m^\star = i_m + i^\star, \qquad (1)$$

where $\lfloor p/q \rfloor$ denotes the integer division of the integers p and q. Note that $i_m - n = i_{m+2n}$, with m, n, integers.

The zonal Hamiltonian is written

$$\mathcal{H} = -\frac{\mu}{2a} - \frac{\mu}{r}\frac{a}{r}\eta \sum_{i\geq 2} V_i \qquad (2)$$

where μ is the earth gravitational parameter, a is the orbit semi-major axis, r is the satellite's radius from the earth's center, $\eta = (1-e^2)^{1/2}$ is the eccentricity function, e is the orbit eccentricity, and, following Kaula developments [4],

$$V_i = \frac{R_\oplus^i}{a^i}\frac{C_{i,0}}{\eta^{2i-1}} \sum_{j=0}^{i} \mathcal{F}_{i,j}(I) \sum_{k=0}^{i-1} \binom{i-1}{k} e^k \cos^k f \cos[(i-2j)(f+\omega) - i_\pi], \qquad (3)$$

in which $R_\oplus$ is the earth's equatorial radius, $C_{i,0}$ is the zonal harmonic coefficient of degree i, I is the orbital inclination, f is the true anomaly of the satellite, ω is the argument of the perigee, and $\mathcal{F}_{i,j}(I)$ are Kaula inclination functions. In the particular case of the zonal problem, these inclination functions are computed from

$$\mathcal{F}_{i,j} = \sum_{l=0}^{\min(j,i_0)} \frac{(-1)^{j-l-i_0}}{2^{2i-2l}} \frac{(2i-2l)!}{l!(i-l)!(i-2l)!} \binom{i-2l}{j-l} s^{i-2l}, \quad i \geq 2l, \qquad (4)$$

in which s is hereafter used to abbreviate the sine of the inclination.

Since we rely on Hamiltonian formalism, all the symbols that enter Eqs. (2)–(4) are assumed to be functions of certain set of canonical variables rather than variables by themselves. In particular, we use Delaunay variables (ℓ, g, h, L, G, H) and hence

$$a = L^2/\mu, \quad \eta = G/L, \quad I = \arccos H/G, \quad \omega = g, \quad f \equiv f(M, G, L),$$

where M is the mean anomaly. The true anomaly is an implicit function of M which involves the solution of the Kepler equation, while the right ascension of the ascending node $h = \Omega$ is absent from the Hamiltonian because of the axial symmetry of the zonal problem.

Analytical solutions to Eq. (2) are not known, but, under certain conditions, useful analytical approximations can be computed by perturbation methods. Thus, a transformation $(\ell, g, h, L, G, H) \rightarrow (\ell', g', h', L', G', H')$ from osculating to mean elements removes, up to some truncation order, short period terms from the original Hamiltonian, yielding

$$\mathcal{H}' = \langle\mathcal{H}\rangle_M + \Delta + \mathcal{O}(C_{2,0}^3) = -\frac{\mu}{2a} - \frac{\mu}{a}\sum_{i\geq 3}\langle V_i\rangle_f + \Delta + \mathcal{O}(C_{2,0}^3), \qquad (5)$$

where terms $\langle V_i \rangle_f$ are taken from Kaula [4], and Δ comprises second order terms of $C_{2,0}$ which are borrowed from [13]. In particular,

$$\langle V_i \rangle_f = \frac{R_\oplus^i}{a^i} C_{i,0} \sum_{j=0}^{i_2} (2 - \delta_{j+i^\star,0}) \mathcal{G}_{i,i_0^\star+j} \mathcal{F}_{i,i_0^\star+j} \cos[(2j+i^\star)\omega + i_\pi], \qquad (6)$$

in which $\delta_{i,j}$ is the usual Kronecker delta function, and

$$\mathcal{G}_{i,j} = \frac{1}{\eta^{2i-1}} \sum_{l=0}^{k-1} \binom{i-1}{q} \binom{q}{l} \frac{e^q}{2^q}, \quad q = 2l+i-2k, \quad \begin{cases} j \le \frac{i}{2} \Rightarrow k = j \\ j > \frac{i}{2} \Rightarrow k = i-j \end{cases},$$

are Kaula's eccentricity functions for the particular case of the zonal problem, cf. Eq. (3.66) of [4]. Besides

$$\Delta = -\frac{1}{2} \frac{\mu}{a} \frac{R_\oplus^4}{a^4} \frac{1}{\eta^7} \left\{ \frac{1+3\eta}{16} (2-3s^2)^2 - q_{0,0} - 2 \left[q_{0,1} + \frac{es^2 \tilde{q}_{0,1}}{(1+\eta)^2} \right] \cos 2\omega \right\}, \qquad (7)$$

with

$$q_{0,0} = \frac{3}{64} e^2 (5s^4 + 8s^2 - 8) - \frac{1}{16} (21s^4 - 42s^2 + 20),$$

$$q_{0,1} = \frac{3}{64} e^2 s^2 (14 - 15s^2),$$

$$\tilde{q}_{0,1} = \frac{3}{16} e(1+2\eta)(4 - 5s^2).$$

Note that all the symbols in Eqs. (5)–(7) are now functions of the Delaunay prime variables.

3 Averaged Flow

The evolution of the system can then be obtained from corresponding Hamilton equations. Since $H' = H = H_0$, for the symmetries of the zonal model, and $L' = L_0'$ because of the averaging over the mean anomaly, the reduced (g', G') system

$$\frac{dG'}{dt} = -\frac{\partial \mathcal{H}'}{\partial g'}, \qquad (8)$$

$$\frac{dg'}{dt} = \frac{\partial \mathcal{H}'}{\partial G'} \qquad (9)$$

decouples from the rest of the flow, which, after solving $g' = g'(t)$, $G' = G'(t)$, will be integrated by quadrature

$$\ell' = \ell'_0 + \int \frac{\partial \mathcal{H}'(g'(t), G'(t); L', H)}{\partial L'} dt, \quad h' = h'_0 + \int \frac{\partial \mathcal{H}'(g'(t), G'(t); L', H)}{\partial H'} dt.$$

In spite of the integrability of the reduced flow, the solution to Eqs. (8)–(9) may involve elliptic and hyper-elliptic functions, thus lacking of physical insight. On the other hand, the equilibria of the reduced flow $dg'/dt = dG'/dt = 0$ correspond to orbits that, on average, have constant eccentricity and argument of periapsis. These orbits are of particular interest in mission designing for artificial satellites and are dubbed "frozen orbits" in the aerospace engineering lingo.

By carrying out the required operations in Eq. (8), we arrive to

$$\frac{dG'}{dt} = C_{2,0}^2 \frac{\mu}{a} \left\{ \sum_{i \geq 3} \frac{\partial \langle V_i \rangle_f}{\partial \omega} + \frac{R_\oplus^4}{a^4} \frac{2}{\eta^7} \left[q_{0,1} + \frac{es^2 \tilde{q}_{0,1}}{(1+\eta)^2} \right] \sin 2\omega \right\},$$

which always vanishes at $g' = \pm \pi/2$, as checked by differentiating Eq. (6) with respect to ω, and noting that $\sin[(2j + i^*)\omega + i_\pi] = \sin 2j\omega$ for even terms whereas $\sin[(2j + i^*)\omega + i_\pi] = \cos(2j + 1)\omega$ in the odd case.

Hence, the condition for orbits with $g' = \pm \pi/2$ to be frozen[1] is obtained by making $dg'/dt = 0$, viz.

$$0 = \dot{\omega}(-, g' = \pm \pi/2, -, L', G', H), \tag{10}$$

where $\dot{\omega}$ is obtained by evaluating the right side of Eq. (9) at either 90 or 270 deg. Computation of this equation requires the computation of different partial derivatives with respect to η, e, and s, and, in particular

$$\frac{\partial \langle V_i \rangle_f}{\partial \eta} = \frac{1 - 2i}{\eta} \langle V_i \rangle_f$$

$$\frac{\partial \langle V_i \rangle_f}{\partial e} = \frac{R_\oplus^i}{a^i} \tilde{C}_{i,0} \sum_{j=0}^{i_2} (2 - \delta_{j+i^*,0}) \frac{\partial \mathcal{G}_{i,i_0^*+j}}{\partial e} \mathcal{F}_{i,i_0^*+j} \cos[(2j + i^*)\omega + i_\pi]$$

$$\frac{\partial \langle V_i \rangle_f}{\partial s} = \frac{R_\oplus^i}{a^i} \tilde{C}_{i,0} \sum_{j=0}^{i_2} (2 - \delta_{j+i^*,0}) \mathcal{G}_{i,i_0^*+j} \frac{\partial \mathcal{F}_{i,i_0^*+j}}{\partial s} \cos[(2j + i^*)\omega + i_\pi]$$

Because of the known singularities of Delaunay variables for circular orbits, the averaged flow is customarily studied in the $(e \cos \omega, e \sin \omega)$ plane. Delaunay variables are also singular for equatorial orbits, a case in which the reduced flow may be studied in the $(s \cos \omega, s \sin \omega)$ plane. Both cases are equivalent to studying the frozen orbits condition from the vanishing of the numerator of $dg'/dt = 0$.

---

[1] I do not discuss the case of other frozen orbits that may exist with $g \neq \pm \pi/2$.

3.1 Local Dynamics: Eccentricity-Vector Diagrams

Alternatively to the direct integration of the reduced flow in Eqs. (8)–(9), for given values of the dynamical parameters H' and L', or $\sigma = H'/L'$ and $a = L'^2/\mu$, the reduced flow can be represented by means of contour plots of the averaged Hamiltonian in mean elements in Eq. (5). Note that $I_{\text{circ}} = \arccos \sigma$ matches the mean inclination of a circular orbit, in which $G' = L'$, but, in general, σ is not representative of the inclination of an elliptic orbit.

The cylindrical map (g, G) misses the case of circular orbits, and hence it is common to use eccentricity-vector diagrams to represent the reduced phase space, although this kind of diagram is not good for very high eccentricities. This is illustrated in Fig. 1 for the simplest case of the J_2 truncation of the zonal potential, which illustrates the typical bifurcation sequence of frozen orbits from the circular case. In order to make the changes more apparent we set $a = R_\oplus$, although it would be obviously unrealistic for actual orbits.

Thus, low inclination circular orbits are stable, whereas elliptic orbits remain with almost constant eccentricity and circulating perigee (top-left plot of Fig. 1). Close to the "critical" inclination of 63.435 deg (or $\sigma = \sqrt{1/5}$) the stability of the circular orbits change in a bifurcation process, and two elliptic frozen orbits emerge with the perigee frozen at either $\omega = 0$ or $\omega = \pi$; the perigee of the elliptic orbits now can oscillate, for orbits close to the elliptic frozen orbits, or circulate (top-right plot of Fig. 1). Smaller values of σ (or higher I_{circ}) result in larger amplitude of perigee libration close to the elliptic frozen orbits (bottom-left plot of Fig. 1). For a slightly higher inclination, circular orbits return to stability in a new bifurcation process, in which two new elliptic frozen orbits emerge, now with the perigee frozen at $\omega = \pm\pi/2$. Further decreasing σ does not introduce qualitative changes in the flow: circular orbits remain stable and the eccentricity of the elliptic frozen orbits grows high with almost constant, critical inclination (bottom-right plot of Fig. 1).

Note that the coupling between the mean eccentricity and inclination $\sigma = H'/L' = \sqrt{1-e^2}\cos I$ can be expressed by the differential relation

$$\frac{e}{1-e^2}\, de = -\tan I \, dI.$$

Therefore, a variation to higher values of the eccentricity of a given orbit implies a concomitant variation of the inclination towards the equatorial plane. This effect is of the same kind as the so-called Kozai resonance (or Lidov-Kozai effect) caused by third-body perturbations [14, 15], in which the eccentricity of librating-perigee orbits can grow high and, in consequence, the higher eccentricity orbits may become very close to the equatorial plane. On the contrary, the second bifurcation of circular orbits which happens in the earth's satellite problem makes that the variations in the eccentricity and inclination remain of the order of J_2, that is very small.

When J_3 is taken into account in Eqs. (8)–(9) the equatorial symmetry of the problem is broken, and the reduced phase space undergoes important qualitative changes. This is illustrated in Fig. 2 for the earth harmonic coefficients: The

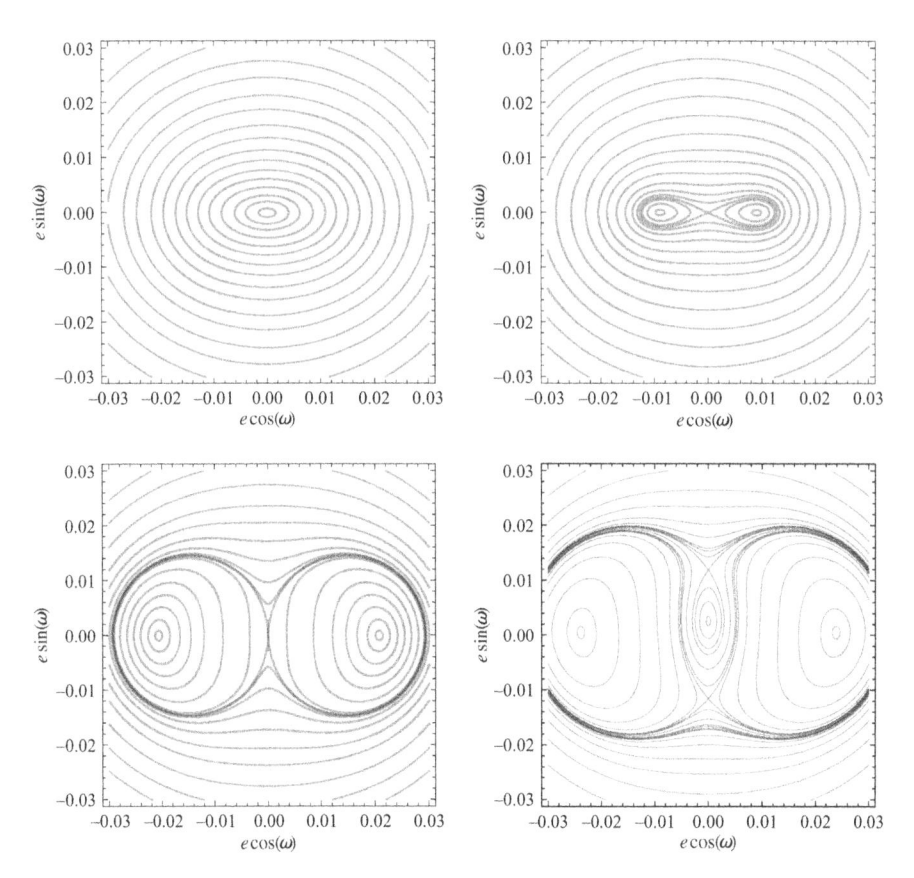

Fig. 1 Eccentricity vector diagrams for the earth's J_2 problem. From left to right and from top to bottom, the (mean) inclination of the circular orbit is $I_{\text{circ}} = 63.43, 63.433, 63.438$, and 63.44 deg, respectively

bifurcation sequence starts close to the critical inclination with slight quantitative differences with respect to the J_2 second order problem. However, the second bifurcation occurs in a completely different way.

Truncation of the Geopotential up to J_4 produces radical changes in the earth's frozen orbits diagram, with the initial bifurcated orbit having non-negligible eccentricity and $\omega = -\pi/2$. Besides, the perigee of the frozen orbits at the second bifurcation is very close to $\omega = -\pi/2$, although for increasing values of σ it approaches to the usual values $\omega = 0$ and $\omega = \pi$, as illustrated in Fig. 3.

Inclusion of other harmonics may still change the frozen orbits portrait. Indeed, as shown in Fig. 4, inclusion of J_5 makes the frozen orbits at $\omega = 0, \pi$ to disappear, whereas truncating the Geopotential up to J_6 make that the elliptic frozen orbit with perigee at $\omega = 90$ deg changes to instability in a bifurcation of two new elliptic frozen orbits. Increasing values of I_{circ} (or decreasing values of σ) make the perigees of these two elliptic orbits to migrate towards $\omega = 0$ and π, respectively, similarly to the J_2–J_4 case.

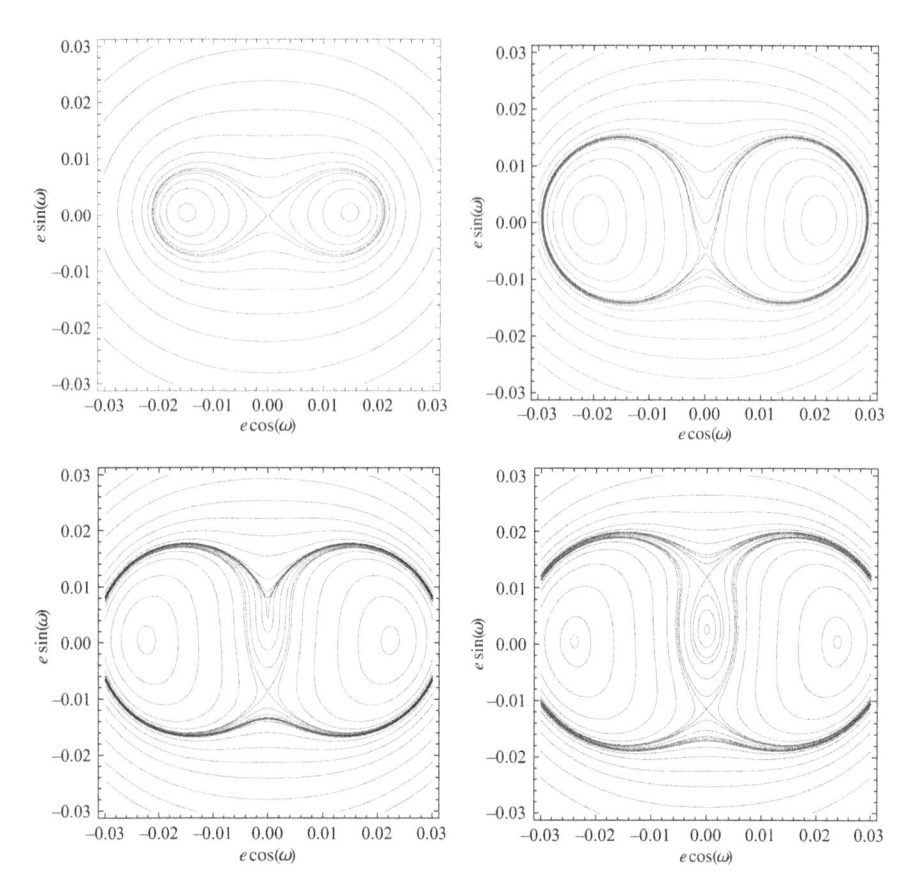

Fig. 2 Typical sequence of bifurcation of frozen orbits for the earth J_2–J_3 problem. From left to right and top to bottom, the (mean) inclination of the circular, orbits is $I_{\mathrm{circ}} = 63.435$, 63.438 63.439 and 63.44 deg

Further changes may be expected for higher degree truncations because the similar magnitude of the zonal harmonic coefficients. However, in practice the scale factor $R_\oplus/a$ is always smaller than unity, a fact that helps in downgrading the influence of the higher degree harmonics on the dynamics, and the situation seems to stabilize when the Geopotential truncation includes the effect of J_7 or higher degree coefficients. Then, the resulting sequence of events is qualitatively the same as the one provided by the J_2–J_5 truncation, but with quantitative variations that may be important, making the bifurcation of elliptic orbits to happen when the parameter I_{circ} takes higher values than in the J_2–J_5 case. This is illustrated in Fig. 4, where the bifurcation phenomenon of the J_2–J_9 truncation is clearly shifted from the J_2–J_5 case.

Hence, eccentricity-vector diagrams in mean elements can be used as a first criterion for deciding which truncation of the Geopotential is representative for the

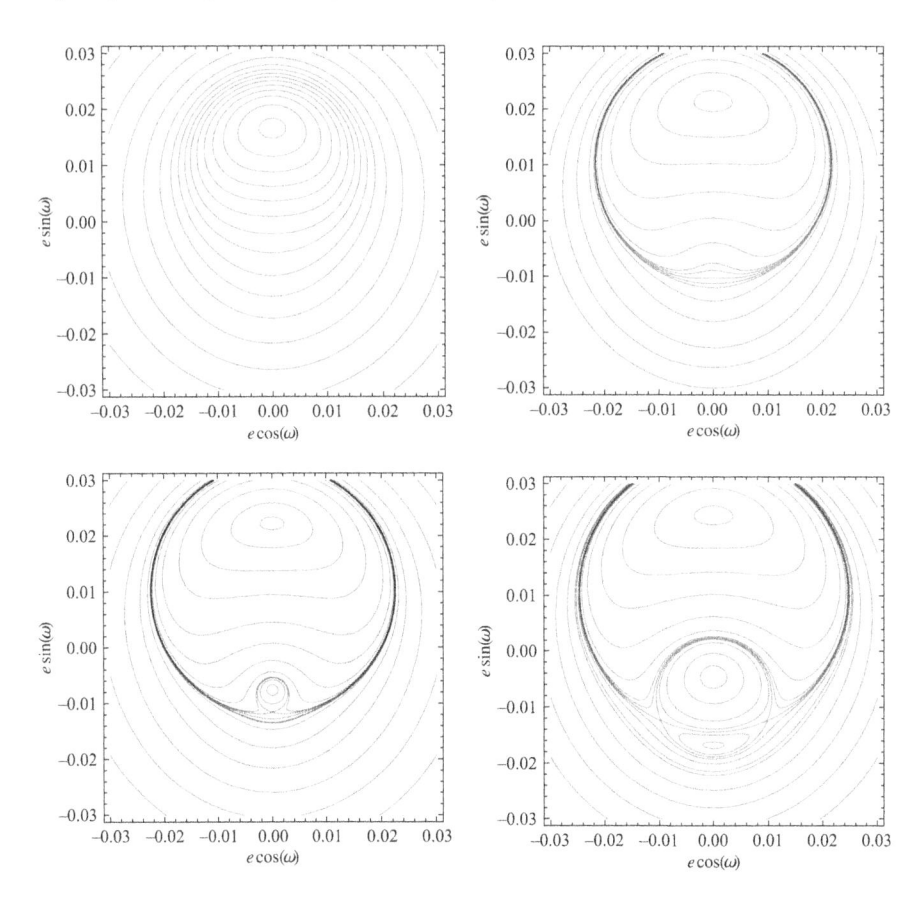

Fig. 3 Typical sequence of bifurcation of frozen orbits for the earth J_2–J_4 problem. From left to right and from top to bottom, the (mean) inclination of the circular, orbits is $I_{\text{circ}} = 63.405$, 63.408, 63.409 and 63.41 deg

class of orbits we want propagate. However, eccentricity-vector diagrams are local in nature, and the insight on the dynamics is only grasped after plotting portraits for different σ values, each of which requires the evaluation of different contour levels of the mean elements Hamiltonian.

3.2 Inclination-Eccentricity Diagrams

The frozen orbits constraint in Eq. (10) can be reformulated in the mean orbital elements as the implicit equation

$$\dot{\omega}(e, I; \omega = \pm \pi/2, a) = 0, \tag{11}$$

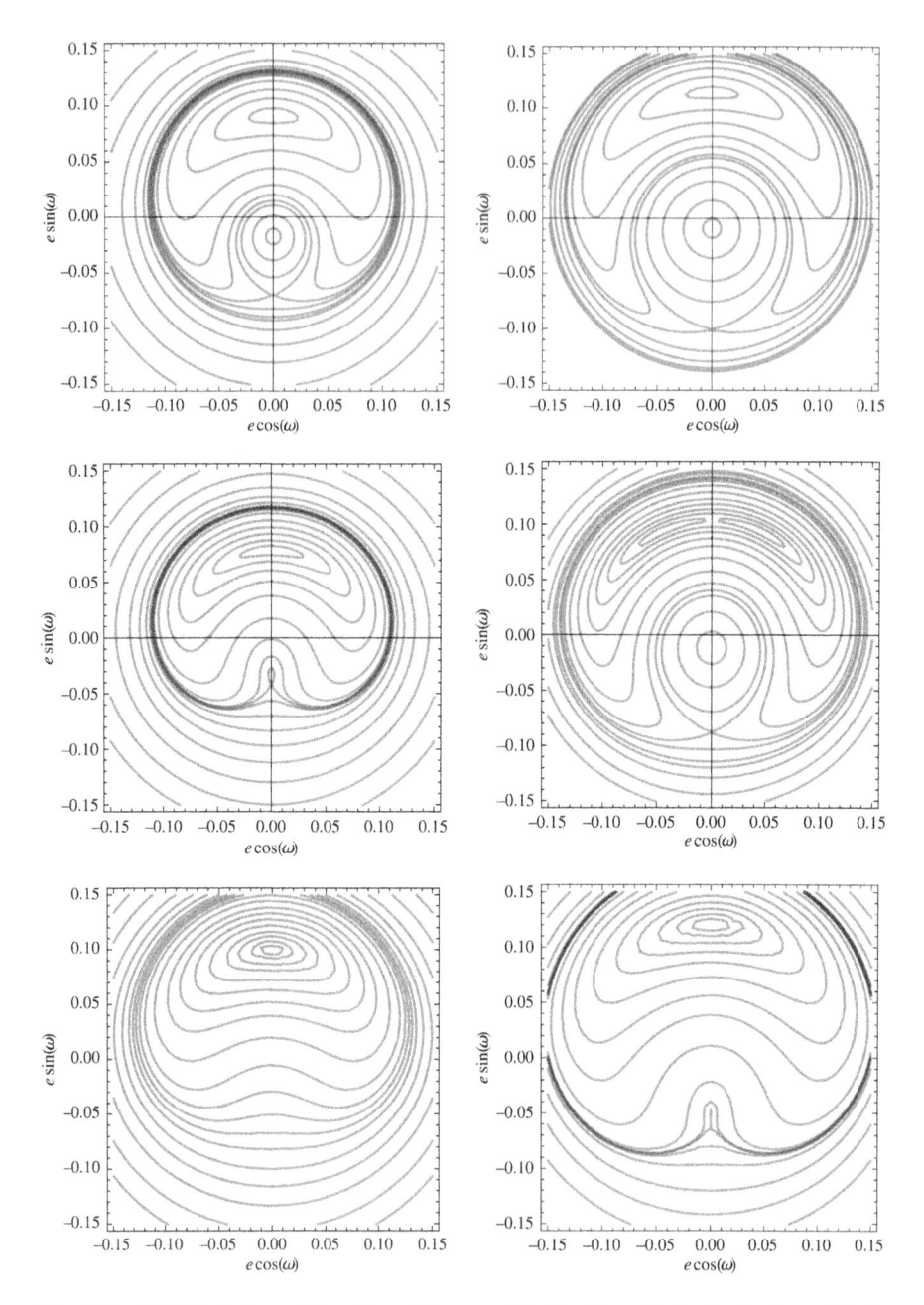

Fig. 4 From top to bottom, bifurcation of frozen orbits for the earth J_2–J_5, J_2–J_6, and J_2–J_9 truncations. Left column: $I_{\text{circ}} = 63.5$; right column $I_{\text{circ}} = 63.57$

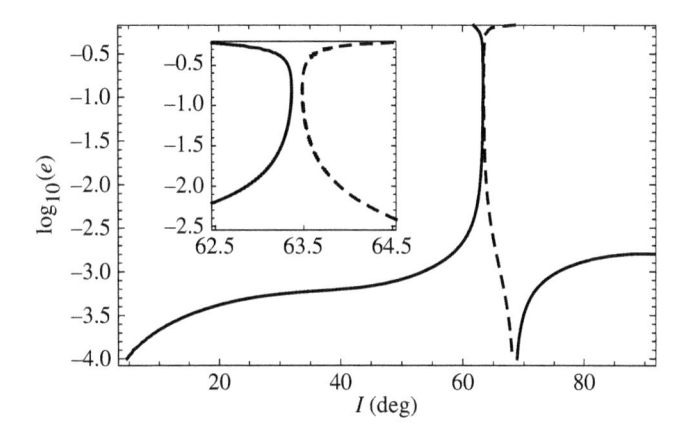

Fig. 5 Frozen orbits' inclination-eccentricity diagram of the earth J_2–J_9 problem for $a = R_\oplus$, with a magnification in the vicinity of the critical inclination. Full line: frozen orbits with $\omega = \pi/2$; dashed line: frozen orbits with $\omega = -\pi/2$

which gives the eccentricity of an orbit with frozen perigee either at 90 or 270 deg as a function of the inclination. Then, for a given (mean) semi-major axis a, the evolution of the frozen orbits with $\omega = \pm\frac{\pi}{2}$ can be depicted in the (e, I) plane by evaluation of Eq. (11). This is illustrated in Fig. 5, which shows an inclination-eccentricity diagram of frozen orbits with $\omega = \pm\pi/2$ for a J_2–J_9 truncation of the Geopotential. The semi-major axis a has been taken equal to the earth's equatorial radius $R_\oplus$ to enhance the effect of the perturbations. As shown in Fig. 5, there exist low eccentricity frozen orbits in all the range of inclinations, except in the vicinity of the critical inclination $I = 63.4$ deg, were the eccentricity of the frozen orbits grows tight. Besides, the diagram discloses the existence of very low eccentricity frozen orbits close to the equator but also in the particular inclinations of 68 deg.

4 Numerical Comparisons

Construction of inclination-eccentricity diagrams of orbits whose mean perigee is frozen at $\pm\frac{\pi}{2}$ is an alternative criterion to asses the sensitivity of the propagation of a given class of orbits with respect to the Geopotential truncation. For a given mean semi-major axis, it only requires depicting a variety of inclination-eccentricity diagrams, each one for a different truncation of the Geopotential. The simple, visual inspection of the different diagrams will provide an important help in the selection of the correct truncation for our particular purposes.

An example is provided in Fig. 6, where the curves are constrained to the case of non-impact frozen orbits with mean semi-major axis equal to 1.2 times the earth equatorial radius. The diagram focus on frozen orbits close to the critical inclination, and shows how the different truncations of the earth's zonal potential

Fig. 6 Inclination-eccentricity curves of frozen orbits for different Geopotential truncations ($a = 1.2 \times R_\oplus$). Full lines: $\omega = \pi/2$; dashed lines: $\omega = -\pi/2$

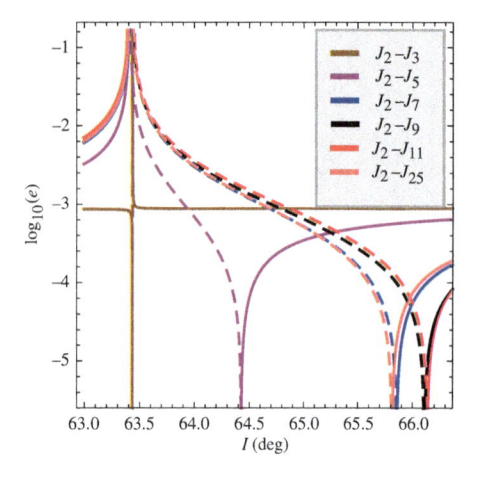

affect the location of the frozen orbits. Differences between the J_2–J_{2n} and J_2–J_{2n-1} truncations are generally negligible in a graphic representation, and the figure is cleared by presenting only curves corresponding to an odd truncation. All models include "centered" second order effects of J_2 [12, 13].

In particular, Fig. 6 shows that truncating the zonal model below the seventh degree may modify the inclination of the lower eccentricity orbits by more than one degree, what would be unacceptable for low earth orbits. Truncation of the Geopotential to higher degrees has the effect of slightly displacing the position of the lower eccentricity frozen orbits to the higher inclinations, but the inclusion of J_{13} in the model, as well as truncations to higher degrees, results in curves that get closer to the J_2–J_7 truncation. Taking higher degree harmonics than J_{23} makes inappreciable changes at the precision of the graphics.

Greater detail on the cases of lower and higher eccentricity frozen orbits is provided in Fig. 7, where the compensation for the effects of the J_9 to J_{12} harmonics when more complete models are used is clearly appreciated in the case of the lower eccentricity orbits (left plot). Orbits with higher eccentricities seem to be less sensitive to the model truncation, as illustrated in the right plot of Fig. 7, where the eccentricity is bounded to the case of non-impact orbits, which happen for $e > 0.06$, and only the curves of frozen orbits with perigee at $\omega = \pi/2$ are displayed. In both cases, truncation to higher degrees than J_{23} introduce negligible quantitative changes in the frozen orbits behavior at the precision of the graphics.

Because the harmonic coefficients of degree m are themselves scaled by the ratio $(R_\oplus/r)^m$, lower order truncations may be acceptable for orbits at higher altitudes. It happens to be the case: as shown in Fig. 8, truncations beyond J_7 do not introduce appreciable variations in the propagation of neither low nor high eccentricity frozen orbits with a mean semi-major axis equal to four times the earth's equatorial radius, which may roughly correspond to GPS and Molniya orbits.

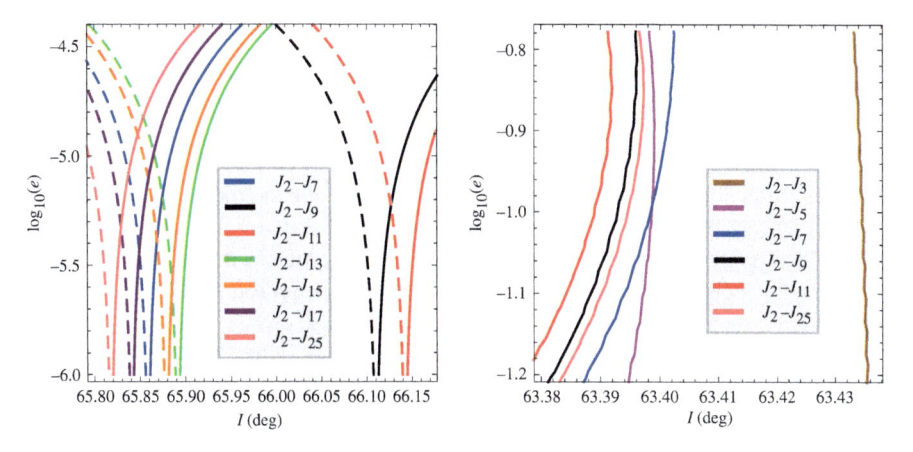

Fig. 7 Magnification of Fig. 6 for the lower (left) and higher eccentricity frozen orbits (right)

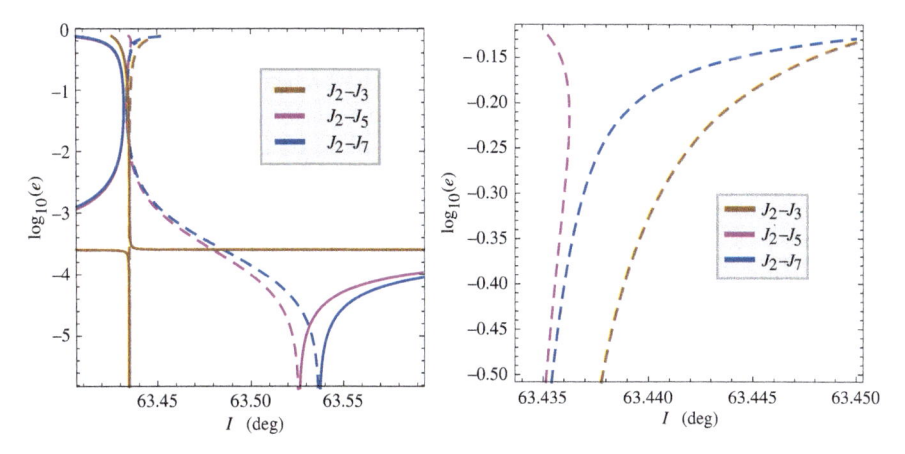

Fig. 8 Inclination-eccentricity curves of frozen orbits for different Geopotential truncations ($a = 4 \times R_\oplus$). Full lines: $\omega = \pi/2$; dashed lines: $\omega = -\pi/2$. Right plot: detail on the region of high eccentricity, unstable frozen orbits

5 Conclusions

A methodology based on the construction of both eccentricity vector diagrams and inclination-eccentricity diagrams of frozen orbits is used to explore the proper dynamical model to use in long-term orbit propagation. This approach allows to disclose the qualitative and quantitative differences introduced in the earth's orbits behavior by the truncation to different degrees of the Legendre polynomials expansion of the zonal Geopotential.

The construction of both kinds of diagrams can be done from direct evaluation of analytical expressions. Indeed, for a given semi-major axis, the frozen orbits constraint for orbits with the argument of the perigee at $\pm\pi/2$ can be written in closed form of the eccentricity, and, therefore, can be efficiently applied to frozen orbits with the higher eccentricities. The use of Kaula recurrences as well as related recurrences derived from them, make the evaluation of the frozen orbits constraint very fast and efficient. Hence, the visual inspection of the diagrams for different truncations of the zonal model permits to decide the required truncation in an inexpensive way. In particular, the construction of eccentricity vector diagrams shows that neglecting the contribution of zonal harmonics of degree higher than 6 may result in spurious frozen orbit solutions. Also, the representation of inclination-eccentricity diagrams shows that, for some kinds of orbits, truncations of the Geopotential to degrees higher than eight may provide worst quantitative results than the J_8 truncation, unless the truncation is extended, at least, up to J_{13}.

Extrapolating the conclusions derived from a particular set of orbits to other orbital configurations cannot be taken for granted; on the contrary, it must be further justified. So future research should explore how a given model truncation may affect to other different orbital configurations. This may require massive numerical propagations using high-fidelity models, to be taken as reference, and progressive truncations, to be compared with the reference. Besides, the influence of tesseral harmonics on resonant orbits should be explored to determine the correct Geopotential truncation to use when propagating these kinds of orbits.

Acknowledgements The author acknowledges partial support by the Spanish State Research Agency and the European Regional Development Fund under Projects ESP2013-41634-P, ESP2014-57071-R and ESP2016-76585-R (AEI/ERDF, EU). Discussions with Hodei Urrutxua, University of Southampton, are acknowledged.

References

1. Vallado, D.A.: An analysis of state vector propagation using differing flight dynamics programs (AAS 05–199). In: Vallado, D.A., Gabor, M.J., Desai, P.N. (eds.) Spaceflight Mechanics 2005. Advances in the Astronautical Sciences, vol. 120, pp. 1563–1592. American Astronautical Society, Univelt, Inc., Escondido, CA (2006)
2. Rosborough, G.W., Ocampo, C.: Influence of higher degree zonals on the frozen orbit geometry. In: Kaufman, B., Alfriend, K.T., Roehrich, R.L., Dasenbrock, R.R. (eds.) Astrodynamics 1991. Advances in the Astronautical Sciences, vol. 76, pp. 1291–1304. American Astronautical Society, Univelt, Inc., Escondido, CA (1992)
3. Cook, G.E.: Perturbations of near-circular orbits by the Earth's gravitational potential. Planet. Space Sci. **14**, 433–444 (1966)
4. Kaula, W.M.: Theory of Satellite Geodesy. Applications of Satellites to Geodesy. Blaisdell, Waltham, MA (1966)
5. Lara, M., de Saedeleer, B., Ferrer, S.: Preliminary design of low lunar orbits. In: Proceedings of the 21st International Symposium on Space Flight Dynamics, Toulouse, pp. 1–15. ISSFD (2009)
6. Lara, M.: Fast computation of inclined, frozen, low lunar orbits in a high degree Selenopotential (AAS 10–220). In: Spaceflight Mechanics 2010. Advances in the Astronautical Sciences, vol. 156, pp. 1769–1781. American Astronautical Society, Univelt, Inc., Escondido, CA (2010)

7. Lara, M., Ferrer, S., Saedeleer, B.D.: Lunar analytical theory for polar orbits in a 50-degree zonal model plus third-body effect. J. Astronaut. Sci. **57**(3), 561–577 (2009)
8. Lara, M.: A Mathematica©-based approach to the frozen orbits problem about arbitrary bodies. The case of a Lunar orbiter. In: Astrodynamics Beyond Borders, Proceedings of the 4th International Conference on Astrodynamics Tools and Techniques (ICATT), WPP-308, pp. 1–8. ESA Publication, Madrid (2010)
9. Lara, M.: Design of long-lifetime lunar orbits: a hybrid approach. Acta Astronaut. **69**(3–4), 186–199 (2011)
10. Coffey, S.L., Deprit, A., Miller, B.R.: The critical inclination in artificial satellite theory. Celest. Mech. **39**(4), 365–406 (1986)
11. Coffey, S.L., Deprit, A., Deprit, E.: Frozen orbits for satellites close to an earth-like planet. Celest. Mech. Dyn. Astron. **59**(1), 37–72 (1994)
12. Exertier, A.: Orbitographie des satellites artificiels sur de grandes periodes de temps. Possibilites d'applications. Ph.D. Thesis. Observatoire de Paris, Paris (1988)
13. Lara, M., San-Juan, J.F., López-Ochoa, L.M.: Proper averaging via parallax elimination (AAS 13-722). In: Astrodynamics 2013. Advances in the Astronautical Sciences, vol. 150, pp. 315–331. American Astronautical Society, Univelt, Inc., Escondido, CA (2014)
14. Lidov, M.L.: The evolution of orbits of artificial satellites of planets under the action of gravitational perturbations of external bodies. Planet. Space Sci. **9**, 719–759 (1962)
15. Kozai, Y.: Secular perturbations of asteroids with high inclination and eccentricity. Astron. J. **67**, 591–598 (1962)

A Parametric Study of the Orbital Lifetime of Super GTO and SSTO Orbits Based on Semi-analytical Integration

Denis Hautesserres, Juan F. San-Juan, and Martin Lara

Abstract A parametric study of the orbital lifetimes of objects in super-geostationary, and super-synchronous transfer orbits (Super GTO and SSTO, respectively) has been carried out based on a fast an efficient semi-analytical orbit propagator ad hoc designed. Results are displayed by means of time-reentry maps, each of which requires the propagation of hundreds of orbits. The maps are displayed in the orbit inclination vs. RAAN plane at the starting epoch of propagation, and provide an important insight in the orbit evolution and final deorbit of Super GTO and SSTO debris, thus showing as a valuable aid to the needs of space surveillance and tracking (SST). The time histories of selected trajectories suggest that the smaller lifetimes are related to resonances of the Kozai type, thus enlarging the scope of the time-reentry maps, which can be used as a general tool to study the dynamics of highly elliptic orbits.

1 Introduction

Geostationary transfer orbits (GTO), Super GTOs, and supersynchronous transfer orbits (SSTO) are highly elliptical orbits (HEO). In these kinds of orbits, the strength of the perturbations switches from perigee, where the earth's spherical harmonics and, possibly, the atmospheric drag are the more important perturbations, to apogee, where the third-body's gravitational pull dominates over other disturbing forces. The

---

D. Hautesserres
CNES, 18 Avenue Edouard Belin, 31401 Toulouse, France
e-mail: denis.hautesserres@gmail.com

J.F. San-Juan
GRUCACI – University of La Rioja, C/Madre de Dios, 26006 Logroño, Spain
e-mail: juanfelix.sanjuan@unirioja.es

M. Lara (✉)
GRUCACI, University of La Rioja, C/Madre de Dios, 26006 Logroño, Spain

Space Dynamics Group, UPM, Pza. del Cardenal Cisneros, 28040 Madrid, Spain
e-mail: mlara0@gmail.com

propagation of HEO orbits is customarily done by means of numerical integration—where regularization is an added bonus for an efficient propagation—with a wealth of methods available, like regularized Cowell [1], Bulirsch-Stoer [2], Dormand and Prince [3], or methods based on the variation of parameters formulation (see, for instance, [4] and references therein). However, when the propagation is extended over long time scales, like in the case of parametric studies, the numerical approach can became prohibitive in terms of computing time.

Alternatively, for these kinds of studies in which, contrary to accurate ephemeris prediction, the true orbit evolution is required one can turn to analytical of semi-analytical integration, where existing software packages like STELA,[1] THEONA [5], DSST,[2] or NADIA [6], perform outstandingly. In this context, and as a result of 1 year Research and Technology contract between the University of la Rioja and CNES, is in which HEOSAT, a Semi-Analytical Theory for HEO propagation, has been developed [7]. HEOSAT includes the main disturbing effects that affect the long-term propagation of HEOs, and takes benefit from a robust implementation in FORTRAN77.

We base on HEOSAT software to carry out a parametric study that allow us to understand the complex behaviors of Super GTOs and SSTOs. In particular, we are interested in uncovering the regions of the phase space of initial conditions that lead to the faster deorbit. We computed hundreds of Super GTOs and SSTOs and found the time spent into orbit. We found that, depending on initial conditions, the lifetime may vary from several months to tens of years. Results are visualized by means of time-reentry maps in the plane of orbit inclination and RAAN at the starting epoch of propagation. The maps provide an important insight in the orbit evolution and final deorbit of Super GTO and SSTO debris, the upper stage of the launcher, thus showing as a valuable aid to the needs of space surveillance and tracking (SST).

2 Super GTO and SSTO Orbits

Super GTO and SSTO orbits are HEO orbits. They have very high apogees and are nowadays interesting orbits to transfer satellites to GEO orbit using electrical propulsion in continuous orbit transfer instead of impulsive chemical propulsion. The altitude evolution of these kinds of orbits is illustrated in Fig. 1, where a full propagation model has been used for the starting epoch given by the CNES Julian Date 24041.0, and with the following initial conditions:

GTO: $a = 24688.1$ km, $e = 0.7, I = 5°, \Omega = \omega = M = 0$
Super GTO: $a = 51528.1$ km, $e = 0.87, I = 30°, \Omega = \omega = M = 0$
SSTO: $a = 56640.6$ km, $e = 0.88, I = 45.3°, \Omega = \omega = M = 0$

where a, e, I, Ω, ω, and M, stand for semi-major axis, eccentricity, inclination, right ascension of the ascending node, argument of the perigee, and mean anomaly,

---
[1]https://logiciels.cnes.fr/content/stela.
[2]https://tastrody.unirioja.es/dsst.

Orbital Lifetime of Super GTO and SSTO Orbits

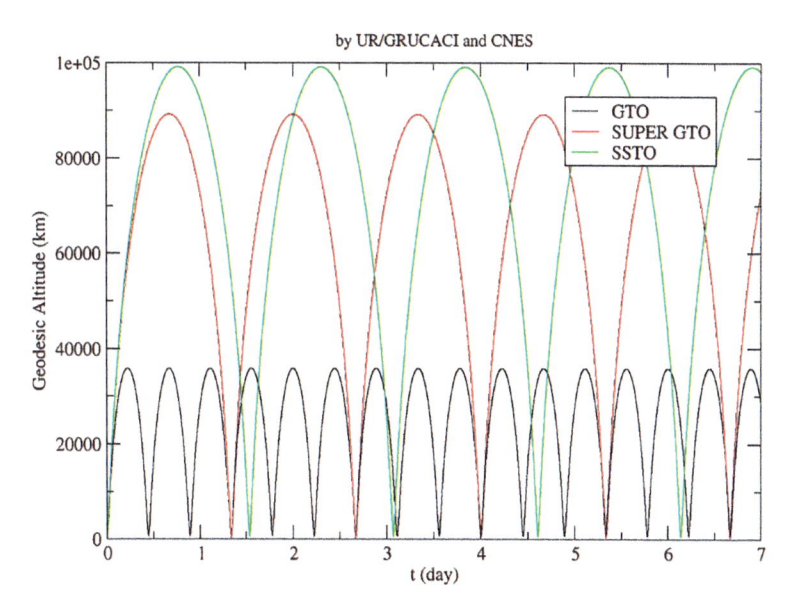

Fig. 1 Typical altitudes of GTO, Super GTO, and SSTO orbits

respectively. Standard values of area, mass, and drag coefficient of geostationary satellites $A = 24\,\text{m}^2$, $m = 3000\,\text{kg}$, and $C_d = 2.2$, have been used in the first two cases, whereas typical values of a Proton launcher were chosen for the SSTO propagation, namely: $A = 25\,\text{m}^2$, $m = 6650\,\text{kg}$, $C_d = 2.2$.

Using electrical propulsion to cover all the thrust needs aboard satellites, including the orbit raising up to the geosynchronous orbit, may result in important savings in GEO transfer. These advantages are not for free, and to reach a super GTO or SSTO orbit requires higher performance of the launcher. Still, the savings can reach the order of 1 ton propellant for a 3 tons satellite in station, and the launch into either Super GTO or SSTO will have the added benefit of reducing the number of perigee crossings through the Van Allen belts [8].

Besides, when compared with low thrust transfer from GTO, one finds that low thrust transfer from Super GTO is achieved with similar propellant mass, but the transfer duration of a Super GTO transfer can be shortened to about 2.6 months, thus providing a notable (∼60%) reduction in the transfer duration [9]. Furthermore, Super GTO and SSTO orbits allow for a faster deorbit of the upper stage of the launcher [10]. We further investigate this claim by constructing the time reentry maps which are described in Sect. 4

3 Semi-analytical Propagation

Modeling of highly elliptic orbits needs to take a variety of perturbations into account, the magnitude of which effects may change radically along the orbit. A sample on how the different effects act on a HEO orbit is illustrated in Fig. 2, in which the different perturbations are scaled with respect to the Keplerian attraction.

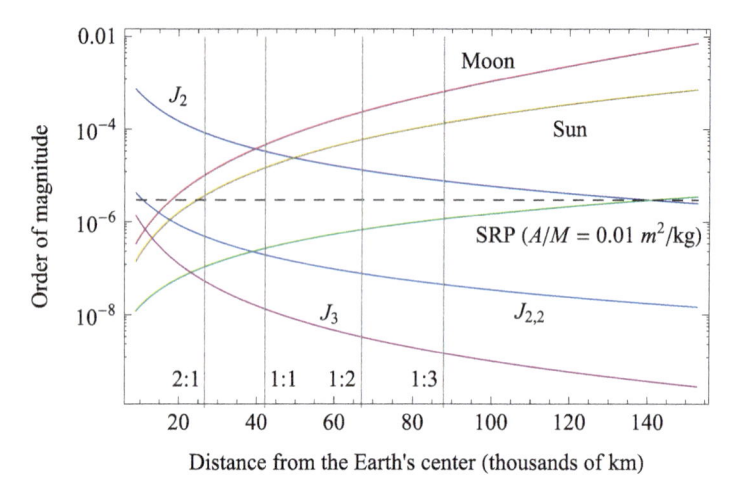

Fig. 2 Relative magnitudes of the different perturbations acting on a HEO orbit (after [7])

The perturbation model takes into account the gravitational effects produced by the first nine zonal harmonics of the Geopotential, including second order effects of J_2, as well as the main tesseral harmonics affecting to the 2:1 and 1:1 resonances. Lunisolar perturbations are modeled in the mass-point approximation by the usual Legendre polynomials expansion of the disturbing function, which is truncated to the second degree in the case of the sun, whereas the accurate modeling of the lunar attraction needs to take up to the sixth degree Legendre polynomial into account. The perturbation model also includes the effects of solar radiation pressure in the cannonball approximation, and the disturbing effects of the atmospheric drag based on the Harris-Priester standard density model [11, 12].

HEOSAT (High Elliptical Orbit Semi-Analytical satellite Theory) is a semi-analytical orbit propagator with a Hamiltonian formalism for the gravitational forces removing the mean anomaly for non-resonant cases and keeping the terms producing long-period effects [7]. The non-gravitational forces are added as generalized forces: based on derivations by [13, 14] for the SRP, and [15, 16] for the atmospheric drag. Thus the method is free of short-period terms and propagates mean elements. More precisely Deprit's perturbation algorithm [17, 18] has been used in the construction of the semi-analytical propagator; it includes second order effects of the $C_{2,0}$ gravitational term, also Kozai-type terms to get (mean) centered elements [19–21]. The semi-analytic theory is in closed form of the eccentricity orbital element except for tesseral resonances [22, 23], and does not consider the coupling between the $C_{2,0}$ gravitational term with the Moon's disturbing effects. In view of the HEO test cases did not experience any eccentricity and/or inclination singularities related troubles, the first version of HEOSAT relies on Delaunay variables; future evolutions of the propagator will be reformulated in nonsingular elements [24].

4 Time Reentry Maps

HEOSAT software is able to propagate hundreds of orbits over long time scales in a reasonable computational time. Thus we can easily compute times to reentry maps, with which we can highlight the complex behaviors and the secular resonances of the orbits.

Indeed, up to 396 super GTO orbits have been propagated with HEOSAT software, as well as the same number of SSTO orbits. They correspond to the nodes of a grid which is constructed by taking 11×36 different initial conditions. The abscisas correspond to inclination values equispaced at intervals of $5°$, starting from $5°$ and ending at $55°$, whereas the ordinates are RAAN values, which are varied each $10°$ from 0 to $350°$. The other initial orbital elements remain the same in all the propagations, which, besides, start at the same initial epoch. In particular we fix $\omega_0 = M_0 = 0$, and take the values $a_0 = 51528.1$ km and $e_0 = 0.87$ for the Super GTO, and $a = 56640.6$ km and $e = 0.88$ for the SSTO.

The propagations were carried out up to a maximum time of 100 years if impact with the earth's surface does not happen earlier. Two maps are presented in Fig. 3, where orbital lifetime is identified by colors. Thus, white regions mean about orbital lifetimes of the order of 1 year, while reddish colors indicate less time in orbit than white, and bluish colors indicate larger times. In particular, dark red means a few months and dark blue goes farther than 50 years in orbit. SSTO map seems to preserve the portrait of super GTO map with minor variations due to the higher importance of third-body perturbations. It seems that the orbits stay in a continuous domain where SSTO amplifies the super GTO effects in a continuous way. Looking at the smallest lifetimes in orbit the SSTO deorbits faster than super GTO. This is due to the stronger effect of the Moon perturbation on the orbit eccentricity.

Analogous maps are presented in Fig. 4, where the time to reentry has been replaced by the eccentricity of the orbit at this particular time. Now, dark red means eccentricities lower than 0.602, red colors mean eccentricities greater than 0.602 and lower than 0.86, while white, blue, and dark blue mean eccentricities close to 0.86, 0.87 and 0.874, respectively. The higher eccentricities seem to be concentrated in the upper right region of the eccentricity map, where high values of the eccentricity are commonly related to low lifetime, thus establishing a correlation with similar regions of the time reentry maps.

5 Secular Resonances

The time to reentry and corresponding eccentricity maps provide an important insight into the orbit evolution and final deorbit of Super GTO and SSTO debris (the upper stage of the launcher). However, these maps show a very complex behavior of these kinds of orbits, which need to be better understood. It is worth noting that Super GTO apogee is about the 25% of the earth-Moon's distance whereas SSTO

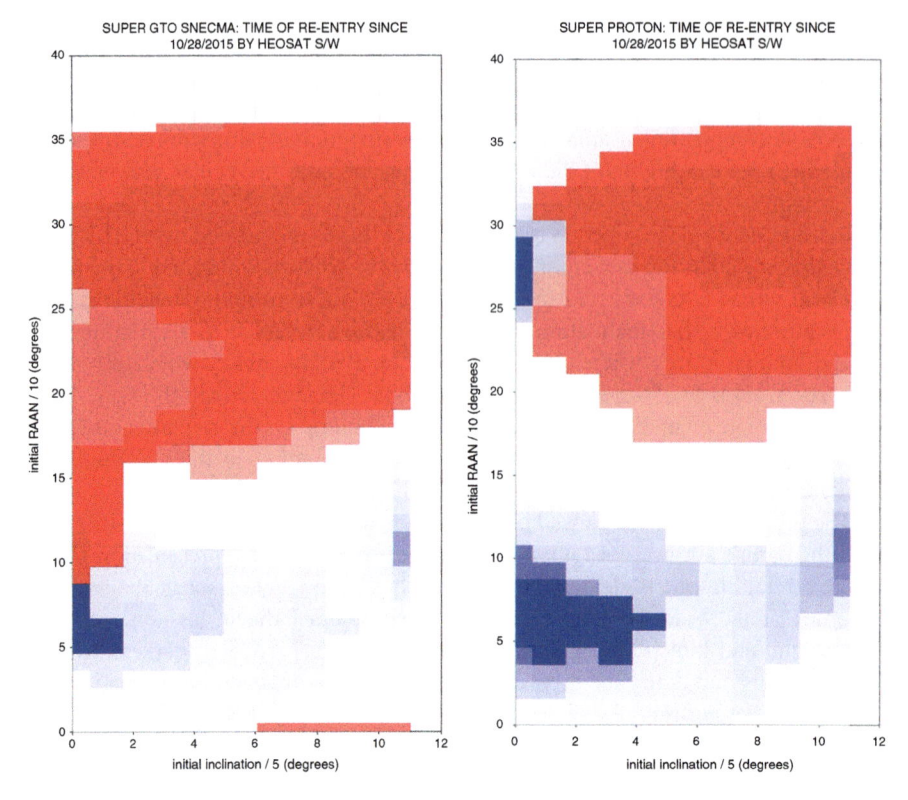

Fig. 3 Orbital lifetime τ; in years, dark red $\tau < 1/2$, red $\tau < 1$, white $\tau = 1$, blue $\tau > 1$, and dark blue $\tau = 50$. Left: Super GTO. Right: SSTO

apogee is about the 28%, and it is well known that this kind of high apogee orbits may latch in resonance with the third-body [25–27]. On the other hand, the growth of the eccentricity may happen also as a consequence of third-body perturbations without need of latching in resonance.

The most common third-body resonances affecting earth orbits are of the secular type. Indeed, in the three-body problem approximation there is a secular interaction between a wide-binary companion and a planet in a triple system or as well between the Moon and an Earth satellite in the Earth-Moon system (Kozai resonance or Lidov-Kozai mechanism). At resonance, there is an exchange between the eccentricity and the inclination and the timescale of the perigee's libration is close to the Kozai's oscillation timescale which is [28]

$$T_{K-I} \approx 70(R_\oplus/a)^{3/2} \approx 3 \text{ years.}$$

But, in extreme cases, resonances between the mean motions of the satellite and the third-body can also happen (Laplace resonances). Furthermore, it has been shown that both kinds of resonances can overlap themselves [29]. Thus, third-body resonances can become very complex and, therefore, difficult to study.

Orbital Lifetime of Super GTO and SSTO Orbits

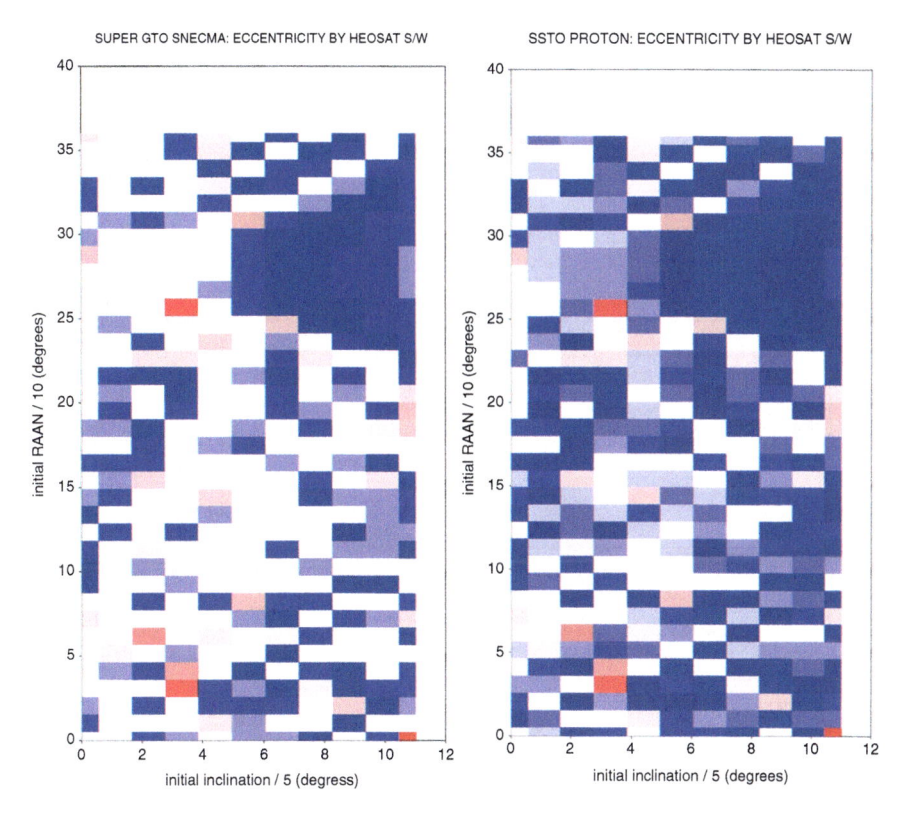

Fig. 4 Eccentricity at reentry: $e < 0.602$ is depicted in dark red, $0.602 \leq e < 0.86$ in red, $e \approx 0.86$ in white, $e \approx 0.87$ in blue and $e \approx 0.874$ in dark blue. Left: Super GTO. Right: SSTO

We investigate how secular resonances can affect Super GEO and SSTO orbits. As shown in Figs. 5 and 6, where $I_0 = 5°$ and $\Omega_0 = 50°$, therefore corresponding to a dark blue region of the time-reentry maps. Both the Super GTO and SSTO orbits undergo secular resonances of the argument of the perigee and the RAAN, which experience temporary captures that make these elements to oscillate. In the case of the RAAN, the oscillations roughly happen about the value 0, whereas the argument of the perigee librates about either 90° or 270°.

To better understand the phenomenon of the secular resonances it is convenient to simplify the perturbations model. As illustrated in Fig. 7, the resonance persists when the perturbation model is simplified to consider only the J_2 zonal term of the Earth potential and the Moon potential up to the degree 2 of its Legendre polynomial expansion. Only one of the orbits of Fig. 7 undergoes a secular resonance because of its low initial inclination, which is the same first secular resonance than appears in the corresponding propagation when using the full perturbation model. In Fig. 8 the time histories of the argument of the perigee and RAAN of the resonant orbit of the simplified model are superimposed to the those of the moon's corresponding

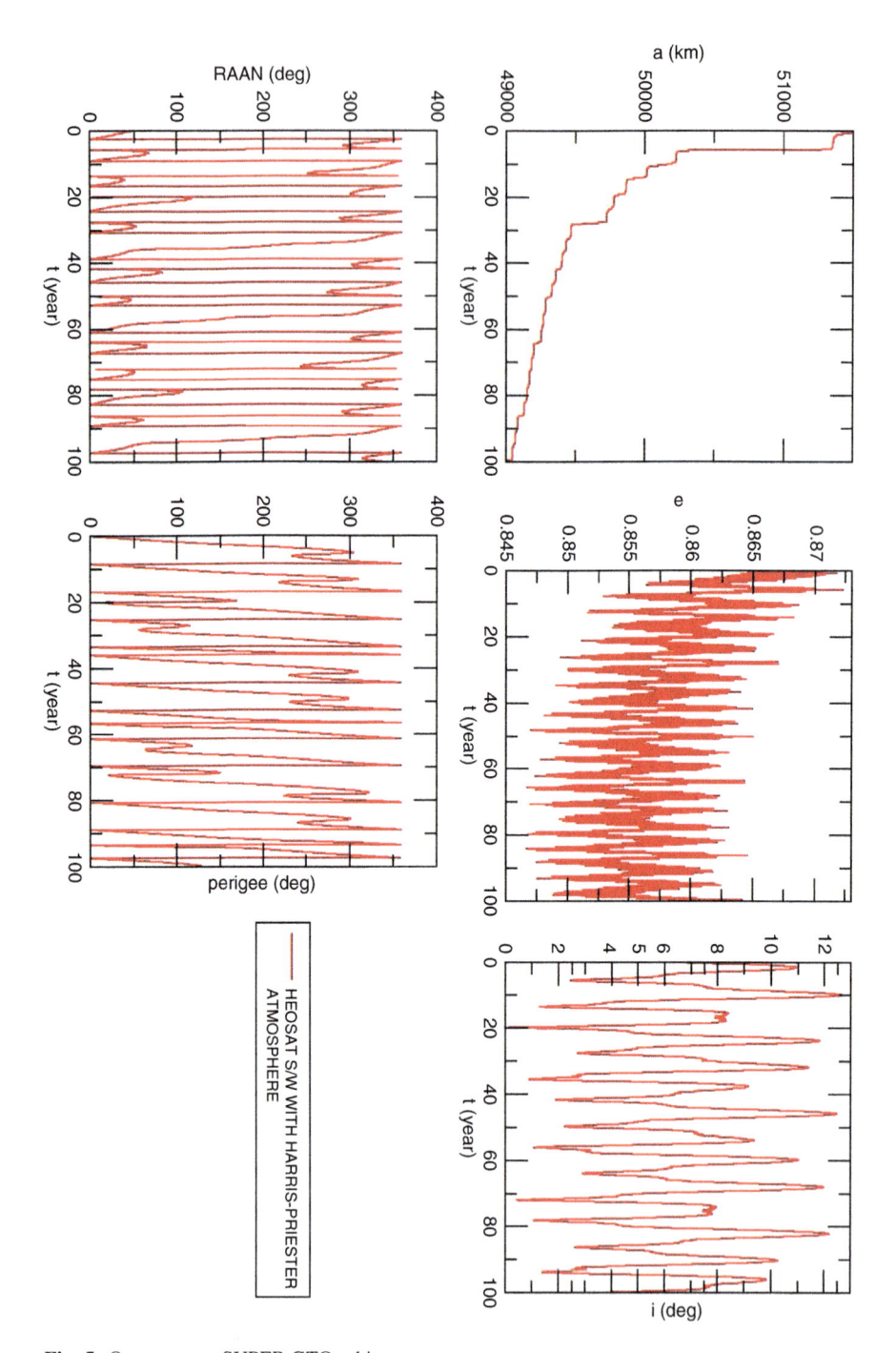

Fig. 5 One resonant SUPER GTO orbit

Orbital Lifetime of Super GTO and SSTO Orbits

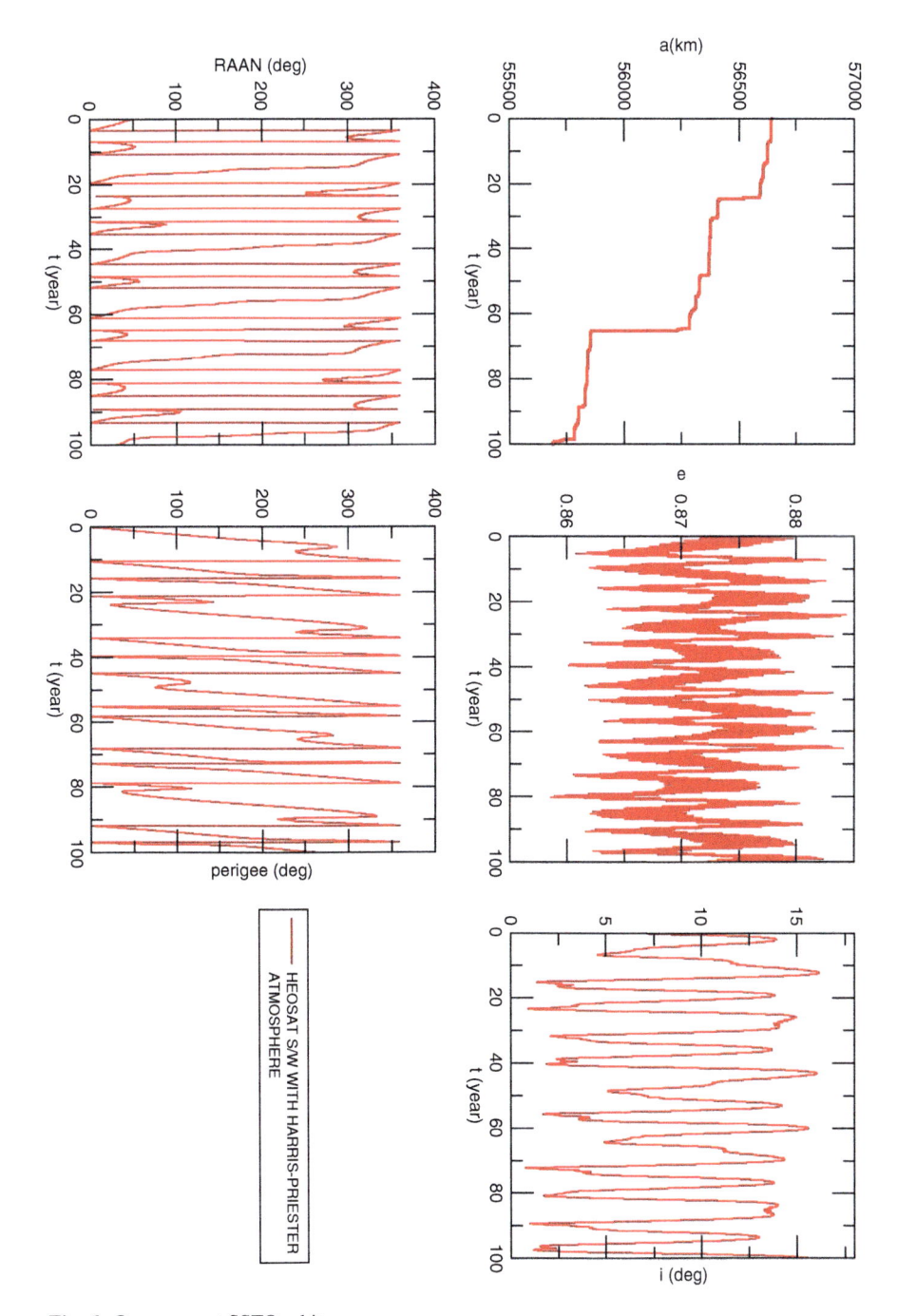

Fig. 6 One resonant SSTO orbit

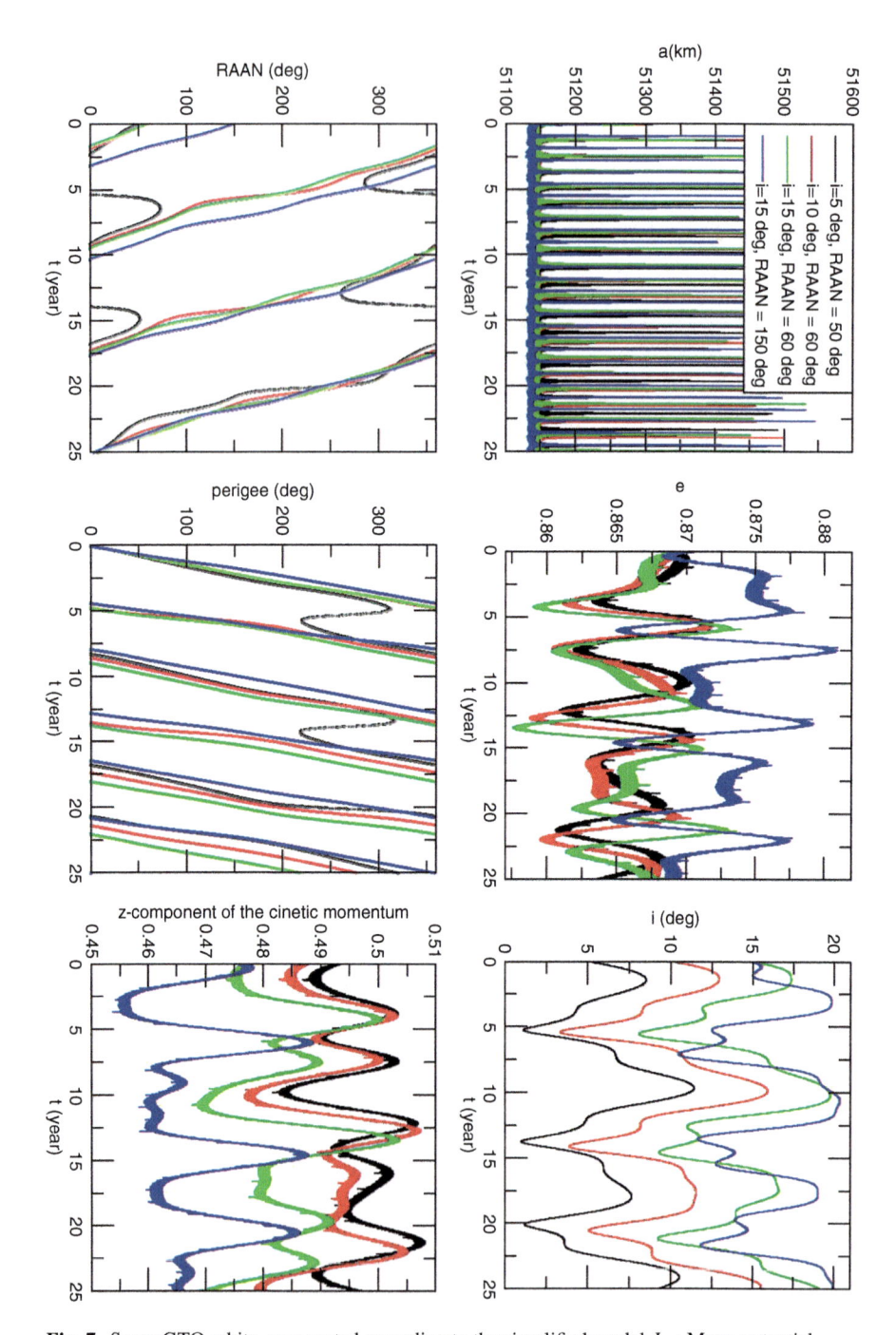

Fig. 7 Super GTO orbits, propagated according to the simplified model J_2 + Moon potential

Orbital Lifetime of Super GTO and SSTO Orbits   95

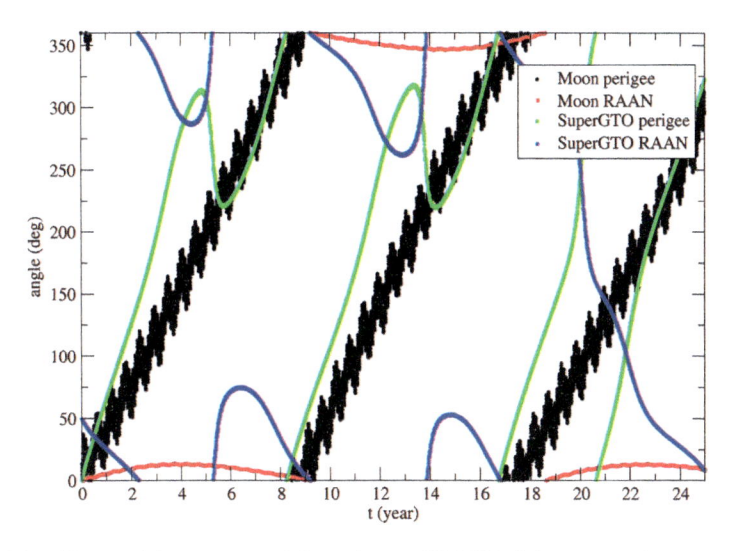

Fig. 8 Time history of the argument of the perigee and RAAN of the moon and a resonant Super GTO orbit of the simplified model (J_2 + Moon potential)

elements. It is clearly apparent in the figure that the variation rates of the Super GTO ω and Ω are synchronized with the rate of the moon's perigee. We recall that the different propagations are carried out in the Earth Centered Inertial (ECI), TOD reference frame for HEOSAT s/w, and Veis' reference frame [30] for the numerical integration which is used to get the Moon's elements and the Cook's resonance relations.

In view of the secular resonances also occur in the simplified model we can use previous results in the literature like Cook's resonance relations [27]. Besides, because the mean motion of the Moon runs much faster than the perigee rates of the super GTO and the SSTO orbits, there is no risk of mean motion resonances and to highlight the resonant angles we only need secular variables. Then we remove the moon rate in Cook's resonance relations. In particular, we get the following Cook's resonances equivalents

$$\dot{\omega} + \dot{\Omega} + \dot{\omega}_d = 0 \quad \text{(No. 7)} \tag{1}$$

$$\dot{\omega} - \dot{\Omega} - \dot{\omega}_d = 0 \quad \text{(No. 9)} \tag{2}$$

$$2\dot{\omega}_d + \dot{\Omega} - 2\dot{\omega} = 0 \quad \text{(No. 10)} \tag{3}$$

$$2\dot{\omega}_d - \dot{\Omega} - 2\dot{\omega} = 0 \quad \text{(No. 11)} \tag{4}$$

$$\dot{\omega} - \dot{\omega}_d = 0 \quad \text{(No. 15)} \tag{5}$$

and plot the left members of these relations using the values of a super GTO, in Fig. 9, and a SSTO, in Fig. 10. In both cases, super GTO and SSTO orbits, the resonance relation No. 7 is close to zero, i.e. a very few hundredth degrees per day.

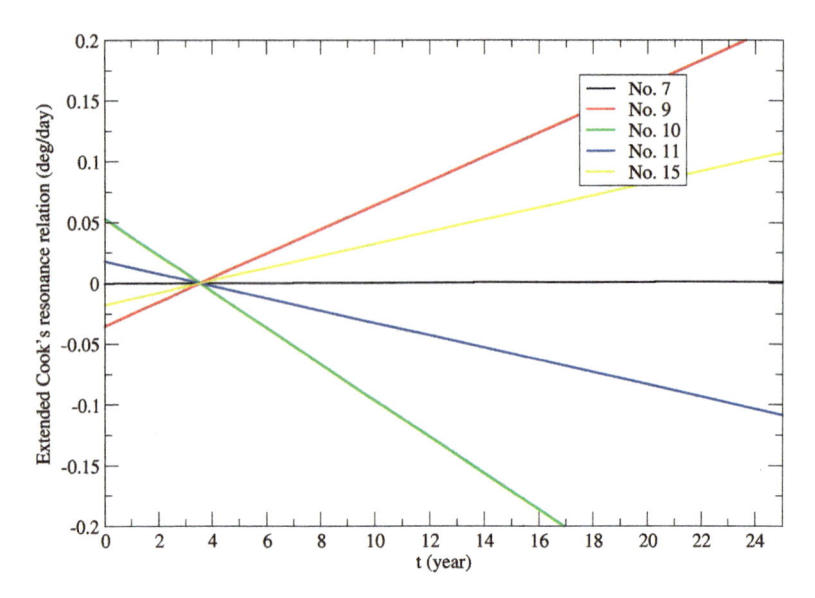

Fig. 9 Cook's resonance relations in Eqs. (1)–(5) for a Super GTO

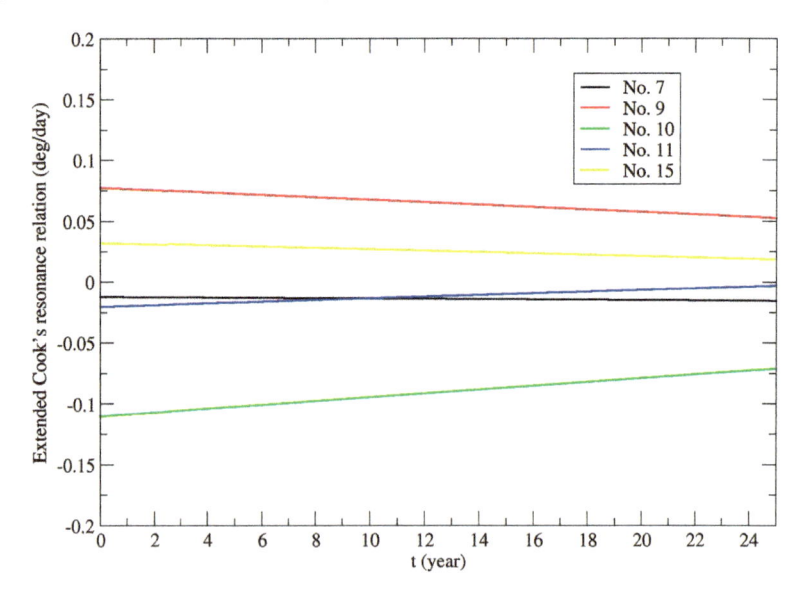

Fig. 10 Cook's resonance relations in Eqs. (1)–(5) for a SSTO

This fact provides a formal explanation of the secular resonance and the observed synchronization between the secular variables of the orbits (argument of the perigee and RAAN) and the Moon's perigee.

Further efforts must be done in better understanding the mechanisms that produce these secular resonances, as, for instance, following Morbidelli's approach [31]. We also guess that there is a resonance function of the inclination and the energy like in Cook's paper [27].

6 Conclusions

We started to study the super GTO and SSTO orbits. To our knowledge this is the first time that this kind of orbits are so far analyzed. Depending on the initial conditions the satellite, or the upper stage of the launcher, in super GTO or SSTO can either stay in orbit for a long time or reenter in a short time because of the influence of the third-body, mainly the moon. We propose to approach the problem by studying time to reentry maps, which are efficiently constructed with HEOSAT, an efficient and powerful software dedicated to the propagation of HEO orbits over long time scales. We also found that Super GTO and SSTO orbits undergo secular resonances of a similar type of the well known Kozai's resonances, and further identified the resonance relations. The maps provide an important insight in the orbit evolution and final deorbit of Super GTO and SSTO debris, the upper stage of the launcher, thus showing as a valuable aid to the needs of the awareness of putting into orbit a satellite, and of space surveillance and tracking. However, this is just an initial approach and the maps show a very complex behaviors of the orbits that needs to be better understood.

Acknowledgements Partial support is acknowledged from project ESP2014-57071-R (JFS and ML) and ESP2013-41634-P (ML) of the Ministry of Economic Affairs and Competitiveness of Spain.

References

1. Borderies, N.: Time regularization of an Adams-Moulton-Cowell algorithm. Celest. Mech. **16**, 291–308 (1977)
2. Gragg, W.B.: On extrapolation algorithms for ordinary initial value problems. J. Soc. Ind. Appl. Math. Ser. B Numer. Anal. **2**(3), 384–403 (1965)
3. Dormand, J.R., Prince, P.J.: A family of embedded Runge-Kutta formulae. J. Comput. Appl. Math. **6**(1), 19–26 (1980)
4. Lara, M.: Note on the ideal frame formulation (2016). arXiv e-prints 1612.08367
5. Golikov, A.R.: THEONA—a numerical-analytical theory of motion of artificial satellites of celestial bodies. Cosm. Res. **50**(6), 449–458 (2012)
6. Hautesserres, D.: Analytical integration of the osculating Lagrange planetary equations in the elliptic orbital motion –J2, 3rd-Body, SRP, + Drag–Software NADIA. In: Presentation at ICDDEA-2015, Amadora (May 18–22, 2015), ICDDEA (2016)
7. Lara, M., San-Juan, J., Hautesserres, D.: A semi-analytical orbit propagator program for highly elliptical orbits. In: Proceedings of the 6th International Conference on Astrodynamics Tools and Techniques, ICATT. ESA (2016)

8. Koppel, C.R.: Advantages of a continuous thrust strategy from a geosynchronous transfer orbit, using high specific impulse thrusters. In: 14th International Symposium on Space Flight Dynamics, Foz do Iguaçu, 8–12 February, 1999. ISSFD (1999)
9. Koppel, C.R.: Low thrust orbit transfer optimiser for a spacecraft simulator (ESPSS–ECOSIMPRO® European Space Propulsion System Simulation). In: Proceedings of the 6th International Conference on Astrodynamics Tools and Techniques, ICATT. ESA (2016)
10. Laursen, E., Waterman, R., Bonner, J.: Proton Launch System Mission Planner's Guide. Technical Report LKEB-9812-1990, Revision 7, International Launch Services, Reston, VA (2009)
11. Harris, I., Priester, W.: Time-dependent structure of the upper atmosphere. J. Atmos. Sci. **19**, 286–301 (1962)
12. Long, A.C., Cappellari, J.O., Velez, C.E., Fluchs, A.J.: Mathematical Theory of the Goddard Trajectory Determination System. Technical Report FDD/552-89/001, National Aeronautics and Space Administration, Goddard Space Flight Center, Greenbelt, MD (1989)
13. Kozai, Y.: Effects of Solar Radiation Pressure on the Motion of an Artificial Satellite. SAO Spec. Rep. **56**, 25–34 (1961)
14. Aksnes, K.: Short-period and long-period perturbations of a spherical satellite due to direct solar radiation. Celest. Mech. **13**, 89–104 (1976)
15. Liu, J.J.F., Alford, R.L.: Semianalytic theory for a close-earth artificial satellite. J. Guid. Control. Dyn. **4**, 576 (1981)
16. Chao, C.C., Platt, M.H.: An accurate and efficient tool for orbit lifetime predictions. In: Spaceflight Mechanics 1991, pp. 11–24 (1991)
17. Deprit, A.: Canonical transformations depending on a small parameter. Celest. Mech. **1**(1), 12–30 (1969)
18. Deprit, A.: The elimination of the parallax in satellite theory. Celest. Mech. **24**(2), 111–153 (1981)
19. Kozai, Y.: Second-order solution of artificial satellite theory without air drag. Astron. J. **67**(7), 446–461 (1962)
20. Lara, M., San-Juan, J.F., López-Ochoa, L.M.: Proper averaging via parallax elimination (AAS 13-722). In: Astrodynamics 2013. Advances in the Astronautical Sciences, vol. 150, pp. 315–331. American Astronautical Society, Univelt, Inc., Escondido, CA (2014)
21. Lara, M., San-Juan, J.F., López, L.M., Cefola, P.J.: On the third-body perturbations of high-altitude orbits. Celest. Mech. Dyn. Astron. **113**, 435–452 (2012)
22. Lara, M., San-Juan, J.F., Folcik, Z.J., Cefola, P.: Deep resonant GPS-dynamics due to the geopotential. J. Astronaut. Sci. **58**(4), 661–676 (2011)
23. Lara, M., San-Juan, J.F., López-Ochoa, L.M.: Averaging tesseral effects: closed form relegation versus expansions of elliptic motion. Math. Probl. Eng. **2013**, 1–11 (2013)
24. Lara, M., San-Juan, J., Hautesserres, D.: Semi-analytical propagator of high eccentricity orbits. Technical Report R-S15/BS-0005-024, Centre National d'Études Spatiales, Toulouse Cedex (2016)
25. Lidov, M.L.: The evolution of orbits of artificial satellites of planets under the action of gravitational perturbations of external bodies. Planet. Space Sci. **9**, 719–759 (1962)
26. Kozai, Y.: Secular perturbations of asteroids with high inclination and eccentricity. Astron. J. **67**, 591–598 (1962)
27. Cook, G.E.: Luni-solar perturbations of the orbit of an earth satellite. Geophys. J. **6**, 271–291 (1962)
28. Malhotra, R.: Orbital resonances in planetary systems. In: CELESTIAL MECHANICS, pp. 380–410. UNESCO-EOLSS (2017)
29. Hautesserres, D.: Extrapolation long terme de l'orbite du satellite SimbolX par la methode de Gragg-Bulirsch-Stoer (GBS). Technical Report DCT/SB/OR/2009-2474, Centre National d'Études Spatiales, Toulouse Cedex (2009)
30. Veis, G.: Precise aspects of terrestrial and celestial reference frames. SAO Special Report, 123, April (1963)
31. Morbidelli, A.: Asteroid secular resonant proper elements. Icarus **105**, 48–66 (1993)

On the Use of Positive Polynomials for the Estimation of Upper and Lower Expectations in Orbital Dynamics

Massimiliano Vasile and Chiara Tardioli

Abstract The paper presents the use of positive polynomials, in particular Bernstein polynomials, to represent families of probability distributions in orbital dynamics. The uncertainty in model parameters and initial conditions is modeled with p-boxes to account for imprecision and lack of knowledge. The resulting uncertainty in the quantity of interest is estimated by representing the upper and lower expectations with positive polynomials with interval coefficients. The impact probability of an asteroid subject to a partially known Yarkovsky effect is used as an illustrative example.

1 Introduction

The treatment of uncertainty in orbit propagation is of fundamental importance to predict the motion of natural and man-made objects. In the specific case of asteroids and space debris a key quantity of interest is the probability of an impact with the Earth or a collision with an operational satellite.

Several methods have been proposed to deal with uncertainty and provide a prediction of the future state of a space object. Most of them start from some assumptions on the probability distribution associated to the uncertain quantities and then model, more or less accurately, the distribution of the quantity of interest. When the nature of uncertainty is epistemic (lack of knowledge), a single probability distribution might not be available. More likely different sources of information may suggest that the probability associated to an uncertain quantity belongs to a finite set for which we can define upper and lower bounds.

M. Vasile (✉)
Aerospace Centre of Excellence, Department of Mechanical and Aerospace Engineering, University of Strathclyde, Glasgow, UK
e-mail: massimiliano.vasile@strath.ac.uk

C. Tardioli
Aerospace Centre of Excellence, University of Strathclyde, 75 Montrose Street, G11XJ Glasgow, UK
e-mail: chiara.tardioli@strath.ac.uk

© Springer International Publishing AG 2018
M. Vasile et al. (eds.), *Stardust Final Conference*, Astrophysics and Space Science Proceedings 52, https://doi.org/10.1007/978-3-319-69956-1_6

However, the fast calculation of these bounds is not a trivial matter. In this chapter an approach based on the use of positive polynomials is proposed to calculate the upper and lower bounds via a simple linear optimisation programme. The uncertain quantities are modeled with p-boxes defined through parametric probability distributions or via positive polynomial expansions [4].

The calculation of the impact probability of an asteroid subject to a poorly known Yarkovsky effect is used as an illustrative example.

2 Worst Case Scenario

The problem under investigation is to evaluate the probability of the following set of events:

$$A_\nu = \{\mathbf{u} \in \mathfrak{U}_0 : f(\mathbf{u}) \leq \nu\} \tag{1}$$

where f is the quantity of interest and $\mathbf{u}$ is a stochastic variable defined in an uncertainty space $\mathfrak{U}_0$ with dimension d. We use the notation $[\mathbf{u}] \in \mathbb{R}^d$ to indicate the convex set of $\mathbf{u}$ such that $\mathbf{u} \in \mathfrak{U}_0 \subseteq \mathbb{R}^d$. If $d = 1$, the uncertainty is an interval and it is also indicated as $[\underline{u}, \overline{u}]$, where $\underline{u}, \overline{u}$ are the lower and upper limits, respectively.

Regardless of the distribution of $\mathbf{u}$ one can define the best and worst case scenarios as follows:

$$\underline{f} = \min_{\mathbf{u} \in \mathfrak{U}_0} f(\mathbf{u}), \qquad \overline{f} = \max_{\mathbf{u} \in \mathfrak{U}_0} f(\mathbf{u}). \tag{2}$$

The solution to (2) gives the limit of variability of f and identifies also two rare events. For any value of $\nu \in [\underline{f}, \overline{f}]$ and a known probability distribution p, the probability associated to A_ν is given by the formula

$$IP(A_\nu) = \int_{A_\nu} p(\mathbf{u}) \, d\mathbf{u}, \tag{3}$$

In the following, the uncertain variables are assumed to be independent and uncorrelated so that the initial uncertainty space is the hyper-rectangle, however, the solution of Eq. (2) does not require $\mathfrak{U}_0$ to be a box and holds true for any generic set. The same is true for Eq. (3).

3 Upper and Lower Expectations

When the uncertainty on the input quantities is epistemic the probability p can belong to a family of parametric distributions or to a set of unknown distributions.

3.1 Representation with Families of Parametric Distributions

Consider the case in which one can reasonably assume that the uncertainty can be quantified with a family of beta distributions with unknown parameters α and β (any other parametric or non-parametric distribution would equally work). Equation (3) then translates into two equations defining the upper and lower probability associated to A_ν:

$$\min_{\alpha,\beta} \int_{A_\nu} p(\mathbf{u})\, d\mathbf{u}, \quad \max_{\alpha,\beta} \int_{A_\nu} p(\mathbf{u})\, d\mathbf{u}, \tag{4}$$

where p is the product of probability $p = \prod_{j=1}^{d} p_j$, where each marginal density mass p_j is a beta distribution function with parameters α_j, β_j.

3.2 Representation with Positive Polynomials

In the general case the integrals in Eq. (4) can be calculated numerically via multidimensional quadrature formula. As an example we can replace the calculation of the exact integrals with an approximation using Halton low discrepancy sequence to generate M sample points (called quasi-Monte Carlo points) in the domain $\mathfrak{U}_0$ and then re-write the integrals in the form:

$$\int_{A_\nu} p(\mathbf{u})\, d\mathbf{u} \approx \frac{1}{M} \sum_{k=1}^{M} I_{A_\nu}(\mathbf{u}_k) p(\mathbf{u}_k) \tag{5}$$

where the samples $\mathbf{u}_k$ are taken from the low discrepancy sequence. Similarly, we can approximate the integrals in Eq. (4):

$$\min_{\alpha,\beta} \sum_{k=1}^{M} I_{A_\nu}(\mathbf{u}_k) \prod_j p_j(\mathbf{u}_k), \quad \max_{\alpha,\beta} \sum_{k=1}^{M} I_{A_\nu}(\mathbf{u}_k) \prod_j p_j(\mathbf{u}_k). \tag{6}$$

subject to the constraint:

$$\frac{1}{M} \sum_{k=1}^{M} p(\mathbf{u}_k) = 1. \tag{7}$$

If the family of distributions is unknown or does not contain only one particular type, one can use an a representation with an expansion in positive polynomials to approximate the extrema of $[p]$ and obtain the upper and lower expectation on A_ν as solutions of a linear problem. In this chapter, in particular, we propose the use of

Bernstein polynomials [4, 7]. The family of probability distributions to which the uncertain variable u_j belongs can be expressed as

$$[p_{c_j}] = \left\{ \sum_{i=1}^{n} c_i^{(j)} B_i(\tau_j(u_j)) \right\}, \qquad (8)$$

where $B_i : [0, 1] \mapsto [0, 1]$ is the ith-univariate Bernstein polynomials of dimension n and τ_j is the change of coordinate from the uncertain interval $[u_j]$ to $[0, 1]$.

Under the independence and non-correlation assumption among the variables, the joint probability distribution is the product of the marginal masses and it is contained in the p-box $[p_{\bar{c}}] = \prod_{j=1}^{d} [p_{c_j}]$ which can be re-written as

$$[p_c] = \left\{ \sum_{\kappa \in \mathcal{K}} c_\kappa \mathcal{B}_\kappa(\tau(\mathbf{u})) \right\}, \qquad (9)$$

with $\mathcal{K} = \{\kappa = (k_1, \ldots, k_d) \in \mathbb{N}^d : 0 \leq k_j \leq n, \forall j\}$, $\mathcal{B}_\kappa$ is a multivariate Bernstein polynomial, $\tau = \prod_{j=1}^{d} \tau_j$, and c is the unknown coefficient vector. Then, the upper and lower expectations are the solutions of the two linear optimisation problems:

$$E_l(A_v) = \min_{c \in \mathscr{C}} \int_{A_v} p_c(\mathbf{u}) \, d\mathbf{u}, \qquad E_u(A_v) = \max_{c \in \mathscr{C}} \int_{A_v} p_c(\mathbf{u}) \, d\mathbf{u}, \qquad (10)$$

The set $\mathscr{C} \in \mathbb{R}^M$ can be assumed to be an hyper-cube, for example, $\mathscr{C} = [0, M]^M$. In discrete form programmes (10) translate into:

$$E_l(A_v) = \min_{c \in \mathscr{C}} \sum_{s=1}^{M} I_{A_v}(\mathbf{u}_s) \sum_{\kappa \in \mathcal{K}} c_\kappa \mathcal{B}_\kappa(\tau(\mathbf{u}_s)), \qquad (11)$$

and

$$E_u(A_v) = \max_{c \in \mathscr{C}} \sum_{s=1}^{M} I_{A_v}(\mathbf{u}_s) \sum_{\kappa \in \mathcal{K}} c_\kappa \mathcal{B}_\kappa(\tau(\mathbf{u}_s)). \qquad (12)$$

subject to the linear constraint:

$$\frac{1}{M} \sum_{s=1}^{M} \sum_{\kappa \in \mathcal{K}} c_\kappa \mathcal{B}_\kappa(\tau(\mathbf{u}_s)) = 1. \qquad (13)$$

3.3 Impact Probability

Positive polynomials are here applied to the estimation of upper and lower impact probabilities of an asteroid subject to the Yarkovsky effect.

We consider a simplified dynamical model of an asteroid under the gravitational force of the Sun and of the Yarkovsky effect. The latter is assumed to be a purely transverse acceleration A_2/r^2, where r is the heliocentric distance and A_2 is a function of the asteroid physical quantities [3]. The dynamical equations, expressed in Keplerian orbital elements, can be reduced to

$$\frac{da}{dt} = \frac{2A_2(1-e^2)}{np^2}, \quad \frac{dM}{dt} = n, \tag{14}$$

where e is the eccentricity, M is the mean anomaly, $n = \sqrt{\mu/a^3}$ is the mean motion of the unperturbed orbit with μ the gravitational parameter, and $p = a(1-e^2)$ is the conic parameter. For $A_2 = 0$ the dynamics (14) reduces to a pure Keplerian motion, while the semi-major axis drifts outwards for $A_2 > 0$, and inwards for $A_2 < 0$.

Although A_2 is unknown, it can be estimated using the available information on the physical model. Following Farnocchia et al. [3], the coefficient A_2 is expressed as

$$A_2 = \frac{4(1-A)}{9} \Phi(1\text{au}) f(\Theta) \cos\gamma, \quad f(\Theta) = \frac{0.5\Theta}{1+\Theta+0.5\Theta^2}, \tag{15}$$

where $\Phi(1\text{au})$ is the standard radiation force factor at 1 astronomical unit, A is the Bond albedo, Θ is the thermal parameter, and γ is the obliquity. The radiation force at 1 W/m² is computed as

$$\Phi(r) = \frac{3L_0}{2c\rho D}, \tag{16}$$

where L_0 is the luminosity of the Sun, i.e., the total power output of the source, R is the mean radius of the asteroid, m_a the mass of the asteroid, and c is the velocity of light.

Using Bowel et al. [1], the Bond albedo can be written as $A = (0.29+0.684G)p_v$, with G the slope parameter and p_v the geometric albedo. Farnocchia et al.[3] related the thermal parameter Θ to the thermal inertia Γ:

$$\Theta = \frac{\Gamma}{\epsilon\sigma T_{ss}^3} \sqrt{\frac{2\pi}{P_{rot}}}, \tag{17}$$

where ϵ is the emissivity coefficient, σ is the Stefan-Boltzmann constant, P_{rot} is the rotation period, and T_{ss} is the subsolar temperature[2]

$$T_{ss} = \left[\frac{(1-A)L_0}{\eta\epsilon\sigma r^2}\right]^{1/4}, \tag{18}$$

where r is the heliocentric distance of the body and η is the so-called beaming parameter, which is equal to one in the case that each point of the surface is in instantaneous thermal equilibrium with solar radiation.

Delbò et al.[2] related the thermal inertia to the diameter D (in km) by the expression

$$\Gamma = d_0 D^{-\psi},$$

with $d_0 = 300 \pm 45 \, \text{Jm}^{-2}\text{s}^{-1/2}\text{K}^{-1}$ and $\psi = 0.36 \pm 0.09$.

Eventually, the diameter can be related to the absolute magnitude H and the geometric albedo by the formula[6]

$$D = 1329 \frac{10^{-H/5}}{\sqrt{p_v}}. \tag{19}$$

The main uncertainty is represented by the obliquity angle: according to La Spina et al. [5] retrograde and direct rotators are in a 2:1 ratio with the NEO population. Therefore, both the inward and the outward drift of the semi-major axis are possible. In addition, other key physical parameters are known with uncertainty. Due to the lack of knowledge in their distributions, they need to be treated as epistemic uncertainty variables.

It is assumed that both the initial conditions and the model parameters are uncertain. The uncertainty space is $\mathfrak{U}_0 = [\mathbf{x}_0] \times [\mathbf{q}] \subset \mathbb{R}^{10}$, where $\mathbf{x}_0 = (a_0, e_0, I_0, \Omega_0, \omega_0, \ell_0)$ is the initial Keplerian orbital element vector, and $\mathbf{q} = (D, G, p_v, \rho, d_0, \psi, P_{rot}, \gamma)$ is the model parameter vector.

The impact risk is computed at the close approach epoch using the projection on the target plane and the impact parameter b. We say that a collision may occur if the b-parameter is less or equal a safety radius: $b \leq R^*$; this threshold is fixed here at 1.5 Earth radii.

Due to the uncertainty in the initial conditions, the final states of the asteroid defined a connected region more or less elongated along its orbit. Therefore, we say that the significant uncertainty of the b-parameter is contained in the interval $[\underline{b}, \overline{b}] \subseteq \mathbb{R}$ given by

$$\underline{b} = \min_{\mathbf{u} \in \mathfrak{U}_0} b(\mathbf{u}), \quad \overline{b} = \max_{\mathbf{u} \in \mathfrak{U}_0} b(\mathbf{u}). \tag{20}$$

Assuming that the orbital elements are uncertain with known distributions (aleatory uncertainty), while the model parameters are uncertain with unknown distributions (epistemic uncertainty), the product of their probabilities is epistemic and it is indicated with the probability box (shortly, p-box) $[p]$. Then the upper and lower impact probabilities are given by the formulas

$$E_u(A_{R^*}) = \max_{p \in [p]} \int_{A_{R^*}} p(\mathbf{u}) \, d\mathbf{u}, \tag{21}$$

$$E_l(A_{R^*}) = \min_{p \in [p]} \int_{A_{R^*}} p(\mathbf{u}) \, d\mathbf{u}, \tag{22}$$

where $A_{R^*} = \{\mathbf{u} \in \mathfrak{U}_0 : b(\mathbf{u}) \leq R^*\}$ is the event of interest.

We can now assume that each uncertainty variables is contained in a probability box (p-box) delimited by two Beta distribution functions:

$$[p_i] = \{\text{cdf}_{\text{Beta}(\alpha,\beta)} : 1 \leq \alpha, \beta \leq 3\},$$

where i is the variable index. This is the situation in which there are two experts with opposite opinions: one believes that the most probable value is the left extrema of the interval (Beta(1, 3)) and the other that most probable value is the right extrema of the interval (Beta(3, 1)); and in the uncertainty analysis we want to take into account both of them. Each p-box can be re-defined as in Eq. (9) with $\mathscr{B}_k, k = 1, \ldots, M$ multivariate Bernstein polynomials of degree 2. This is due to the fact that each Beta function can be approximated with a positive polynomials series. The degree 2 is because Beta(3,1), Beta(3,1) are approximated exactly with a Bernstein polynomial of degree 2.

Figure 1 shows the cumulative distribution function of the b-parameter corresponding to an aleatory and epistemic case. Curve P represents the case when all variables are aleatory with known Beta distributions with parameters $\alpha = 3, \beta = 3$ for the orbital elements and Beta functions $\alpha = 1, \beta = 1$ (uniform distribution) for the Yarkovsky parameters. On the contrary when uncertainty on the model parameters and initial conditions is epistemic one obtains the upper and lower expectations (curves E_u and E_l, respectively). For all the possible values of the uncertain parameters the impact probability in Eq. (22) is 1 since $b \leq 1.5R_\oplus$, with $R_\oplus$ the Earth radius, for every b.

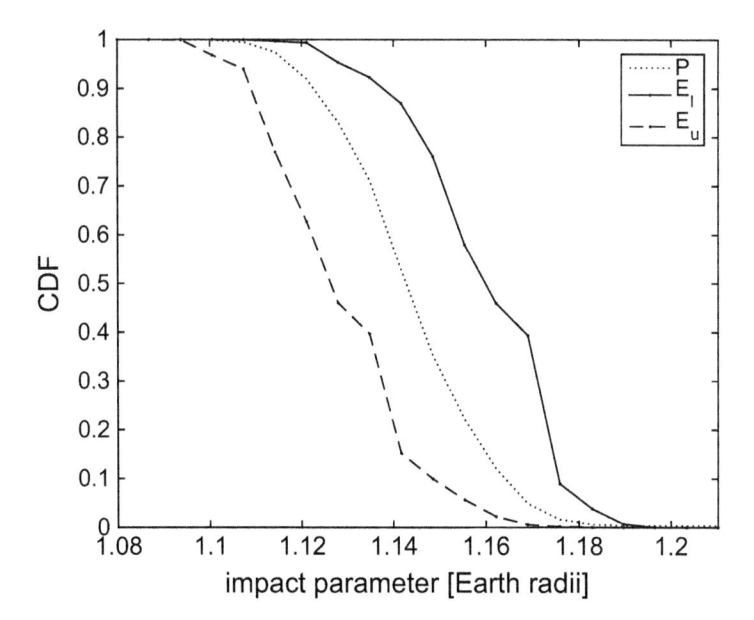

Fig. 1 Impact probability before deflection

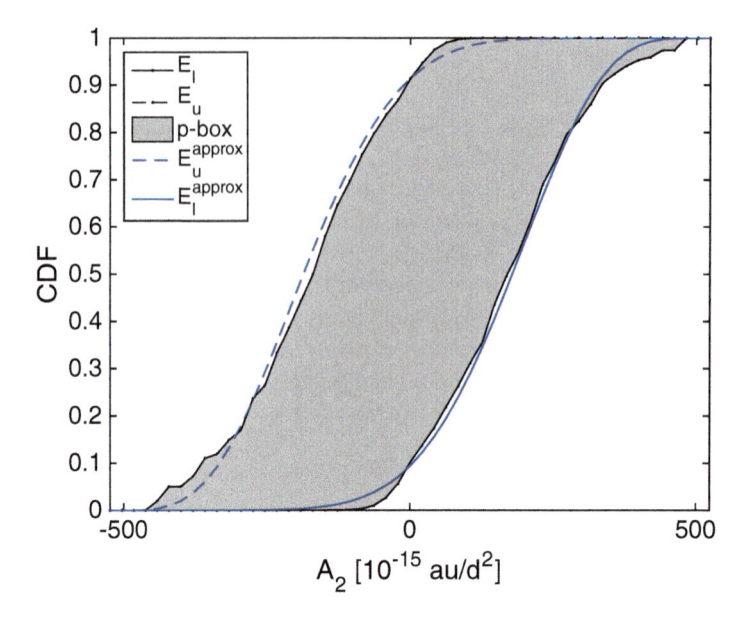

Fig. 2 Upper and lower expectation of the Yarkovsky parameter A_2

From the same analysis one can estimate an upper and lower expectation of the Yarkovsky parameter A_2. Following Sect. 3.2, we can solve Eq. (10) with $\mathscr{C} = [0, M]^M$, $M = 6561$, and the integral approximation given by Eq. (6) on $2 \cdot 10^5$ quasi-Monte Carlo samples. The upper and lower expectation delimiting the p-box are computed on 50 bins in the interval $[-524, 524]$ au/d^2, using Bernstein polynomials as described in Sect. 3.2 on 10^4 quasi-Monte Carlo points. The p-box of A_2 is shown in Fig. 2. The Yarkovsky parameter is computed at a fixed distance of $a\sqrt{1 - e^2}$. In the dynamical model we will sample A_2 from distributions P such that $E_l \leq P \leq E_u$. To simplify the problem the upper and lower expectation have been approximated by Beta functions (Fig. 2):

$$E_u \approx \text{cdf}_{\text{Beta}(5,10)} \qquad E_l \approx \text{cdf}_{\text{Beta}(10,5)}.$$

4 Conclusion

The paper demonstrates the use of positive polynomial expansions to approximate upper and lower expectations on the impact probability of an asteroid with the Earth. The proposed approach leads to the solution of a simple linear programme with a single linear constraint. The use of Bernstein polynomials, as proposed in this paper, allows for the representation of any set of probability distributions with finite support. The main limitation is the exponential growth of the number of polynomial coefficients with the number of dimensions. However, this problem is equally present in Gaussian mixture models although in this case no parameters, appearing nonlinear in the mixture model, need to be defined. The number of terms in the expansion can be calibrated to achieve the desired representation. Furthermore, if uncertain variables are independent, one can define the multivariate Bernstein polynomial simply as the product of univariate polynomials and then calculate the coefficients of each each univariate Bernstein polynomial. In this way the number of coefficients grows linearly with the number of dimensions. The resulting optimisation problem is not linear anymore but has a special form and can be solved very efficiently with a nonlinear solver.

Acknowledgement The work in this paper was partially supported by the Marie Curie FP7-PEOPLE-2012-ITN Stardust, grant agreement 317185.

References

1. Bowell, E., Hapke, B., Domingue, D., Lumme, K., Peltoniemi, J., Harris, A.W.: Application of photometric models to asteroids. In: Binzel, R.P., Gehrels, T., Matthews, M.S. (eds.) Asteroids II, pp. 524–556 (1989)
2. Delbò, M., dell'Oro, A., Harris, A.W., Mottola, S., Mueller, M.: Thermal inertia of near-Earth asteroids and implications for the magnitude of the Yarkovsky effect. Icarus **190**, 236–249 (2007). https://doi.org/10.1016/j.icarus.2007.03.007
3. Farnocchia, D., Chesley, S., Vokrouhlický, D., Milani, A., Spoto, F., Bottke, W.: Near earth asteroids with measurable Yarkovsky effect. Icarus **224**, 1–13 (2013)
4. Ghosal, S.: Convergence rates for density estimation with Bernstein polynomials. Ann. Stat. **10**(3), 1264–1280 (2001)
5. La Spina, A., Paolicchi, P., Kryszczynska, A., Pravec, P.: Retrograde spins of near-earth asteroids from the Yarkovsky effect. Nature **428**, 400–401 (2004)
6. Pravec, P., Harris, A.W.: Binary asteroid population. 1. angular momentum content. Icarus **190**, 250–259 (2007)
7. Zhao, Y., Ausín, C., Wiper, M.P.: Bayesian multivariate Bernstein polynomial density estimation. In: UC3M Working Papers. Statistics and Econometrics. Universidad Carlos III de Madrid. Departamento de Estadística (2013)

Part III
Space Debris Monitoring, Mitigation, and Removal

Trajectory Generation Method for Robotic Free-Floating Capture of a Non-cooperative, Tumbling Target

Marko Jankovic and Frank Kirchner

Abstract The paper illustrates a trajectory generation method for a free-floating robot to capture a non-cooperative, tumbling target. The goal of the method is to generate an optimal trajectory for the manipulator to approach a non-cooperative target while minimizing the overall angular momentum of the entire system (chaser plus target). The method is formulated as an optimal control problem (OCP) and solved via an orthogonal collocation method that transforms the OCP into a nonlinear programming problem (NLP). This way the dynamical coupling between the base and manipulator is actively used to reach the optimum capturing conditions. No synchronization of the relative motion between the target and chaser is necessary prior to the maneuver. Therefore, there is an inherent propellant advantage of the method when compared with the standard ones. The method is applied in 2D simulation using representative targets, such as a Vega 3rd stage rocket body, in a flat spin. The results of simulations prove that the developed method could be a viable alternative or a complement to existing free-flying methods, within the mechanical limitations of the considered space manipulator. The study of the capture and stabilization phases was outside the scope of the present paper and represents future work that needs to be performed to analyze the operational applicability of the developed method.

1 Introduction

Recent studies of the space debris population in Low Earth Orbit (LEO) have concluded that its certain regions have reached a critical density of objects, which will eventually lead to a cascading process called the *Kessler syndrome* [14]. Thus, there is a consensus among researchers that the active debris removal (ADR) should be performed in the near future if we are to preserve the space environment for future

---

M. Jankovic (✉) · F. Kirchner
DFKI-Robotics Innovation Center, Robert-Hooke-Str 1, 28359 Bremen, Germany
e-mail: marko.jankovic@dfki.de; frank.kirchner@dfki.de

generations [14]. Among the proposed ADR capture technologies, those involving orbital robotics are at the moment the most mature ones since they have been successfully tested in-orbit in more than one occasion. Moreover, these technologies are among the most versatile ones given the high number of degrees of freedom they can generally control. Furthermore, once developed they can easily be re-purposed for other in-orbit tasks, such as the on-orbit servicing. However, robotic based solutions have been until today confined only to objects with very low level of non-cooperativeness (i.e. low attitude rates). Moreover, no robotic spacecraft has ever performed a capture of a non-cooperative, tumbling object,[1] especially in the free-floating mode. The former mode is here defined as a control mode of a space robot during which the attitude control system (ACS) of the base spacecraft is non-active during the operation of the manipulator as opposed to the free-flying mode during which the ACS actively counteracts the motion of the base spacecraft due to the operation of the manipulator. The free-floating mode is especially interesting for robotic multi-target ADR missions or on-orbit servicing since it would lead to fuel savings and therefore extension of the overall duration of missions.

Grasping a target that has a residual angular momentum without considering it in the approach phase could pose difficulties to the ACS in the capture and post-capture phases of a mission and most probably would result in a failed maneuver [5, 6]. In fact, the actual capture phase involves physical contact between two bodies and transfer of forces and momenta. For as long as the capturing is not completed, these contact forces will have a random character. Therefore, it is advisable to have a free-floating spacecraft during the maneuver to avoid random effects triggering the activation of an ACS that would lead to shocks and damage of the spacecraft and target [9]. Furthermore, in the post-capture phase the system will need to accommodate a wide variety of angular momenta, therefore requiring heavier reaction wheels and high powered actuators. This directly influences the total spacecraft mass and power consumption which is generally to be avoided. To overcome these limitations, the robotic control subsystem should be developed in such a way that is capable of performing the capture maneuver autonomously, taking into consideration its free-floating dynamics, as well as the angular momentum of the target object during the approach phase of the manipulator.

In this context, the following paper illustrates a method for trajectory generation of the approach phase of a spacecraft mounted manipulator that takes the advantage of its free-floating dynamics to facilitate the capture of a non-cooperative, tumbling target. This is done by pre-loading a desired angular momentum onto the base spacecraft and using the manipulator to transfer it from the base to the arm itself. The method is formulated as an optimal control problem (OCP) and solved as a nonlinear programming problem (NLP).

The structure of the remainder of the paper is as follows: Sect. 2 is dedicated to a brief literature survey where major differences of the existing methods are pointed out with respect to (w. r. t.) the developed method. In Sect. 3 the trajectory

---

[1]To best of our knowledge.

generation method is defined by introducing the main notation, equations of motion, objectives and assumptions of the method. Section 4 presents the implementation of the method as an OCP. In Sect. 5 the results of numerical simulations are illustrated, considering representative targets, such as a Vega 3rd stage rocket body, in a flat spin. Finally, Sect. 6 provides the concluding remarks of the paper and envisioned future work that will improve the developed method.

2 State-of-the-Art

The capture of a target by means of a manipulator mounted on a spacecraft is a well known problem dating back to the early 1980s. Since then there has been a great variety of fundamental research performed on this topic. However, most of the times the dynamical coupling between the manipulator and its base has been regarded as a disturbance to be suppressed by either actively controlling the base, by means of an ACS [1, 2], or passively, by means of an optimized path of the manipulator [8] in order to maintain a fixed attitude of the base spacecraft in the inertial system for communication purposes. Furthermore, the stabilization and detumbling of the compound, once the target was captured, has been, most of the times, relegated entirely to an ACS of the base spacecraft [2, 13]. This could lead to a failed maneuver if a high angular momentum of the target is considered due to the limitations of an ACS, such as the amount of propellant or dimensions of reaction wheels (RWs).

More specifically, in [5], Dimitrov et al. describe a trajectory generation method formulated as an NLP which facilitates the post-capture maneuver by pre-loading an angular momentum in the RWs of the ACS and using the manipulator to transfer it from the base to the manipulator. However, the authors focus mainly on minimizing the attitude disturbance of the base during maneuvers and assume an active ACS through the contact phase, which might lead to unexpected behavior of the overall system as evidenced in [9]. Furthermore, while they do minimize the angular momentum to be transferred to the base, they do not deal with the management of the angular momentum of the overall system (i.e. chaser plus target) relegating this task to the RWs and thus ACS of the base which might be a problem in case of angular velocities of targets higher then those considered by the authors, i.e. $> \pm 1°/s$.

In [13], Lampariello et al. on the other hand, describe a motion planning for the on-orbit grasping of a non-cooperative target. The motion planning includes the whole capture maneuver (i.e. from approach to the stabilization) for typical target tumbling motions, i.e. flat spin of $\pm 4°/s$. The method is formulated as an OCP and solved as an NLP. The collision avoidance is included in the method as an inequality constraint. However, the angular momentum management of the stack is relegated entirely to the ACS of the base spacecraft during the post-capture phase. The approach of the manipulator is performed in the free-floating mode and considered cost functions include the mechanical energy of the manipulator which allows reduced joint torques and velocities of the manipulator.

In [8], Flores-Abad et al. consider the capture of a tumbling target as an OCP trying to minimize joint torques and the attitude disturbance of the base spacecraft during the contact phase. This is done by directing the contact force through the center of mass of the compound. However, the method formulated in this way is found hard to be accomplished, due to the generally unpredictable contact dynamics. Moreover, management of the angular momentum of the target has not been addressed and the collision avoidance has not been mentioned. Furthermore, only a 2 degrees of freedom (DOF) manipulator is considered in the study.

In [1, 2], Aghilli addressed the capture of a tumbling target as an OCP of the pre- and post-capture phases of a space robot. In the pre-grasping phase, an optimal trajectory is planned to intercept a grasping point on the target with zero relative velocity, subject to acceleration limit and adequate target alignment. In the post-grasping phase, the manipulator is used to damp out the angular and linear momenta of a target as quickly possible subject to the constraints of the manipulator. However, both phases are performed in the free-flying mode, thus assuming usage of a coordinated control between the ACS and the manipulator. Moreover, the collision avoidance problem was not tackled in these studies.

The method presented in this paper builds upon the mentioned studies and tries to solve their issues. Mainly, the method is based on the Bias Momentum Approach (BMA) developed by Dimitrov et al. in [5]. However, with respect to the mentioned method, the novelty of the following work consists mainly in: (a) free-floating mode of the chaser (i.e., the ACS is completely switched-off during the approach maneuver), (b) limited kinetic energy of the chaser system at the end of the maneuver, (c) management of the overall angular momentum of the stack, (d) collision avoidance. This way, a spacecraft mounted robot would be able to capture a non-cooperative, tumbling target without the need to synchronize the relative attitude motion and with limited need to de-tumble the stack after the capture.

3 Method Definition

The capture of a tumbling target may be described essentially in four phases as illustrated in Fig. 1:

1. a chaser spacecraft performs the observation and pose[2] estimation of a target object,
2. a chaser approaches a target using its ACS to place the target in a predefined berthing box,
3. the manipulator of the spacecraft approaches the designated berthing feature in either free-flying or free-floating mode,

---

[2]Defined as position and orientation of an object.

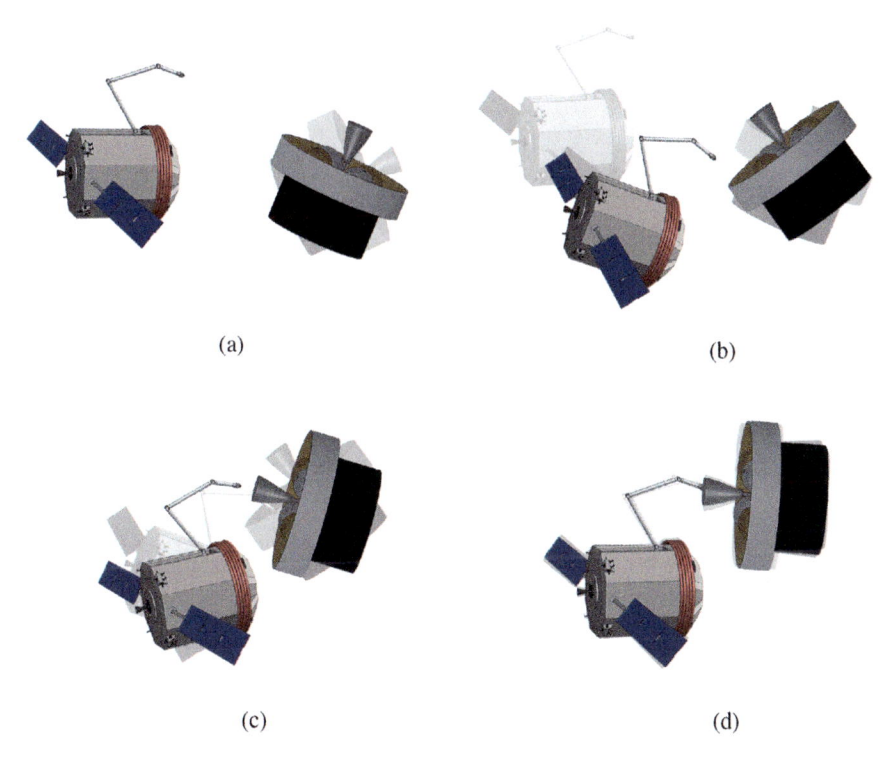

Fig. 1 Mission phases of a robotic capturing of a non-cooperative target. (**a**) Observation. (**b**) Spacecraft approach. (**c**) Manipulator approach. (**d**) Capture and stabilization

4. the end-effector of the manipulator captures the berthing feature and the relative motion between the two objects is stabilized.

In case of angular speeds of a target higher then 5°/s, the current doctrine [3, 4, 16] dictates a synchronization maneuver that should precede the mentioned phases in order to match the relative attitudes of the robotic chaser spacecraft and tumbling target. However, this maneuver requires non negligible fuel requirements[3] which is why it should be minimized as much as possible especially in case of multi-target ADR or on-orbit servicing (OOS) missions, where the spacecraft would need to capture or service multiple targets during one mission. Thus, the method described in this paper aims at minimizing this requirement by exploring the possibility of generating an optimal trajectory of the manipulator to approach a target that would reduce the need for the follow-up de-tumbling phase. This phase is illustrated in Fig. 1c. It ends with the manipulator able to make the contact with a berthing feature

---

[3]Especially considering that during one capture it will need to be performed twice, once to synchronize the motion and the second time to de-tumble the chaser-target stack.

of the target. It is assumed that only one useful berthing feature on the target exist and that it is pre-selected by a user beforehand for simplicity.

3.1 Equations of Motion and Main Notation

The dynamics of a generic rigid free-floating manipulator composed of a free-floating base and an m-link serial manipulator with no external forces/moments acting on the spacecraft, can be expressed as follows [22]:

$$\begin{bmatrix} H_b & H_{bm} \\ H_{bm}^\top & H_m \end{bmatrix} \begin{bmatrix} \ddot{x}_b \\ \ddot{\phi}_m \end{bmatrix} + \begin{bmatrix} c_b \\ c_m \end{bmatrix} = \begin{bmatrix} 0 \\ \tau \end{bmatrix} \tag{1}$$

where x_b expresses the pose of the base spacecraft, ϕ_m are the joint angles of the manipulator and τ are the joint torques. H_b, H_m and H_{bm} are the inertia matrices of the base body, manipulator and the coupling between the base and arm, respectively. c_b and c_m are the velocity dependent non-linear terms of the base and arm, respectively [22].

All the variables are expressed w. r. t. an inertial reference frame, Σ_I, assumed to be moving in an orbital plane of the space robot and thus translating with it for the short duration of the capture maneuver [22].

Integrating the upper part of Eq. (1) w. r. t. time it is possible to obtain the equation of the total momentum of the free-floating system around the center of mass (COM) of the base spacecraft as [22]:

$$\mathscr{L} = \begin{bmatrix} P \\ L \end{bmatrix} = H_b \dot{x}_b + H_{bm} \dot{\phi}_m \tag{2}$$

where P and L are the linear and angular momenta of the robotic system around the center of mass of the base.

Alternatively, the total momentum equation can be re-written w. r. t. Σ_I as [5]:

$$\begin{bmatrix} P \\ L \end{bmatrix} = H_b \dot{x}_b + H_{bm} \dot{\phi}_m + \begin{bmatrix} 0 \\ r_b \times P \end{bmatrix} \tag{3}$$

where P and L are the linear and angular momenta of the robotic system around the inertial reference frame, Σ_I.

Therefore, the angular momentum equation can be expressed in a shorter form as [5]:

$$L = \tilde{H}_b \omega_b + \tilde{H}_{bm} \dot{\phi}_m + r_g \times P \qquad (4)$$

where $\tilde{H}_b$ and $\tilde{H}_{bm}$ are the modified inertia matrices as defined in [5, 21] and r_g is the vector from Σ_I to the COM of the whole system.

Alternatively, Eq. (4) can be written in even a shorter form as [5]:

$$L = L_b + L_{bm} + L_p \qquad (5)$$

where the individual contributions to the overall angular momentum of the space robot, by the partial angular momenta of the base, L_b, robot, L_{bm} and linear momentum, L_p, are clearly evidenced and give us some insight on how the overall angular momentum of the space robot can be managed if no external forces are applied to the centroid of the system. In fact, Eqs. (4) and (5) express the first order non-holonomic constraint of a free-floating robot [22] which couples the motion of the manipulator and its base spacecraft. Therefore, assuming no external forces applied to the system and no linear motion of the center of mass of the overall system, i.e. $L_p = 0$, it is evident from Eq. (4) that because of the conservation of momentum any movement of the manipulator will only results in a redistribution of the already present angular momentum in the system. Thus, if we are to optimize the distribution of the angular momentum of the system during the approach phase, where the ACS of the base spacecraft is switched off, the only way to achieve it would be by optimizing the motion of the manipulator, i.e. its joint angular velocities or more concretely the joint torques.

3.2 Method Objectives and Assumptions

The objective of the developed method is to provide an optimal approach strategy of the manipulator, based on a proper distribution of the angular momentum within the system, to facilitate the control of the manipulator and ACS in the post-capture phase and thus lower the overall propellant needs of the mission.

The assumptions made in this study are the following:

1. the target undergoes a constant flat spin motion, i.e. a pure rotational motion around its principal axis of inertia, assumed to be known in advance and being $\geqslant 5°/s$ in magnitude,
2. the inertial characteristics of the target are well known,
3. there are no external forces acting on the entire system (chaser plus target),
4. there is no relative linear motion between the bodies at the beginning of the approach maneuver,

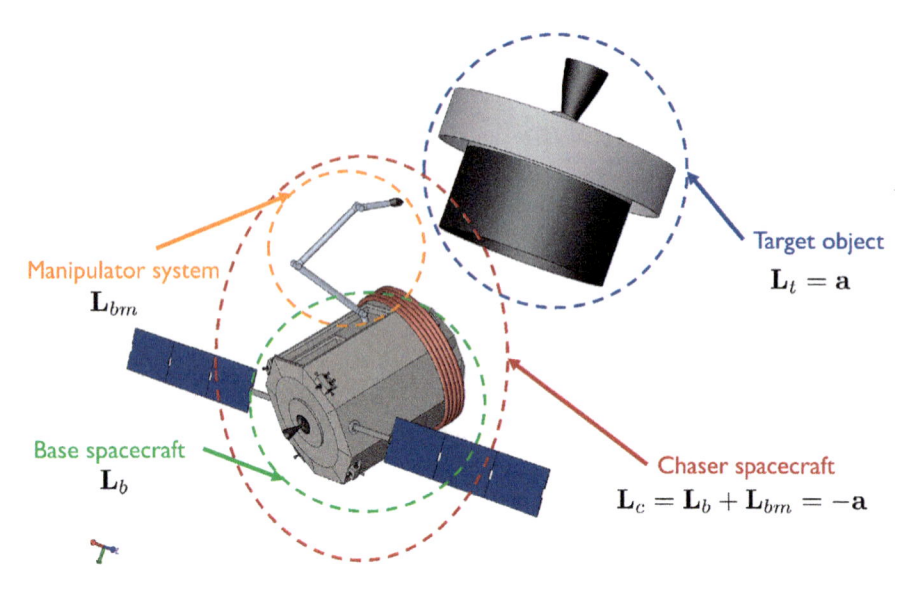

Fig. 2 Distribution of partial angular momenta of the system

5. the system is composed only of rigid bodies,
6. the ACS of the chaser spacecraft is switched-off during the approach phase,
7. the manipulator is redundant with respect to the imposed task constraints,
8. the contact and de-tumbling phases are not considered at the moment by the optimization method.

With those assumptions in mind and referring to Eq. (5) and Fig. 2, the angular momentum of the entire system (chaser plus target) can be sufficiently defined by two variables: L (or better L_c), and L_t, representing the angular momentum of the chaser and target, respectively.

To facilitate the post-capture (i.e. the stabilization) phase of a capture mission, the management of the angular momentum of the entire system has to be taken into consideration during the approach phase. To this end, the developed method considers as the most favorable condition the following one: $\boldsymbol{L}_c = \boldsymbol{L}_b + \boldsymbol{L}_{bm} = -\boldsymbol{L}_t$. Should this condition be satisfied, no post-capture angular momentum management should be needed since the overall momentum of the system (chaser plus target), at the contact of the manipulator with the target, would be zero. To reach this condition while taking into consideration the made assumptions, $-\boldsymbol{L}_t$ needs to be initially pre-loaded onto the base spacecraft by means of an ACS, prior to the manipulator maneuver and then re-distributed via the manipulator. The amount of momentum to be re-distributed from the base to the manipulator will depend on the desired pre-contact strategy, as outlined in [5]. In our case, we would like to off-load the base spacecraft as much as possible to minimize the transfer of angular momentum from the target to the base spacecraft in the post-capture phase of the mission. To reach such a state the following condition needs to be imposed onto the trajectory of the robotic system at the end time of an approach maneuver [5]:

$$\begin{cases} \|L_{bm}(t_f)\| \leq \|L_t\| \\ L_{bm}(t_f) \cdot L_t < 0 \end{cases} \qquad (6)$$

The condition indicates that the optimal amount of angular momentum to be transferred onto the manipulator prior to a contact phase should be in any case smaller or equal in magnitude and opposite in direction to that of the target such that only a minimal residual angular momentum will be transferred to the base spacecraft after the contact phase. Ideally that condition would be:

$$\begin{cases} L_{bm}(t_f) = -L_t \\ L_b(t_f) = 0 \end{cases} \qquad (7)$$

where the angular momentum initially pre-loaded onto the base is completely transferred to the manipulator. However, this might not always be possible due to mechanical limitations of manipulators which is why the condition expressed in Eq. (6) is used and not the one expressed in Eq. (7).

4 Method Implementation

The trajectory generation method developed to optimize the distribution of partial angular momenta of the chaser spacecraft has been implemented as a constrained nonlinear optimization problem (OCP). The OCP has been transformed, via a direct collocation method, into a nonlinear programming problem (NLP) and solved using the MATLAB's optimization toolbox.

Given the complexity of the OCP at hand and the need for a good initial guess, the optimization is performed in two steps. At first a reasonable guess of the trajectory is obtained with: a simple objective function, limited number of collocation points and relaxed error tolerances. Then, the solution is refined using a more complex objective function, larger number of collocation points and smaller error tolerances.

Morse specific information about the implementation of the method can be found in what follows.

4.1 Mathematical Formulation of the Optimization Problem

The optimization problem of the manipulator approach maneuver is formulated as a single-phase continuous-time trajectory optimization problem with an objective to minimize in the first step a path integral:

$$\min_{t_0, t_f, x(t), u(t)} \int (\tau^\top \tau) dt \qquad (8)$$

where its minimization entails smaller joint torques during the maneuver. t_0 and t_f are the initial and final time of the approach phase, respectively; x are the state variables of the system (i.e. $x \triangleq [x_b; \phi_m; \dot{x}_b; \dot{\phi}_m]$) and u are the control variables of the system (i.e. $u \triangleq \tau$ joint torques).

In the second step the objective function has been expressed in Bolza from, thus containing both a boundary objective and a path integral. The objective function defined for this second step has the following expression:

$$\min_{t_0, t_f, x(t), u(t)} \kappa_1\left(\sqrt{\omega_b^2(t_f)}\right) + \int_{t_0}^{t_f} \left[\kappa_2(\dot{\phi}_m^T \dot{\phi}_m) + \kappa_3(\tau^\top \tau)\right] d\tau \qquad (9)$$

where the Mayer term of the objective function, i.e. the boundary objective, expresses the magnitude of the angular velocity of the base, ω_b, at t_f and the Lagrange term of the objective function, i.e. the path integral, is a sum of the squared values of joint velocities and joint torques. k_1, k_2 and k_3 represent the weights of individual terms of the objective function and are custom to each problem.

The minimization of the Mayer term is meant to contain the magnitude of the angular velocity of the base to achieve a desired distribution of partial momenta of the robot (as described in Eq. (6)), while the minimization of the Lagrangian term should decrease the angular velocities of the joints. The torque squared term represents only a regularization term of the objective function, as suggested in [11].

Both objective functions are subject to:

- the free-floating dynamics of a space robot (see Eq. (1));
- kinematic constraints:

$$\begin{aligned} x_h(t_f) - x_t(t_f) &= 0 \\ \dot{x}_h(t_f) - \dot{x}_t(t_f) &= 0 \end{aligned} \qquad (10)$$

which define the final pose and velocities of the end-effector of the manipulator, i.e. x_h and $\dot{x}_h$, respectively;

- state, control, time and collision avoidance constraints:

$$\begin{aligned} \phi_m^- &\leq \phi_m(t) \leq \phi_m^+ \\ \dot{\phi}_m^- &\leq \dot{\phi}_m(t) \leq \dot{\phi}_m^+ \\ \tau^- &\leq \tau \leq \tau^+ \\ t^- &\leq t_0 < t_f \leq t^+ \\ r_{coll}(\phi_m(t)) &\leq 0 \end{aligned} \qquad (11)$$

where r_{coll} represents the collision avoidance constrains;

- initial and final bounds:
$$\begin{aligned} x_b(t_0) &= \mathbf{0} \\ \boldsymbol{\phi}_m(t_0) &= \boldsymbol{\phi}_{m0} \\ \dot{\boldsymbol{\phi}}_m(t_0) &= \mathbf{0} \\ \dot{x}_b(t_0) &= -\boldsymbol{H}_b^{-1}\boldsymbol{M}_t\dot{x}_t \\ t_o &= 0 \\ 0 &\le t_f \le 60 \end{aligned} \quad (12)$$

where M_t is the mass matrix of the target as defined in [17] and x_t is the pose of the target object. The expression $\dot{x}_b(t_0) = -\boldsymbol{H}_b^{-1}\boldsymbol{M}_t\dot{x}_t$ reflects the desired initial conditions, defined in Sect. 3.2 on page 117, i.e. $L_c(t_0) = L_b(t_0) = -L_t$.

4.2 Implementation of the Optimization Problem

The described OCP has been implemented as an NLP using an open-source MATLAB library OptimTraj [12] designed for solving continuous-time, single-phase trajectory optimization problems. The library implements several direct collocation methods, such as the trapezoidal and Hermit-Simpson collocations, multiple shooting and orthogonal Chebyshev-Lobatto collocation.[4] The latter was chosen in this study as a collocation method of choice given its robustness and efficiency when compared to lower order collocation methods implemented in OptimTraj library.

The solution of an NLP was then found with the MATLAB's fmincon closed-source, solver which is part of the Optimization Toolbox.

The kinematics and dynamics of the free-floating platform were developed using the Recursive Dynamics Simulator (ReDySim)[18, 19] and SPAcecraft Robotics Toolkit (SPART) [20] MATLAB libraries.

The collision avoidance of bodies was implemented via the MATLAB function *distLinSeg* developed by Sluciak, O., using an algorithm for fast computation of the shortest distance between two line segments developed by Lumelsky [15].

5 Method Evaluation

The trajectory generation method outlined in the previous section has been evaluated in 2D simulation studies performed with a 4 DOF manipulator whose mass properties were chosen to be similar to those of the robotic arm used in the Japanese ETS-VII robotic mission [22], as visible in Table 1.

The considered target objects were: a spacecraft (S/C), having similar physical characteristics to those of the robotic base spacecraft, and a Vega 3rd stage rocket

---
[4]Implemented in the library using an additional open-source MATLAB library Chebfun [7].

Table 1 Inertial parameters of the chaser and targets

Body	Mass (kg)	I_{zz} (kg m^2)
Link 1	57.46	0.2281
Link 2	38.43	0.1525
Link 3	26	0.1032
Link 4	18.49	0.0734
Base S/C	868.92	579.28
Generic target S/C	868.92	579.28
Vega 3rd stage R/B	999.6	1228.47

body (R/B). The inertia tensor of the latter was approximated for simplicity to that of a cylinder having: a mass, diameter and length of: ~1000 kg,[5] 1.907 m and 3.467 m, respectively. The inertia tensor of the target S/C on the other hand was calculated considering a cube having an edge length of 2 m and a mass of 868.92 kg. More details can be found in Table 1.

For the purposes of simulation, the targets were assumed to be in a flat-spin attitude motion around their principal axis of inertia (i.e. z axis in simulations) with a constant rate of $-5°/s$.

The initial configuration of the robot was assumed in all simulations to be the following:

$$\begin{aligned} \boldsymbol{x}_b(t_0) &= \boldsymbol{0}_{3\times 1} & (\text{m},°) \\ \boldsymbol{\phi}_m(t_0) &= [-60, 45, 45, -30]^T & (°) \\ \dot{\boldsymbol{\phi}}_m(t_0) &= \boldsymbol{0}_{4\times 1} & (°/s) \\ \dot{\boldsymbol{x}}_b(t_0) &= -\boldsymbol{H}_b^{-1}\boldsymbol{M}_t\dot{\boldsymbol{x}}_t & (\text{m/s},°) \end{aligned} \quad (13)$$

The grasping features/points on the targets were assumed to be the nozzles of their main propulsion systems, placed at a distance of 0.6 and 0.9 m from the main bodies of the S/C and Vega 3rd stage R/B, respectively. Their inertial 2D position was therefore assumed to be at $\boldsymbol{r}_{gf} = [0.3149, 3.6615]^T$ meters. The initial position of the centroid of the base spacecraft was instead assumed to be at the origin of the inertial frame, Σ_I, therefore at $\boldsymbol{r}_b = [0, 0]^T$ meters.

The results of the conducted simulation studies, based on the previous constraints and conditions, are presented in Figs. 3 and 4. The optimized trajectories were obtained using 30 collocation points and objective function weights $\kappa_1 = 1$, $\kappa_2 = 0.02$, $\kappa_3 = 0.001$. In all studies, the initial guess of a trajectory was always generated with a lower-order, trapezoidal collocation method, in order to find a first coarse solution to an easy to solve problem, thus reducing the amount of computational effort needed for the overall optimization.

From the results it can be concluded that, with the developed method it is possible to obtain a locally optimal approach trajectory of a free-floating space robot to capture a tumbling target under the imposed assumptions and conditions.

---

[5]833 kg + 20% of margin.

Trajectory Generation Method for Robotic Free-Floating Capture... 123

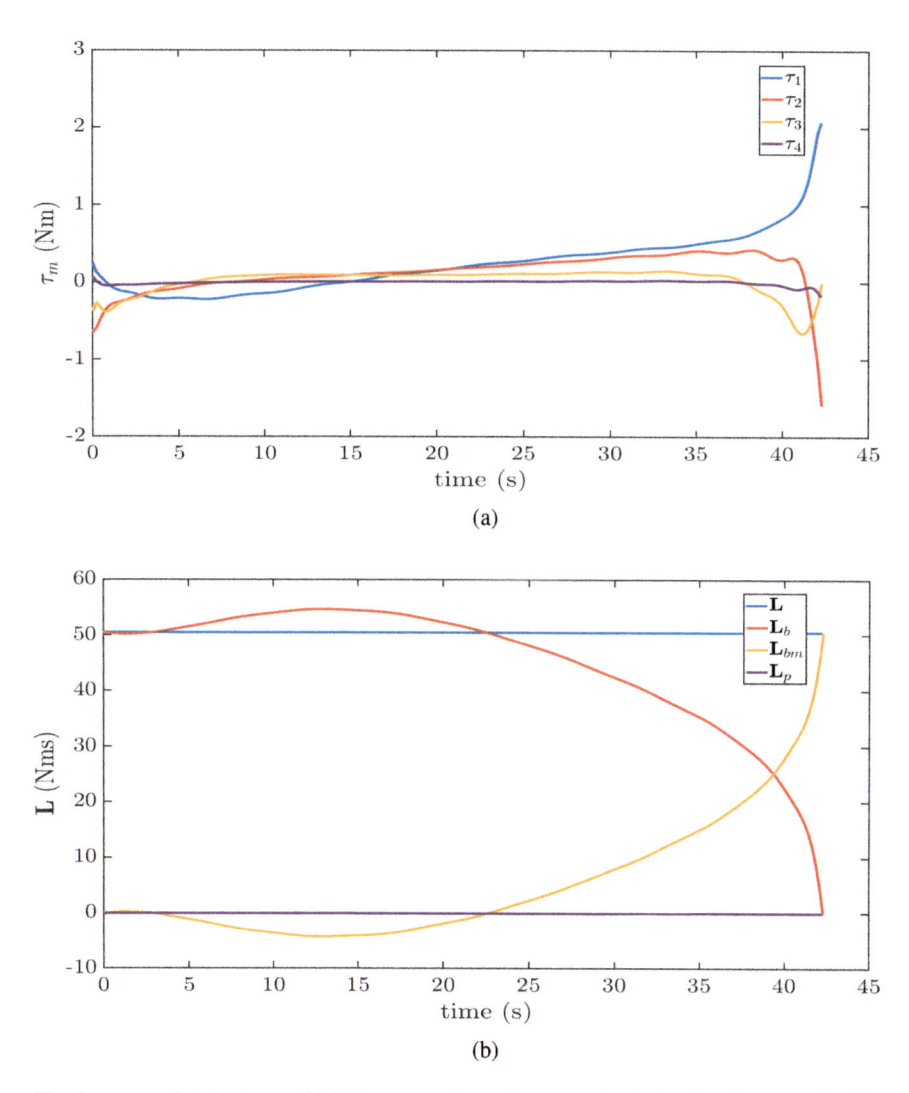

Fig. 3 Approach trajectories of joint torques and angular momenta of the robot for a generic S/C target having $\omega_t = -5°/s$. (**a**) Joint torques of the manipulator. (**b**) Angular momenta of the chaser S/C

Specifically, in case of a target having mass properties similar to those of the chaser base S/C, an ideal momentum distribution, as expressed in Eq. (7), is possible, as visible in Fig. 3b. This condition assures that no angular momentum will be transferred through the manipulator to the base during a contact phase. Therefore, the post-capture manipulator and ACS control would be significantly simplified since the overall capture operation could be ideally achieved simply by servo locking the manipulator joints.

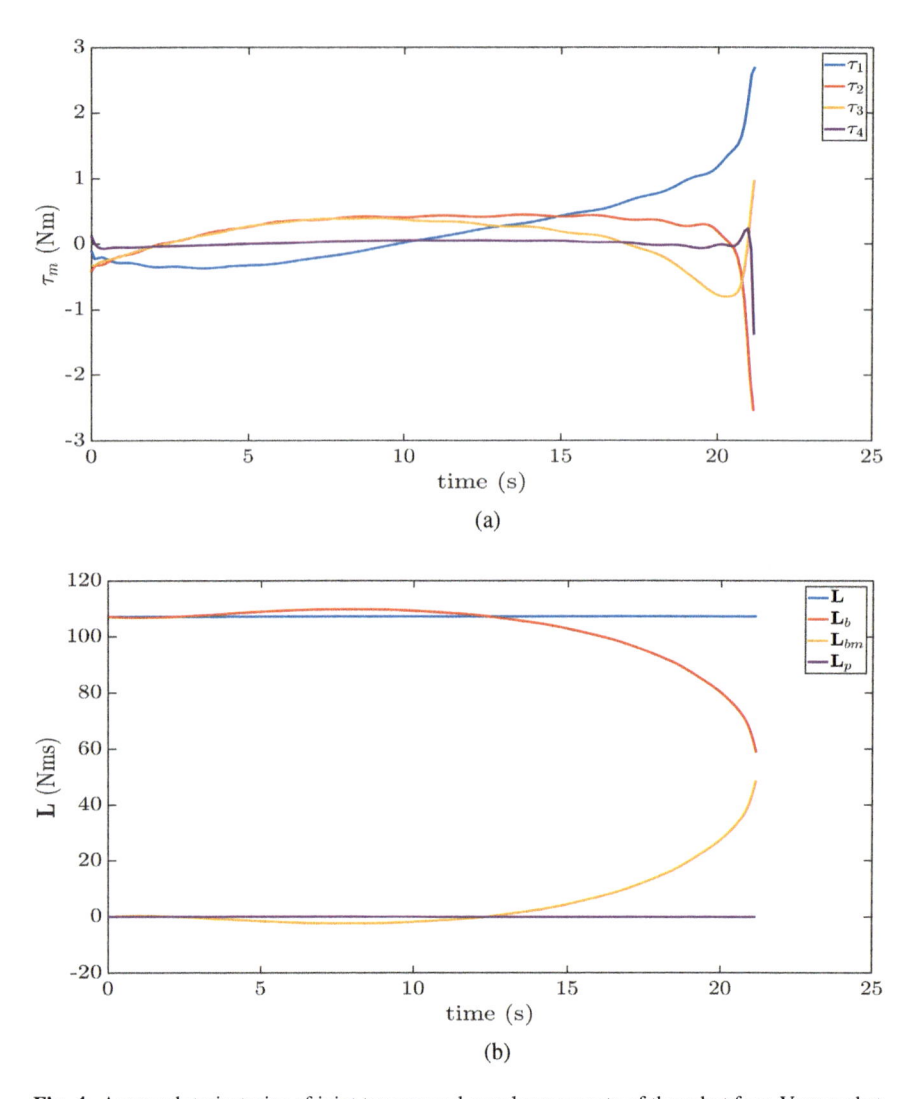

Fig. 4 Approach trajectories of joint torques and angular momenta of the robot for a Vega rocket body target having $\omega_t = -5°/s$. (**a**) Joint torques of the manipulator. (**b**) Angular momenta of the chaser S/C

In case of a target with bigger inertial properties and/or higher angular velocity, only partial momentum distribution transfer is achievable, as visible in Fig. 4b. This distribution would cause the remaining angular momentum, resulting from the difference between the target angular momentum and that of the manipulator, i.e. $\boldsymbol{L}_t - \boldsymbol{L}_{bm}$, to transfer to the base through the manipulator, with a transfer rate depending on a variety of conditions, such as the pre-contact configuration of the manipulator, contact forces, post-capture control, etc.. Therefore, the post-capture manipulator control and the overall capture procedure would be more complex with respect to the ideal case. Nevertheless, this is not found to be a major drawback of the implemented method since it could be overcome by either changing the mass ratio between the base S/C and manipulator or by using this method in combination with more standard free-flying methods where only a residual angular velocity of a target would be compensated with the developed method.

The fuel requirements (i.e. the required change in velocity or Δv) of the proposed method were found to amount to $\Delta v = 0.1$ m/s, which is four times less then that of a standard syncing maneuver requiring $\Delta v = 0.4016$ m/s, for the maneuver performed at the same distance of the grasping feature from the COM of the chaser S/C.

Based on the previous results it is possible to asses that the advantages of the method are: (a) minimum angular motion of the compound after the capture (w.r.t. the inertial reference frame), (b) minimum redistribution of L_t within the chaser upon the contact, thus implying easier post-capture manipulator and ACS control, (d) no need for an ACS during the approach and post-capture phases of the manipulator. The disadvantages of the current implementation of the method are on the other hand assessed to be: (a) inability to transfer very high magnitudes of L_b or better L_t onto the manipulator, (b) need for a pre-loaded angular momentum onto the base S/C, (c) nonexistence of tracking phase that would permit uncertainties in the position of the grasping feature.

6 Conclusions

A method for the trajectory generation of the approach phase of a robotic ADR or OOS mission has been described. The method has been formulated as an OCP from the point of view of redistribution of the angular momentum between the manipulator and its base spacecraft. The objective of the optimization is to limit the transfer of the angular momentum from the target to the base spacecraft at the time of contact. The method allows to deal with tumbling targets without the need for relative synchronization and usage of the ACS during the de-tumbling phases. The method has proven to be a viable option to more traditional methods since: it would allow an easier post-capture/stabilization control of the manipulator and base S/C, and introduce fuel savings especially in case of multi-target ADR or OOS missions. Nevertheless, the current implementation of the method presents some limitations that will need to be tackled in the near future to prove its operational applicability in a real-world scenario. Those limitations include the nonexistence of a tracking phase during the manipulator approach and lack of analysis of the impact of the developed method onto the subsequent capture and stabilization phases.

Acknowledgements The research work here presented was supported by the Marie Curie Initial Training Network Stardust, FP7-PEOPLE-2012-ITN, Grant Agreement 317185. Thus, the authors would like to thank the European Commission and the Research Executive Agency for their support and funding. Moreover, we would like to thank researchers Dr. Dimitrov, D. (developer of the Bias Momentum Approach (BMA) [5] that inspired our method) and Dr. Shah, S. V. (developer of the ReDySim MATLAB toolbox that we used extensively in our method) for their help, time and advice.

References

1. Aghili, F.: Coordination control of a free-flying manipulator and its base attitude to capture and detumble a noncooperative satellite. In: 2009 IEEE/RSJ International Conference on Intelligent Robots and Systems, pp. 2365–2372. IEEE, New York (2009). https://doi.org/10.1109/IROS.2009.5353968
2. Aghili, F.: Pre- and post-grasping robot motion planning to capture and stabilize a tumbling/drifting free-floater with uncertain dynamics. In: 2013 IEEE International Conference on Robotics and Automation, pp. 5461–5468. IEEE, New York (2013). https://doi.org/10.1109/ICRA.2013.6631360
3. Bonnal, C., Ruault, J.M., Desjean, M.C.: Active debris removal: recent progress and current trends. Acta Astronaut. **85**, 51–60 (2013). https://doi.org/10.1016/j.actaastro.2012.11.009
4. Castronuovo, M.: Active space debris removal-a preliminary mission analysis and design. Acta Astronaut. **69**(9–10), 848–859 (2011). https://doi.org/10.1016/j.actaastro.2011.04.017
5. Dimitrov, D.: Dynamics and control of space manipulators during a satellite capturing operation. PhD Thesis, Tohoku University (2005). http://drdv.net/pdf/drdv_thesis.pdf
6. Dimitrov, D.N., Yoshida, K.: Utilization of distributed momentum control for planning approaching trajectories of a space manipulator to a target satellite. In: Proceedings of 'The 8th International Symposium on Artificial Intelligence, Robotics and Automation in Space - iSAIRAS', 603, pp. 1–8. ESA Publications Division-ESTEC, Munich, Germany (2005)
7. Driscoll, T.A., Hale, N., Trefethen, L.N.: Chebfun Guide (2014). http://www.chebfun.org/
8. Flores-Abad, A., Wei, Z., Ma, O., Pham, K.: Optimal control of space robots for capturing a tumbling object with uncertainties. J. Guid. Control. Dyn. **37**(6), 2014–2017 (2014). https://doi.org/10.2514/1.G000003
9. Innocenti, L.: CDF Study Report: e.Deorbit, e.Deorbit Assessment. Technical Report, ESTEC - ESA, Noordwijk, The Netherlands (2012)
10. Jankovic, M., Paul, J., Kirchner, F.: GNC architecture for autonomous robotic capture of a non-cooperative target: preliminary concept design. Adv. Space Res. **57**(8), 1715–1736 (2016). https://doi.org/10.1016/j.asr.2015.05.018
11. Kelly, M.: An introduction to trajectory optimization: how to do your own direct collocation. SIAM Rev. (under pub, 2017). https://goo.gl/GH2PHy
12. Kelly, M.: OptimTraj - Trajectory Optimization for Matlab (2017). https://github.com/MatthewPeterKelly/OptimTraj
13. Lampariello, R., Hirzinger, G.: Generating feasible trajectories for autonomous on-orbit grasping of spinning debris in a useful time. In: IEEE International Conference on Intelligent Robots and Systems, pp. 5652–5659. DLR, Tokyo, Japan (2013). https://doi.org/10.1109/IROS.2013.6697175
14. Liou, J.C.: An active debris removal parametric study for LEO environment remediation. Adv. Space Res. **47**(11), 1865–1876 (2011). https://doi.org/10.1016/j.asr.2011.02.003
15. Lumelsky, V.J.: On fast computation of distance between line segments. Inf. Process. Lett. **21**(2), 55–61 (1985). https://doi.org/10.1016/0020-0190(85)90032-8

16. Matsumoto, S., Ohkami, Y., Wakabayashi, Y., Oda, M., Ueno, H.: Satellite capturing strategy using agile Orbital Servicing Vehicle, Hyper-OSV. In: Proceedings 2002 IEEE International Conference on Robotics and Automation (Cat. No.02CH37292), vol. 3, pp. 2309–2314. IEEE, New York (2002). https://doi.org/10.1109/ROBOT.2002.1013576
17. Oki, T., Abiko, S., Nakanishi, H., Yoshida, K.: Time-optimal detumbling maneuver along an arbitrary arm motion during the capture of a target satellite. In: 2011 IEEE/RSJ International Conference on Intelligent Robots and Systems, pp. 625–630. IEEE, New York (2011). https://doi.org/10.1109/IROS.2011.6095159
18. Shah, S.V., Nandihal, P.V., Saha, S.K.: Recursive dynamics simulator (ReDySim): a multibody dynamics solver. Theor. Appl. Mech. Lett. **2**(6), 063,011 (2012). https://doi.org/10.1063/2.1206311
19. Shah, S.V., Saha, S.K., Dutt, J.K.: Recursive Dynamics Simulator (ReDySim) (2016). http://redysim.weebly.com/
20. Virgili-Llop, J.: SPART: SPAcecraft Robotics Toolkit (2016). https://github.com/NPS-SRL/SPART
21. Xu, Y., Kanade, T.: Space Robotics: Dynamics and Control, p. 285. Springer, New York (1993)
22. Yoshida, K., Wilcox, B.: Space robots and systems. In: Siciliano, B., Khatib, O. (eds.) Springer Handbook of Robotics, chap. 45, pp. 1031–1063. Springer, Berlin (2008). https://doi.org/10.1007/978-3-540-30301-5_46

Taxonomy of LEO Space Debris Population for ADR Capture Methods Selection

Marko Jankovic and Frank Kirchner

Abstract This paper illustrates a novel taxonomy method for LEO space debris population whose aim is to classify the LEO space debris objects such that it is possible to identify, safety wise, for each one the most suitable Active Debris Removal (ADR) capture method. The method is formulated in two distinct layers. In the first layer, a main class of an object is identified, based on its most prominent dynamical and physical traits. At this stage identifying the most suited ADR method is still a difficult task, due to the crude nature of the parameters used for the classification. The second taxonomic layer is thus performed on top of the first one. In it the break-up risk index and levels of non-cooperativeness of an object are identified and the ADR association is refined. Examples of application of the developed taxonomy are presented at the end of the paper and conclusions are drawn regarding the best methods to be used for the main categories of LEO space debris under investigation for future ADR missions.

1 Introduction

With the start of the human space activities in 1957, the LEO, once pristine and void, started showing signs of congestion which will lead to a critical density of objects in orbit and eventually to a cascading problem predicted by Kessler and Cour-Palais in 1978, called the Kessler syndrome. The space community is trying, up-to-date, to counteract this syndrome only with a set of non-binding mitigation measures that would, if applied correctly, assure no future sources of space debris [8]. Despite these efforts, studies on the subject showed that the population of objects bigger then 10 cm (considered to be lethal for any active satellite) is expected to rise by

---

M. Jankovic (✉) · F. Kirchner
DFKI-Robotics Innovation Center, Robert-Hooke-Str 1, 28359 Bremen, Germany
e-mail: marko.jankovic@dfki.de; frank.kirchner@dfki.de

© Springer International Publishing AG 2018
M. Vasile et al. (eds.), *Stardust Final Conference*, Astrophysics and Space Science Proceedings 52, https://doi.org/10.1007/978-3-319-69956-1_8

75 % in LEO in the next 200 years, despite those measures [9]. Thus, to stabilize the LEO environment and reduce the population of space debris the in-orbit mass needs to be actively removed [8].

Among all the phases of an ADR mission, the capture appears to be the most challenging one, since it generally involves close-range maneuvering and contact with a target. Moreover, no spacecraft has ever performed a capture of a completely non-cooperative target. Furthermore, the design of the "capture mechanism" drives the design of the whole chaser spacecraft which is why it is considered in this paper as the most distinctive and difficult phase of an ADR mission.

Targets considered as relevant in this study are those larger then 10 cm, since they are considered as lethal for any active satellite and are capable of generating more lethal fragments when impacting an operation spacecraft [7, 11]. Moreover any removal of objects smaller or equal to 10 cm is as of today considered non practical [5, 7]. With this in mind, the number of suitable ADR technologies to tackle those targets can be restricted and essentially divided into two categories: *contact* and *contactless* [4].

The *contact* methods considered in this paper include technologies based on: robotics (e.g. clamps, manipulators) and tethers (e.g. nets, harpoons). The *contactless* methods considered in this paper include technologies based on: plume impingement (e.g. chemical, electrical thrusters), ablation (e.g. lasers, solar concentrators) and electromagnetic forces (e.g. eddy brakes, electrostatic tractors).

They all have advantages and disadvantages but none of them can be applied to every type of target. Therefore, choosing one ADR method over another, in the initial stages of the mission planning, is essential. Nevertheless, this is generally a difficult and time consuming task, especially in the initial stages of the mission planning, mainly due to the dimensions of the parameter space describing each method and target object. Moreover, there is no easy way to express the degree of hazard that an object represents for an ADR mission.

One way of solving the first part of the problem would be to provide a means to compare the listed capture devices. This was attempted by creating a survey where experts in the field of ADR were able to evaluate (to best of their knowledge) ADR technologies in few categories, such as: technological availability, safety, reusability, versatility, etc. The total number of experts that agreed to participate to the developed survey was 35. Their professional status ranges from university professors and senior researchers in leading European and American academic research institutions to project managers in leading European aerospace companies.

The result of the survey concerning the capture technologies can be seen in Fig. 1. A higher bar represents a better overall weighted score of a method and different color of a bar indicates the weighted score of each category. For the scores, median values of the answers were used in order to take into consideration the overall distribution of the answers since not all of the technologies were evaluated by the same number of experts.

Analyzing more closely Fig. 1, it is evident that manipulator-based capture technique has the highest score, as it was expected, since it is the most mature technology among them all. However, the scores of other technologies are not that

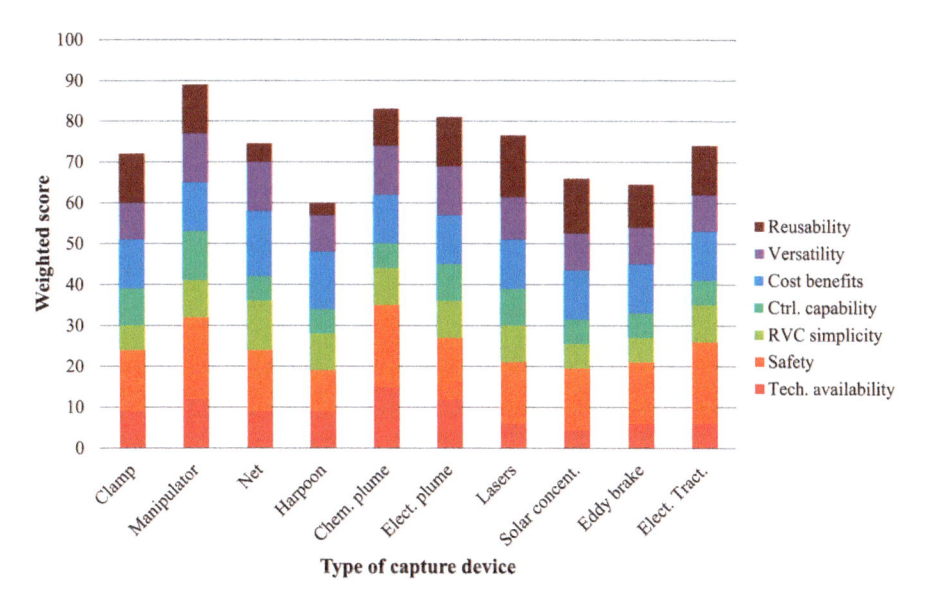

Fig. 1 Capture devices classification based on answers from 35 experts

different, evidencing once again the difficulty in raking them and consequently choosing one method over another. This is especially true if other factors need to be taken into consideration and not only the overall score, e.g. bigger safety requirements, versatility, etc. Therefore, another way is needed to solve more readily the mentioned problem. In this paper, this has been identified as providing a proper scientific classification of the space debris population that is able to point out the most suitable ADR method safety wise via a taxonomic method. This way the parameter space describing each object would be reduced to few significant quantities which would be used to properly identify, group, and discriminate space objects while at the same time providing the information about the most suitable ADR method, safety wise, that could be used to capture it.

In this context, the following paper presents a taxonomy of LEO space debris population, based on the taxonomic scheme developed by Früh et al. in [2], to support ADR decision making and classification of the space debris. The outcome of this research is a method for the classification of LEO space debris population and selection of the most suited ADR method, safety wise, for the selected target [4].

The structure of the remainder of the paper is as follows: Section 2 is dedicated to a brief review of previous studies of taxonomy of space debris. Section 3 presents the developed taxonomic method which is divided into: the formulation of the main LEO space debris classes and degree of hazard that an object poses to an ADR effort. Section 4 illustrates an application of the proposed taxonomy to some of the most representative objects of the main categories of space debris under investigation for ADR, i.e. intact LEO rocket bodies and spacecrafts. Section 5 provides the concluding remarks of the paper and the envisioned future work that will improve the developed method.

2 State-of-the-Art

In case of space debris, the ancestral decent of an object is generally known in advance which is why the taxonomy of space debris is done in reveres with respect to a biological taxonomy [2]. Despite this advantage, the taxonomy of space debris is still an unexplored field of research due to the large dimensions of the parameter space and immaturity of almost all ADR technologies. Nevertheless, there have been in the past some attempts to develop a taxonomy of space debris objects and the most relevant ones are outlined in what follows.

Wilkins et al. in [14] describe a basis for a resident space objects (RSO) taxonomy, based on the structure of the Linnaean taxonomy. Moreover, they also illustrate an algorithmic approach to the satellite taxonomy based on the open source probabilistic programming language, Figaro. The goal of the framework is to classify and identify without ambiguity the class of an RSO based on observation data, while providing the probability of the correct association [14]. However, the purpose of the framework was not to aid ADR therefore it falls short in classifying the objects according to their principal physical and dynamic characteristics that would be most useful for that purpose. Furthermore, it does not deal with the hazard that objects would pose to an ADR mission. Thus, although the framework could be extended and modified to include those properties, it was determined that it would require quite an effort and therefore was not considered as a basis for our approach [4].

Früh et al. in [2], on the other hand, describe a phylogenetic taxonomy based on more specific physical and dynamic traits of LEO objects with the goal of identifying their main classes and sources of origin. Moreover, they provide a way of visualizing the main traits of object by means of a concise acronym. However, this framework was also not explicitly developed to aid future ADR missions planning, therefore some of the discerning traits were missing (e.g. the break-up risk index or the existence of a berthing feature), while others were not defined in a rigorous manner, thus leaving space for individual interpretation (e.g. the material parameter). However, it does includes a hazard scale of objects based on their size, velocity and area-to-mass ration (AMR), thus indicating how dangerous an object is for the surrounding population. Therefore, it was considered as a good basis for our own taxonomic method and was refined and extended to include more specific traits (e.g. the risk that an object poses to the mission and its level of non-cooperativeness) [4].

3 Method Formulation

The taxonomy described in this paper consists of two layers developed to aid the initial mission planning of future ADR missions and provide an easy way to visualize, with an acronym (see Fig. 2), the main characteristics of an object and its

URXLlo-9-1L

Debris class — Debris hazard

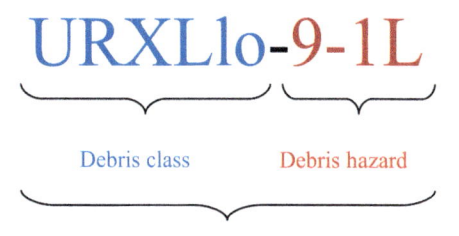

Most suited ADR method: Net-based

Fig. 2 Example of application of the taxonomic method to the 1967-045B Cosmos-3M 2nd stage. The acronym identifies: an uncontrolled (U), regularly rotating (R), convex (X), large (L) object with low AMR (lo), having a criticality number (CN) equal to 9 and level of non-cooperativeness equal to 1L (see Table 7 for more details) [4]

hazard. The first part of the acronym, defined in Fig. 2 with a label "*Debris class*", refers to the first layer of taxonomy and it indicates the class of a debris based on its most prominent physical and dynamical characteristics. In fact, every letter in this group refers to a specific characteristic of an object, i.e. **U** stands for *uncontrolled*, **R** for *regularly rotating*, **X** for *regular convex*, **L** for *large* and **lo** for *low area-to-mass ration* (AMR). Already at this stage some conclusions about the most suitable ADR capture method for that class can be made. However, an uncertain result is to be expected due to the crude nature of the traits used for the formulation of the classes [4].

To eliminate the mentioned uncertainty and narrow down the ADR association, a second layer of taxonomy is to be performed, on per object basis, and is indicated in Fig. 2 with a label "*Debris hazard*". It consists of individuating the break-up risk index of an object (indicated in the figure with the number **9**) as well as its level of non-cooperativeness (indicated in the figure with the symbol **1L**), which essentially highlights the hazard that the target represents for its capture based on its: passivation state, age, probability of spontaneous break-up, angular rate, properties of the capturing interface (if any), etc. [4].

3.1 Debris Classes

In general, defining a taxonomy of any kind involves the following steps: (a) *collection* of data, (b) *identification* of groups and (c) *classification* of groups [10]. This means that at first all the relevant data about the objects that we would like to classify should be collected. Then, objects should be sorted in groups, based on their most relevant and distinguishing features. Finally, *taxa* should be ranked and ordered to make a taxonomic tree which delineates its ancestral decent and minimum amount of information necessary to positively identify an object [2]. Therefore, in this paper the first layer of the taxonomy consists of: (a) defining the main characteristics of LEO objects, (b) building of the taxonomic tree and (c) formulating the classes of LEO objects [4].

Table 1 Main characteristics of the first layer of taxonomy [4]

Characteristics	Definitions
Object type	Artificial: *man-made object*
	Natural: *non-man made object*
Orbit type	LEO: 80–2000 km
	MEO: 2000–35,786 km
	GEO: at 35,786 km
	HEO: > 35,786 km
Orbital state	Controlled (C): *actively controlled*
	Uncontrolled (U): *self-explanatory*
Attitude state	Actively stabilized (S): *three axis stabilized*
	Regularly rotating (R): *passively controlled/uncontrolled stable (no precession)*
	Tumbling (T): *irregular attitude motion*
External shape	Regular convex (without appendages) (X): *cylindrical or spherical shapes*
	Regular polyhedral (with appendages) (P): *regular cubic shapes of spacecrafts*
	Irregular (I): *self-explanatory*
Size	Small (S): < 10 cm (up to 5 cm)
	Medium (M): 10 cm–1 m
	Large (L): > 1 m
Area-to-Mass Ratio (AMR)	Low (lo): < 0.8 m^2/kg
	Medium (me): 0.8–2 m^2/kg
	High (hi): > 2 m^2/kg

The principle of taxonomic distinction dictates that placing an objects into a taxa must be performed without ambiguity [2]. Therefore, it must be done based on their most relevant and distinguishing features. The main characteristics identified as sufficient to classify space debris objects without ambiguity in this layer are: the orbital state, attitude state, shape, size and area-to-mass ratio (AMR). The definition of those characteristics can be seen in Table 1. A more detailed description of the listed characteristics can be found in our previous publication on the topic [4].

Using the previously defined characteristics it is possible to build a taxonomic tree of space debris objects (see [4]) which allows us to identify the LEO classes of space debris objects. The result of this step are 18 classes, out of which only four (see Table 2) are considered as relevant for any future ADR effort, given their mass and size. Therefore, only those classes were considered for the next taxonomic layer.

Table 2 Main classes of the taxonomic tree

Acronym of the class	Example object
Passively stable, intact objects	
URXLlo	Upper stages/decommissioned spacecrafts with momentum bias
URPLlo	Decommissioned spacecrafts with momentum bias
Intact uncontrolled objects	
UTPLlo	Generic decommissioned tumbling spacecrafts
UTXLlo	Tumbling upper stages

3.2 Debris Hazard

The second layer of taxonomy is to be performed on per object basis and consists of individuating: (a) a break-up risk index of an object and (b) its level of non-cooperativeness. The goal of the layer is to identify the hazard and difficulty that an object poses to its capture in order to pin-point the most suited ADR capture method for that object, safety wise. This requires a more specific knowledge of physical and dynamical traits of objects, not all of which are available in publicly accessible databases. Therefore, to overcome this limitation and restrict the number of possible permutations, only a limited amount of decisive traits was considered in this layer despite the fact that a bigger parameter space would yield a more precise results [4].

The break-up risk index of an object is defined as the highest criticality number (CN) calculated, in accordance with the ESA's standard on Failure modes, effects (and criticality) analysis (FMEA/FMECA) (see [1]), as a product between the severity number (SN) and probability number (PN) of possible failure modes of space debris objects [4].

According to [1], the FMEA shall be performed mainly by: (a) describing the product to be analyzed, (b) identifying all potential failure modes and their effects on the product, (c) evaluating each failure mode in terms of the worst potential consequences and assigning a severity category, (d) identifying preventing measures for each failure mode and (e) documenting the analysis. Based on this methodology and data from the ESA's Database and Information System Characterising Objects in Space (DISCOS[1]) it was possible to identify (for large LEO non-passivated objects) two types of possible failure modes of which we have documented information: explosions or malfunctions of propulsion/attitude systems and explosions of battery packs. The distribution of those two events varies between spacecrafts and rocket bodies and is 33 and zero events due to propulsion and batteries, respectively, in case of rocket bodies while it is 3 and 10 events due to propulsion and batteries, respectively, in case of spacecrafts. Collisions are excluded from this study since a

---
[1]https://goo.gl/e279ln.

only a total of 6 events occurred versus 46 due to a malfunction of on-board systems. Future studies might overcome this current limitation of the method.

For passivated objects, no break-up risk exists from on-board stored energy, however the embrittlement of external surfaces due to the thermal cycling and erosion is to be expected and is considered in this study as a possible failure mode.

The estimated consequences of each identified failure mode and thus the associated SN of each mode are the following. For explosions or malfunctions of propulsion/attitude systems the severity of that failure mode depends greatly on the stored fuel type. This study distinguishes between the cold gas, solid, cryogenic and hypergolic fuel. Based on the median number of fragments generated from 44 LEO explosions, it is assumed that the severity number associated with those fuels is equal to: 1 for modes involving cold gas, 2 for those involving cryogenic and solid fuels and 3 for those involving hypergolic fuels. SN 4 is reserved for objects having hypergolic type of fuel in large quantities and liquid form. This situation is typically to be expected in case of objects that have been decommissioned early in the mission due to an irrecoverable in-orbit failure or wrong orbital insertion.

For explosions of battery packs an SN of 2 was assumed due to a median number of generated debris from historical data (i.e. 65.5 fragments).

In case of ruptures of external surfaces, due to the embrittlement, the assumed SN is equal to 1, which corresponds to a minor or negligible mission degradation mainly due to the cracking of the external paint.

The next step towards the definition of the break-up risk index is the identification of the probability of occurrence of assumed failure modes. This was done by using the probability of occurrence levels visible in Table 3 and the data, extrapolated from the ESA's DISCOS and US Air Force's Space Track[2] databases.

A distinction was made between payloads/spacecrafts and rocket bodies since a significant difference in the number and distribution of break-up events between these two types of objects (see Fig. 3) was determined. Moreover, for non-passivated objects the non-parametric Kaplan-Meier estimator was used to determine their cumulative failure probability instead of the probability calculated as a normalization of the number of events with respect to the number of launches, as done in our first paper on this topic [4]. The mentioned estimator was originally developed in medicine to estimate a survival curve from a population sample, including incomplete observations [6], which is exactly the case of space debris since the

Table 3 Probability levels, limits and numbers

Probability Level	Limits	Number
Probable	$P > 10^{-1}$	4
Occasional	$10^{-3} < P \leq 10^{-1}$	3
Remote	$10^{-5} < P \leq 10^{-3}$	2
Extremely remote	$P \leq 10^{-5}$	1

---

[2]https://www.space-track.org.

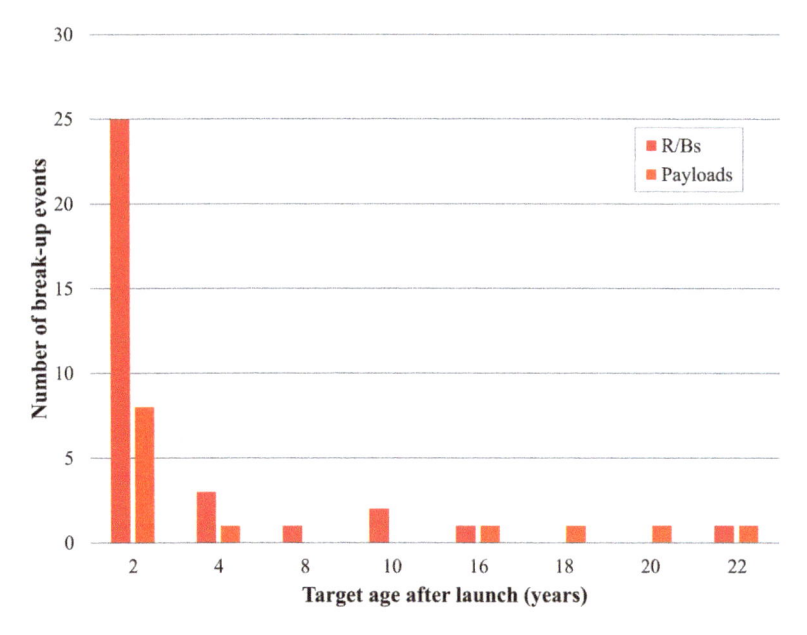

Fig. 3 Distribution of LEO break-up events (source: DISCOS database)

estimation of the cumulative failure probability needs to take into consideration re-entered objects.

Due to a vary small number of samples (i.e. 46) the calculation of a cumulative probability distribution was made without any distinction between the failure modes. A future study might tackle this point in more depth and overcome this current limitation of the method. In total, a population of 2304 and 2442 large, defunct, LEO spacecrafts and rocket bodies, respectively, was analyzed using the data from the US Air Force's Space Track and ESA's DISCOS databases. The result of the analysis are the cumulative failure probability distributions visible in Figs. 4 and 5, for spacecrafts and rocket bodies, respectively, which can be used to obtain PNs of desired non-passivated objects if the data is paired with Table 3.

For passivated objects the PN, is determined by calculating the following linear functions, $PN = 2 \times 10^{-4} \times age$, for spacecrafts, and $PN = 4 \times 10^{-4} \times age$, for rocket bodies, to reflect the maximum values of *PNs* obtained with the Kaplan-Meier estimations (see Figs. 4 and 5).

Using the previously defined severity and probability numbers it is possible to assign the criticality numbers to each failure mode of each object by either using the formula: $CN = SN \cdot PN$ or the criticality matrix represented in Table 4.

With this in mind, an object is to be considered as critical for capture if:

- the consequences of the failure mode are to be considered **catastrophic**, i.e. the SN of the failure mode is 4 (see Table 4), or
- the failure mode is greater or equal to **8** (see Table 4).

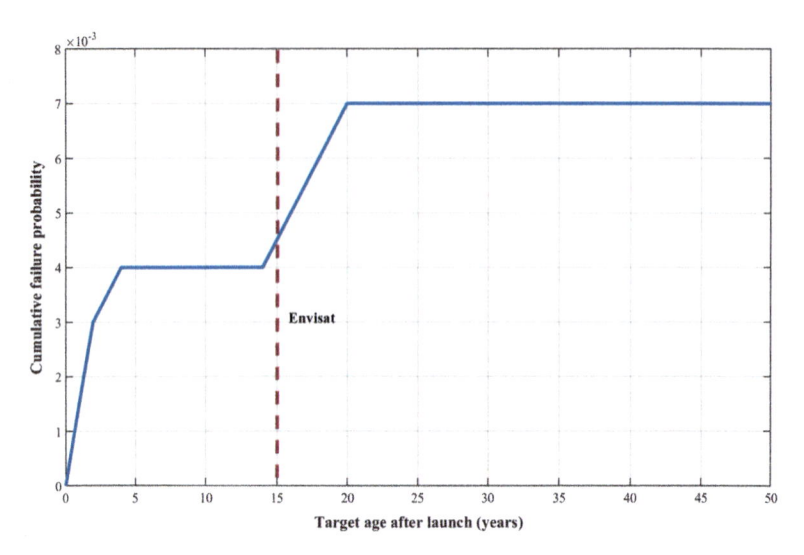

Fig. 4 Cumulative failure probability distribution for spacecrafts with Envisat as an example object

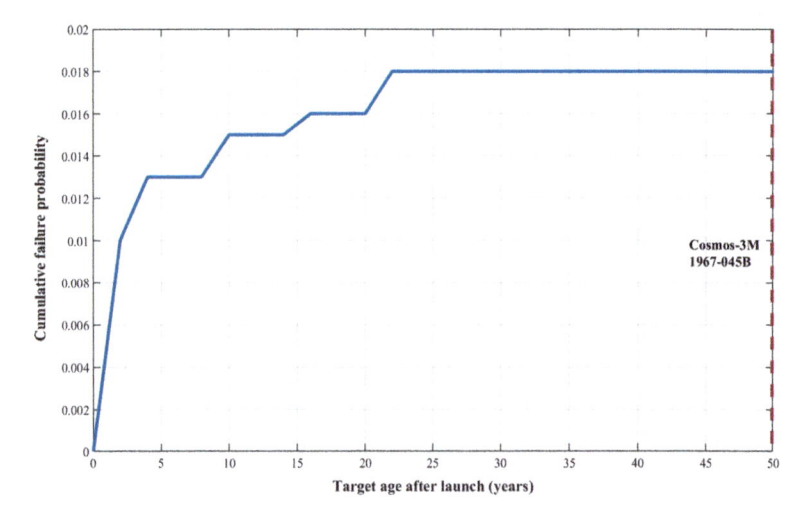

Fig. 5 Cumulative failure probability distribution for rocket bodies with Cosmos-3M 1967-045B as an example object

In these cases, any close contact with the target is to be avoided by using **only net-based methods** that can perform a capture from a considerable stand-off distance. Moreover, a special care should be exerted during the capture and stabilization of these objects to avoid sources of sparks. Therefore, harpoon-based methods are to be avoided since they assume a penetration of a target and thus would only add more hazard to the mission.

Taxonomy of LEO Space Debris

Table 4 Criticality matrix

		Probability			
		10^{-5}	10^{-3}	10^{-1}	1
		PN			
Severity	SN	1	2	3	4
Catastrophic	4	4	8	12	16
Critical	3	3	6	9	12
Major	2	2	4	6	8
Negligible	1	1	2	3	4

Table 5 Levels of non-cooperativeness of a target [4]

	Capture interface & ADR association							Mechanical clearance & ADR	
	Rate			Berth.		Material			
Level	Low	Med	High	Y	N	Iso	An	L	S
1	x			x		x		Manipulator	
2	x				x	x		Clamp w sync./Tether	Tether
3	x				x		x	Clamp w sync./Net	Net
4		x		x		x		Manipulator w sync.	
5		x			x	x		Clamp w sync./Tether	Tether
6		x			x		x	Clamp w sync./Net	Net
7			x					Contactless	

For the remaining CNs, the associated ADR capture methods were chosen as follows:

- **robotic/tether-based** methods for CNs **1–4** or for failure modes classified as **negligible,**
- **net/contactless** methods for the CN equal to **6.**

The association was carried out based on the engineering judgment regarding the maturity of a technology and distance that the chaser spacecraft needs to maintain during the capture. Thus, it was performed only with safety in mind given the many uncertainties surrounding most of the currently considered ADR targets.

With the identification of the break-up risk index it is possible to estimate how dangerous the capture of a specific target would be but at this stage there is no indication on how difficult it would be. Moreover, even at this point of the method the uncertainty in choosing the best ADR capture methods is still to be expected. To solve this, a level of non-cooperativeness of an object needs to be identified and is performed using Table 5. In total, 14 levels of non-cooperativeness were identified and each level is expressed as a combination of an Arabic numeral (from *1* to *7*, with *1* being the *least non-cooperative* and *7* the *most non-cooperative* level) and a letter indicating the dimensions of the mechanical clearance of the capturing interface (i.e. *large (L)* or *small (S)*) (see Fig. 2 on page 133) [4].

Table 6 Main characteristics/taxa of the first layer of taxonomy

Characteristics	Definitions
Angular rate	Low: < 5 deg/s
	Medium (Med): 5–18 deg/s
	High: $\geq$ 18 deg/s
Berthing feature existence	True (Y): *dedicated berthing feature exists*
	False (N): *dedicated berthing feature does not exist*
Capturing interface material	Isotropic (Iso): *e.g. metal, ceramics or polymer*
	An-isotropic (An): *other*
Mechanical clearance	Small (S): < 0.28 m^2
	Large (L): $\geq$ 0.28 m^2

The definition of the traits used to define the levels of non-cooperativeness can be seen in Table 6. A more detailed description and definition of the listed characteristics can be found in our previous publication on the topic (i.e. [4]).

The ADR association performed in Table 5 was done using a qualitative approach based on the engineering judgment of the capabilities of the considered ADR methods. Therefore, a manipulator was considered as the first choice in case of an object having a dedicated berthing feature since it is the most mature one among the considered capture technologies. A non-existing berthing feature precludes the usage of the manipulator thus, it was assumed that these cases should be tackled by methods capable of capturing a surface rather then a particular feature of a target. Hence, clamp and tethered methods were considered in these cases, based on the mechanical clearance available on the target. However, this association is only to be used as a complement to the one already performed with the break-up risk analysis and as an additional filter to identify the most suited capture ADR method with respect to (w.r.t) the overall safety of a mission. For example, if the CN of an object dictates that the associated capture methods are robotic/tether-based and its level of non-cooperativeness is equal to 1L, the most suited capture method for that target would be a manipulator-based method due to its level of non-cooperativeness. Should there ever arise a conflict between the associated capture methods identified during the break-up risk analysis and definition of the levels of non-cooperativeness of an object, the most suited method or methods identified with the former analysis are always to have the priority. For example, should the identified ADR class be net-based, due to a high CN, and the level of non-cooperativeness equal to 1L, a net-based system is to be considered as the most suitable to capture method for that object, instead of the manipulator-based system, due to the high criticality number of the possible failure mode [4].

4 Method Application

To study the practicality of the developed taxonomic method, it was applied to representative objects of the most attractive families of space debris for future ADR missions, such as the European Envisat, Soviet/Russian SL-16 and SL-8 rocket bodies [8, 13]. The results are visible in Table 7.

Most of the physical data about the objects was obtained from the ESA's DISCOS database. However, other traits were obtained from on-line resources such as: Encyclopedia Astronautica,[3] Gunter's Space Page,[4] Earth Observation Portal[5] and RussianSpaceWeb.com.[6] Others, evidenced in table with an asterisk (*), were defined based on the engineering judgment using available resources (e.g. assuming that old objects are subject to a regular slow rotation around one axis was based on

Table 7 Examples of taxonomy application

DISCOS name	Envisat	Zenit-2 II stage	Cosmos-3M II stage
COSPARID	2002-009A	1985-097B	1967-045B
Mass [kg]	8110	8225.970	1434
DISCOS classification	Payload	Rocket body	Rocket body
Shape	Box + 1 panel	Cylinder	Cylinder
Mean size [m]	13.5	7.15	3.3
Mean area [m^2]	74.39	33.426	10.179
Orbital state	Uncontrolled	Uncontrolled	Uncontrolled
Attitude state	Tumbling	Reg. rotating*	Reg. rotating*
External shape	Reg. polyhedral	Reg. convex	Reg. convex
Size	Large	Large	Large
AMR	Low	Low	Low
Debris class	UTPLlo	URXLlo	URXLlo
Severity number	3	2	3
Probability number	1	3	3
Risk index	3	6	9
Angular rate	Low	Low*	Low*
Berthing feature	Yes	No*	Yes*
Interface material	Isotropic	Isotropic*	Isotropic*
Mech. clearance	Small	Large*	Large*
Non-coop. level	1S	2L	1L
Taxonomic acronym	URPLlo-3-1S	URXLlo-6-2L	URXLlo-9-1L
ADR capture methods	Manipulator-based	Net-based	Net-based

---

[3] http://goo.gl/iVOgvS.
[4] http://goo.gl/f21ATh.
[5] https://goo.gl/SWwGSl.
[6] http://goo.gl/XR6JK.

the conclusions of [12] and not actual data). In fact, currently there are no publicly available databases[7] that contain these kind of information. We acknowledge that this might undermine the precision of the identified ADR capture methods for the use-case objects. However, we are convinced that this does not undermine the validity of the developed taxonomic method and its main characteristic to concisely describe the main properties of objects and the hazard that they represents for an ADR effort. Therefore, the results of these applications are to be considered at the moment only as indicative. Moreover, please note that the probability number of the Envisat has been added based on the results of the e.Deorbit CDF Study Report [3]. Should have we estimated its PN using Fig. 4, that number would have been different (i.e. 3 instead of 1). Therefore, the recommended capture methods would have been different, i.e. they would have been based on net/contactless technologies. This is to indicate a conservative nature of the developed method and break-up probability numbers. However, anytime that a deeper study on the cumulative probability of a break-up of an object has been performed it is advisable to be used in the developed taxonomy to obtain a more precise association.

From these examples it is possible to make a conclusion that objects having a hypergolic type of fuel on-board are most likely to be tackled by net-based methods, due to their high criticality number. For targets having a non-hypergolic type of fuel on-board the most suited ADR capture method depends greatly on the identification of their levels of non-cooperativeness. Which in turn dictate that if we are to peruse the ADR in the near future the traits identified in Table 6 will need to be identified for the most appealing ADR targets.

5 Conclusions

A method for the classification of space debris and ADR association has been described. The outcome of the taxonomy is an easy to interpret acronym, which describes at a glance the most prominent features of objects and the hazard they pose to an ADR effort. The method has been formulated in two layers identifying a debris class and its capture hazard. For the latter a statistical analysis has been performed on the available data using the Kaplan-Meier estimator to estimate the cumulative probability distribution of failure modes. The application of the method to representative objects of three families of space debris has been also illustrated. The results of that application indicate that objects having a hypergolic type of fuel (e.g. Soviet/Russian SL-8 rocket bodies) are most likely to be tackled by net-based methods due to their high break-up risk index. For all the other targets, the most suited ADR capture method depends greatly on the identification of their levels of non-cooperativeness which was performed in this study based on the engineering judgment using limited resources. Therefore, the results are to be considered at

---

[7] At least to best of our knowledge.

the moment only as indicative, at least until the necessary data is made available to the public. Nevertheless, this does not undermine the validity of the developed method and its immediacy when it comes to identifying the most suited ADR capture method necessary in the initial phases of mission planning.

However, the method does not include collisions as possible source of breakup and the cumulative probability distribution is calculated using the data of both propulsion/attitude and battery failure modes, without distinction. Furthermore, only a non-parametric analysis of the probability is implemented making any future predictions of PNs impossible. Therefore, these three issues are the shortcomings of the presented method and will be tackled in our future research.

Acknowledgements The research work here presented was supported by the Marie Curie Initial Training Network Stardust, FP7-PEOPLE-2012-ITN, Grant Agreement 317185. Thus, the authors would like to thank the European Commission and the Research Executive Agency for their support and funding.

References

1. ECSS Secretariat: Space product assurance: failure modes, effects (and criticality) analysis (FMEA/FMECA) (2009). http://www.ecss.nl/
2. Früh, C., Jah, M., Valdez, E., Kervin, P., Kelecy, T.: Taxonomy and classification scheme for artificial space objects. In: 2013 AMOS (Advanced Maui Optical and Space Surveillance) Technical Conference. Maui Economic Development Board (2013)
3. Innocenti, L.: CDF study report: e.Deorbit, e.Deorbit assessment. Technical Report, ESTEC - ESA, Noordwijk (2012)
4. Jankovic, M., Kirchner, F.: Taxonomy of LEO space debris population for ADR selection. In: Proceedings of the 67th International Astronautical Congress (IAC-2016), pp. 1–15. International Astronautical Federation (IAF) (2016)
5. Kaplan, M.H.: Space debris realities and removal. On-line (2010). https://goo.gl/kYhtU4
6. Kaplan, E.L., Meier, P.: Nonparametric estimation from incomplete observations. J. Am. Stat. Assoc. **53**(282), 457–481 (1958). https://doi.org/10.2307/2281868
7. Kaplan, M., Boone, B., Brown, R., Criss, T., Tunstel, E.: Engineering issues for all major modes of in situ space debris capture. In: AIAA SPACE 2010 Conference & Exposition, September, pp. 1–20. Space Department Applied Physics Laboratory, AIAA, Anaheim, CA (2010). https://doi.org/10.2514/6.2010-8863
8. Liou, J.C.: An active debris removal parametric study for LEO environment remediation. Adv. Space Res. **47**(11), 1865–1876 (2011). https://doi.org/10.1016/j.asr.2011.02.003
9. Liou, J.C., Johnson, N.L., Hill, N.M.: Controlling the growth of future LEO debris populations with active debris removal. Acta Astronaut. **66**(5–6), 648–653 (2010). https://doi.org/10.1016/j.actaastro.2009.08.005
10. Mayr, E.: Systematics and the Origin of Species, from the Viewpoint of a Zoologist. Harvard University Press, Harvard (1999)
11. Mcknight, D.: Pay me now or pay me more later: start the development of active orbital debris removal now. In: Proceedings of the 2010 AMOS Conference, pp. 1–21. Maui Economic Development Board, Maui, Hawaii (2010)

12. Praly, N., Hillion, M., Bonnal, C., Laurent-Varin, J., Petit, N.: Study on the eddy current damping of the spin dynamics of space debris from the Ariane launcher upper stages. Acta Astronaut. **76**, 145–153 (2012). https://doi.org/10.1016/j.actaastro.2012.03.004
13. Rossi, A., Valsecchi, G.B., Alessi, E.M.: The criticality of spacecraft index. Adv. Space Res. **56**(3), 449–460 (2015). https://doi.org/10.1016/j.asr.2015.02.027
14. Wilkins, M.P., Pfeffer, A., Schumacher, P.W., Jah, M.K.: Towards an artificial space object taxonomy. In: 2013 AMOS (Advanced Maui Optical and Space Surveillance) Technical Conference. Maui Economic Development Board (2013)

Remote Sensing for Planar Electrostatic Characterization Using the Multi-Sphere Method

Heiko J. A. Engwerda, Joseph Hughes, and Hanspeter Schaub

Abstract Applications like the Electrostatic Tractor (ET), remote sensing of space debris objects, or planetary science investigating asteroid charging, benefit from a touchless method to assess the electrostatic potential and charge distribution of another body. In the ET, accurate predictions of the force and torque between a passive space object and tug spacecraft are critical to ensure a robust closed loop control. This paper presents a novel, touchless method for determining both the voltage and a Multi-Sphere-Method (MSM) model which can be used to determine the charge distribution, force, and torque on a nearby space object. By means of potential probes, Remote Sensing for Electrostatic Characterization (RSEC) can be performed. Here the space tug shape and electrostatic potential is assumed to be known. The probes measure the departure from the expected potential field about the tug and determine an MSM model of the passive object's potential distribution. This paper outlines a method for estimating the voltage and charge distribution of a neighboring charged spacecraft undergoing a planar rotation given measurements of voltage over a full rotation. Assuming knowledge of the tug spacecraft's voltage and charge distribution, the rotation rate and distance to the debris, numerical simulation results illustrate that the constructed model of the debris can be characterized within a few percent error.

H. J. A. Engwerda
Faculty of Aerospace Engineering, Delft University of Technology, Kluyverweg 1, Delft, 2629 HS, The Netherlands
e-mail: h.j.a.engwerda@student.tudelft.nl

J. Hughes · H. Schaub (✉)
Aerospace Engineering Sciences Department, University of Colorado, ECEE 275, 431 UCB, 80309-0431 Boulder, CO, USA
e-mail: joseph.hughes@colorado.edu; hanspeter.schaub@colorado.edu

© Springer International Publishing AG 2018
M. Vasile et al. (eds.), *Stardust Final Conference*, Astrophysics and Space Science Proceedings 52, https://doi.org/10.1007/978-3-319-69956-1_9

1 Introduction

Spacecraft formation flying is a popular topic within the aerospace community and offers many benefits. Swarms of satellites can provide a low cost solution to many space operations and allow for scientific studies that can not be performed with single spacecraft platforms [6]. Coulomb formation flying allows for small corrections within a satellite constellation without the use of propellant, but rather through the electrostatic force [35]. Between multiple charged bodies a Coulomb force exists which is inversely proportional to the square of the separation distance and product of the charges. Measuring the potential of the bodies is of importance to predict this force and the torques resulting from the charge distribution. Especially if one of the bodies is uncooperative, measurements of electric potential are crucial in maintaining operational safety. Over-prediction of the potentials may limit formation flying performance, while under-prediction can lead to collisions [14]. This paper presents a possible solution for estimating the potential and charge distribution of a nearby charged spacecraft with a known planar rotation rate using measurements of voltage in the vicinity of the debris. The force and torque on both bodies can be found from the voltage and charge distribution on the debris assuming the voltage and charge distribution are known for the tug.

One application of Coulomb formation flying with uncooperative spacecraft is the Electrostatic Tractor (ET), as proposed by Moorer and Schaub [24, 29], which applies the Coulomb interaction to reorbit space debris. Debris removal has been an increasingly popular topic in the recent years. The total amount of officially catalogued objects in Earth orbit exceeds 17,000, while only 23% of these objects is a payload [26]. Especially for regimes with many assets such as the Geostationary orbit (GEO), the development of the debris population is anxiously monitored since this orbit offers unique Earth monitoring and sensing possibilities. Of the total market of USD 20 billion, the majority of insured satellites resides in this orbit at 35,786 km altitude [7]. Since there is no passive clean-up mechanism, such as there is atmospheric drag in Low Earth Orbit, objects have to be manually removed. The most cost efficient option is to move the objects to a graveyard orbit, which is typically around 300 km above GEO [16, 20]. At this altitude the decay time back to GEO exceeds 200 years, offering a (temporary) mitigation solution [2]. Over the years, several concepts have been proposed for moving the debris to such an orbit. The Ion Beam Shepherd proposed by Bombardelli and Pelaez [4] and the ET offer contact-less removal opportunities decreasing risks associated with debris removal, such as break-up events. Concepts such as nets, harpoons and grappling devices do introduce such risks as they require an established contact [21]. This paper focusses specifically on sensing potentials and estimating force and torque for the ET concept, but touchless electrostatic characterization has a broad applicability.

For example, there is interest in knowing the local potentials of asteroids and the moon. These measurements could help scientists better understand dust transport across the lunar surface [10, 13] and asses risks encountered by spacecraft and astronauts during asteroid rendezvous missions [17]. It is of interest to measure these voltages without making electrical contact and thereby corrupting these measurements through discharge. It is also very difficult to make contact with a foreign body safely, as the Rosetta and Hayabusa missions have shown [34]. In order to apply the method described in this paper to such a mission, dielectric characteristics have to be incorporated in the model, which is beyond the scope of this paper.

In the ET concept, the tug irradiates the debris with electrons using an on-board electron gun. This causes the tug to become positively charged while the debris charges negatively. An electrostatic force and torque are felt on both craft, which can be used for touchless actuation [28]. Establishing the ET force is feasible in GEO due to the locally large Debye lengths of 200 m and more [9]. The attractive Coulomb force resulting from the potential difference can be utilized to create a link between the bodies, while thrusting can be performed from the tug to move the multiple-body system to a graveyard orbit. As demonstrated by Albuja, inactive satellites can have very large rotation rates and depending on the symmetry of the body these rates evolve over time [1]. If the charge distribution on tug and debris is non-symmetric, the Coulomb force will generate a torque which can be used to detumble the debris in the span of a couple of days [3].

The Coulomb force can be determined from the accelerations inferred from Light Detection and Ranging (LiDAR) measurements or other ranging methods over a long time. As discussed previously, Coulomb formation flying is inherently open-loop unstable and therefore estimation of the potential of debris in a feed-forward procedure is crucial. By means of measurements, the potential of the debris and thus the Coulomb force can be determined in real time. Possible sensing methods include placement of a probe on the debris, surface measurements, evaluation of charged particles [23, 27] and contact-less electric field (E-field) or potential probing [25]. Because this paper focuses on remote and real-time sensing applications, only potential and E-field probes are considered.

By means of measuring the electric potential, Remote Sensing for Electrostatic Characterization (RSEC) of the debris can be performed. One or multiple probes are extended from the tug spacecraft using booms and register the electric potential field over time. This measurement is combined with the known debris spin rate and passed into a numerical solver to determine the potential of the debris and a Multi-Sphere Method (MSM) model for it. An artist's impression of this method can be found in Fig. 1. This paper describes a touchless method for determining the voltage of a nearby space object and constructing an MSM model for it, which can be used to predict the force and torque on the passive object. Additionally, the sensitivity of the model with respect to input parameters and their errors is evaluated.

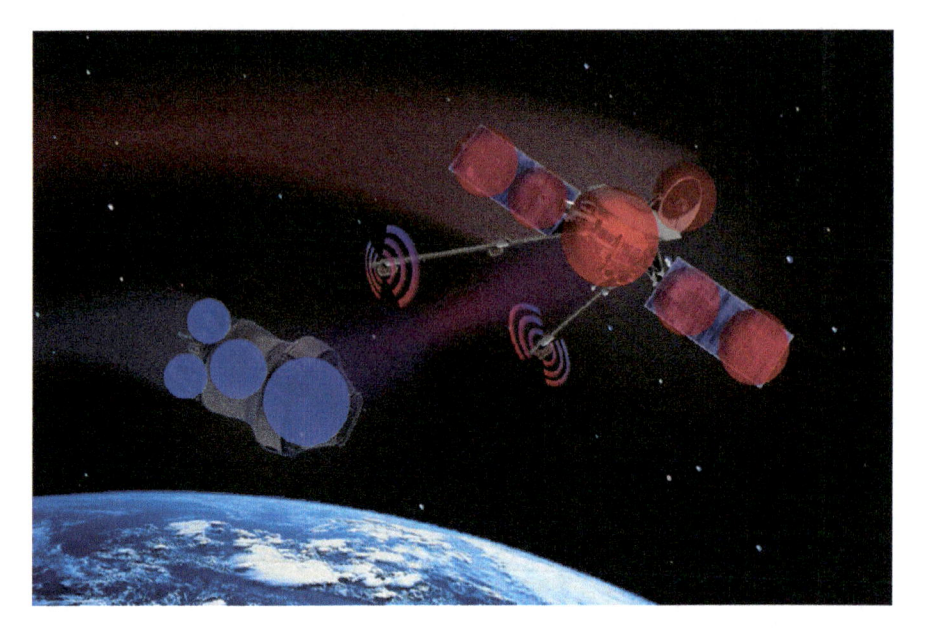

Fig. 1 By means of potential probes, the electrostatic interaction between space objects can be characterized

2 Sensing Electric Potential

The RSEC model requires measurement of kilovolt potential at specific locations between tug and debris. Measuring potentials of spacecraft in space is a proven technology and can be performed even for kilovolt magnitudes. In its first year of operations, instruments on board of the Spacecraft Charging AT High Altitudes (SCATHA) satellite measured potentials as high as -14 kV during eclipse [25]. These measurements were obtained from plasma detectors, which consisted of electron- and ion detectors, as well as an electric field detector.

The most common methods to sense the electric potential in space is by means of Langmuir probes and emissive probes. Emissive probes are electrically heated which causes electron emission. This results in the probe reaching a floating potential and allows measurement of the plasma potential without requiring voltage sweeps [22]. Langmuir probes on the other hand do require voltage sweeps over a bias voltage. The measurements obtained from a sweep are related to the characteristic I–V curve and can be used to determine many plasma parameters [5]. Due to the high potential of the tug and sparse electron density in GEO, obtaining unambiguous measurements is expected to be difficult.

On Earth, electric field mills are commonly used to determine the electric field strength of thunderstorm clouds [8]. As E-fields are often more conveniently measured than potential fields in atmospheric conditions [30], field mills are an appropriate choice for validation of the RSEC method in an atmospheric environment.

Most potential measurements that have been flown were designed to measure the ambient plasma potential or spacecraft potential with respect to the ambient (low potential) plasma. In this study, the spacecraft will be charged to tens of kV which requires adaptation of current available measurement methods. Around the tug, a large electron deficit exists, reducing the effectiveness of Langmuir probe measurements. Measurement of the space potential or electric field which results from being nearby a highly charged object is expected to be a difficult but possible task in space plasma. The authors are not aware of measurements currently being performed for mapping the spatial dependence of such high voltages (kV) or strong fields (kV/m).

Apart from the influence of the tug and debris on the sensors, there are a number of sources which may corrupt the measurements, such as interference of the booms and fluctuating space weather. Due to the high potential of the tug, not all of the emitted photo-electrons will be able to escape from the influence of the tug. When a probe is located in this photo-electron cloud, the measurements will be biased. In order to determine the radius r_{max} of this cloud, consider the following energy balance.

$$\underbrace{E_k + E_p}_{\text{Initial}} = \underbrace{E_k + E_p}_{\text{Final}} \frac{1}{2} m_e v_e^2 + q \frac{V_s R_s}{R_s} = q \frac{V_s R_s}{r_{max}} \quad (1)$$

In these equations, the electrons with mass m_e are emitted from the surface of a sphere of radius R_s and potential V_s with an electron-velocity v_e. The initial velocity (left hand side of Eq. (1)) can be determined with Eq. (2).

$$\frac{1}{2} m_e v_e^2 = q \phi_0 \quad (2)$$

Considering an average kinetic energy of 2 eV [15], the velocity with which the photo-electrons are emitted is $v_e = 840$ km/s. Substituting values in Eq. (1) gives a maximum distance travelled by the emitted electrons of 0.3 mm. Since close to the sphere the field deviates from that of a point source, the flat plate approximation as discussed by Grard may be more representative [12], although application of this more realistic model is not expected to change the results drastically. The calculated value shows that the photo-electron cloud has a negligible influence on the probe measurements as long as they are placed on booms at distances on the order of multiple meters such as considered in the rest of this paper.

Another disturbance on the potential and electric field of two spheres is due to the presence of the electron beam. As an approximation, the beam can be represented as a line charge, for which the following equation holds.

$$V = \int \frac{k_c dq}{r} = k_c \lambda \Psi \qquad k_c = \frac{1}{4\pi \epsilon_0} \quad (3)$$

The charge density λ can be determined from the kinetic energy of electrons in the beam (Eq. (4)), where the integral over the beam length is represented by the parameter Ψ. The relation for Coulomb's constant k_c is given in the right hand side equation with ϵ_0 the permittivity of vacuum.

$$\lambda = \frac{I}{v_e}\frac{1}{2}mv_e^2 = eV \qquad (4)$$

Assuming an electron beam current of $I = 10$ mA and a potential V of 20 kV over a 10 m beam, the potential in the neighborhood of the beam is in the 0–8 V regime. Such a beam current is 10 times higher than the minimum current required for charge transfer with a 4 m radius debris, as derived by Hogan et al. [15]. Compared to the kilovolt potentials around the two bodies, this potential has minor significance and is neglected in the further analysis.

3 Electrostatic Characterization Model

The following section describes the RSEC model and the equations and assumptions required to construct this model. A graphical representation of the application of RSEC is given in Fig. 2.

The potential field resulting from the tug-debris system is measured by probes. The numerical RSEC model is then applied to find the parameters in an MSM model of the debris and it's voltage which best matches this measured potential field. This MSM model and potential are used to estimate the force and torque on the debris. This allows time-varying measurements of the voltage field to be used to estimate for the force and torque on both bodies faster-than-realtime.

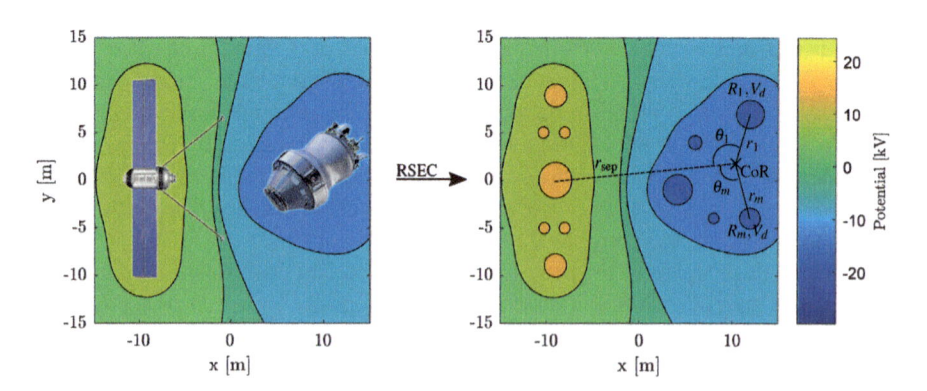

Fig. 2 Angles and distances from the center of Rotation (CoR) towards spheres on the debris

3.1 Force, Torque and Potential Equations

Multiple methods have been developed which can be used to model the electrostatic characteristics of bodies in space. The easiest approximation is a 2-sphere model, but for small separation distances the tug and debris can not be accurately represented with single spheres. The non-homogeneous charge distribution and shape of the bodies affects not only the Coulomb force, but also introduces a torque on the bodies. Where the charge distribution can be very accurately predicted with Finite Element Method (FEM) [18] or the Method of Moments (MoM) [11], the computation time required often precludes real time simulations, in particular with FEM. An alternative is the MSM, which reduces computation time for a simple sphere-cylinder system from over an hour with FEM to a fraction of a second [33].

If an MSM model of a tug consisting of n and debris of m-spheres is considered, the Coulomb force is written as a summation over all spheres [32].

$$\vec{F}_c = k_c \sum_{j=1}^{m} \sum_{i=1}^{n} \frac{q_i q_j}{r_{i,j}^3} \vec{r}_{i,j} \qquad (5)$$

In this equation the Debye shielding effect is neglected since the Debye length at GEO is much larger than the tug-debris separation distance (≈ 200 m vs 20 m) [31]. The torque on the debris is given by

$$\vec{T}_d = k_c \sum_{j=1}^{m} \sum_{i=1}^{n} \frac{q_i q_j}{r_{i,j}^3} \vec{r}_{i,j} \times \vec{r}_j \qquad (6)$$

which is effectively the cross product of the forces and distance from the debris center of the rotation to the spheres. In order to compute the torque on the tug, r_j can be substituted by r_i.

Introducing multiple spheres in the MSM model also introduces a multiple of parameters which have to be solved for. In order to obtain as many unique equations as there are unknowns, electric potential measurements are taken over time for a rotating debris object. While the charge of all spheres varies over this tumbling motion, it is assumed that the electric potential of both bodies stays approximately constant during application of the tractor [32].

Again considering a tug represented by n and debris of m-spheres, the charge on any of the total k spheres can be calculated from:

$$\begin{bmatrix} q_1 \\ q_2 \\ \vdots \\ q_k \end{bmatrix} = \frac{1}{k_c} \begin{bmatrix} 1/R_1 & 1/r_{2,1} & \cdots & 1/r_{k,1} \\ 1/r_{1,2} & 1/R_2 & \cdots & 1/r_{k,2} \\ \vdots & \vdots & \ddots & \vdots \\ 1/r_{1,k} & 1/r_{2,k} & \cdots & 1/R_k \end{bmatrix}^{-1} \begin{bmatrix} \phi_1 \\ \phi_2 \\ \vdots \\ \phi_k \end{bmatrix} \qquad (7)$$

The center matrix is known as the inverse capacitance matrix and it consists of the inverse self-capacitance C_t^{-1} of the tug and debris C_d^{-1} as well as the inverse mutual capacitance $C_{t,d}^{-1}$ and $C_{d,t}^{-1}$. In the MSM, inverse self capacitance is given analytically for a sphere, and the inverse mutual capacitance is found by treating both spheres as point charges:

$$[C^{-1}] = \begin{bmatrix} C_t^{-1} & C_{t,d}^{-1} \\ C_{d,t}^{-1} & C_d^{-1} \end{bmatrix} \qquad C_{d,t}^{-1} = C_{t,d}^{-1\mathrm{T}} \qquad (8)$$

The inverse capacitance matrix describes the relationship between voltage and charge on both bodies at an instance of time. Since the position of the spheres is fixed within the bodies, the rotation rate determines the position at any instance of time. The only entries in the matrix that vary with this rate, and therefore have to be determined at each timestep, are $C_{t,d}^{-1}$ and its transpose. Pre-computing the other entries enables quick evaluation of this matrix at each time step. The inverse of the capacitance matrix can then be found from a block matrix inversion [19]. The potential curve that has to be matched against potential measurements follows from Eq. (7). The potential at an instance of time at the probe location Φ_p is determined from the summation of all charges over their respective distance to the probe $r_{i,p}$.

$$\Phi_p = k_c \sum_{i=1}^{k} \frac{q_i}{r_{i,p}} \qquad (9)$$

In this equation, the influence of the charge of the probe itself is neglected as this is expected to be calibrated by circuitry.

3.2 Obtaining the Best Fit MSM Model

Now consider the model given in the right-hand side of Fig. 2. The location of the spheres on the debris is specified with respect to the center of rotation (CoR) which is at a constant and known distance from the tug center of mass. A constant rotation about the axis through the CoR, orthogonal to the plane of the spheres, is assumed to be determined by means of LiDAR. Alternatively this rotation rate could be estimated by waiting for the probe measurements to repeat. By doing so the potential measured by the probes at any time is a function of the state vector $\vec{x}$, which includes full planar MSM model and the voltage of the debris:

$$\vec{x} = [R_1, r_1, \theta_1, \ldots R_m, r_m, \theta_m, V_d] \qquad (10)$$

Note that the characteristics of the tug are not defined in the state vector as they are assumed to be known a priori. These variables can be introduced to create a more general model, but for now they are assumed constant. Furthermore, the distance r_3

and rotation rate are not included as they are expected to be determined from LiDAR measurements. By using a numerical solver such as Matlab's *fmincon* function, the following cost function J is minimized in order to find the best fitting debris characterization.

$$J = \left\| \vec{\phi}_P(\vec{x}, \vec{t}) - \vec{\phi}_M(\vec{t}) \right\| \tag{11}$$

Here $\vec{\phi}_M$ and $\vec{\phi}_P$ are vectors containing the measured and predicted probe potential at all measurement instances $\vec{t}$. In order to find a better solution, some non-linear inequality constraints are applied. As a first constraint, all of the spheres have to be located within a rectangular box of which the contours have a 10 cm offset from most outward surface of the truth model. The second constraint makes sure that all sphere locations are unique by calculating the distance between spheres and requiring a minimal separation of 50 cm. This constraint is enforced because closely placed spheres make the inverse capacitance matrix difficult to invert numerically. Additionally, upper and lower bounds are applied on the state vector to restrict the radii of the spheres and their potential, such that no non-zero radii and potentials exist. The initial state vector is constructed as a spiral of spheres such that no sphere locations are identical, and all have a -15 kV potential.

The obtained solution state vector describes the electrostatic characteristics of the debris. After obtaining this characterization, potential measurements can be used to determine the force and torque at any instance of time, assuming the rotation rate and potential stay constant. As the potential of the tug and debris is expected to vary with the space weather, the orbit does play a role. From an analysis performed by Denton et al., it follows that except for a few hours after local midnight, the ambient plasma temperature and density varies over the scale of hours [9]. Considering that force and torque estimates could be made multiple times per minute, assuming the potential to be constant is deemed a good assumption.

4 Numerical Simulation Example

In the following section, the results from an example numerical simulation are given to demonstrate the applicability and performance of the RSEC method. Furthermore, a parameter sweep and sensitivity study are performed to examine the robustness of the method.

4.1 Set-Up and RSEC Results

The simulation is performed for a truth model consisting of 18 spheres on the debris and 4 spheres on the tug. By means of the RSEC method, the debris is approximated

Table 1 Truth model input parameters

Sphere	Location (x y) [m]	Radius [cm]
Tug		
1	(−5 1)	10
2	(−5 −1)	20
3	(−7 1)	5
4	(−7 −1)	3
Probe	Location (x y) [m]	Measurements
Probes		
1	(−1 5)	15
2	(−1 −5)	15
Tug potential: +20 kV	Debris potential: −20 kV	
Sphere	Location (x y) [m]	Radius [cm]
Debris		
1	(5 1)	10
2	(5 −1)	8
3	(7 1)	7
4	(7 −1)	4
5	(8 0)	3
6	(6 0)	7
7	(4 0)	9
8	(8 1)	2
9	(8 −1)	3
10	(6 1)	5
11	(6 −1)	4
12	(4 1)	9
13	(4 −1)	8
14	(6 2)	4
15	(6 3)	5
16	(6 4)	1
17	(5 −2)	4
18	(9 0)	2

with a solution consisting of only 10 spheres. Two probes acquire the potential measurements, being located approximately 5 m from the tug under a 45° angle with the debris. In order to obtain the measured potential curves, 15 measurements equally distributed over a full rotation of the debris are used. An overview of these parameters can be found in Table 1.

The potential measured by the probes in both the truth (18 sphere) and solution (10 sphere) model are plotted over rotation angle in Fig. 3. From these curves, it can be concluded that the numerical solver is able to quite accurately determine the potential curve of the solution. The error in potential between truth and solution is at most 0.01%. The corresponding potential field of truth and solution can be found in Figs. 4 and 5 respectively.

Fig. 3 Measured potential by two probes, potential of the solution, and difference between the two

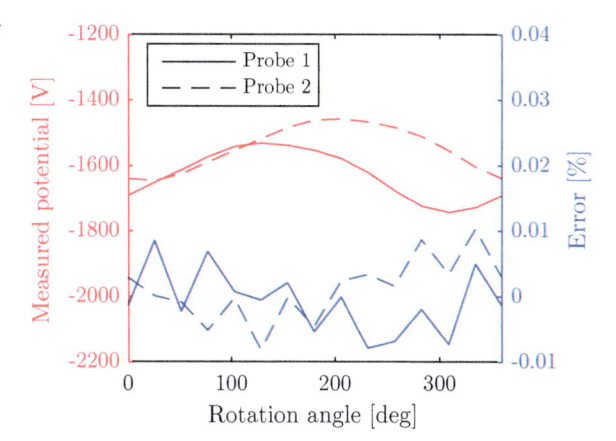

Fig. 4 Surface plot showing the 2D potential field of the truth model tug-debris system

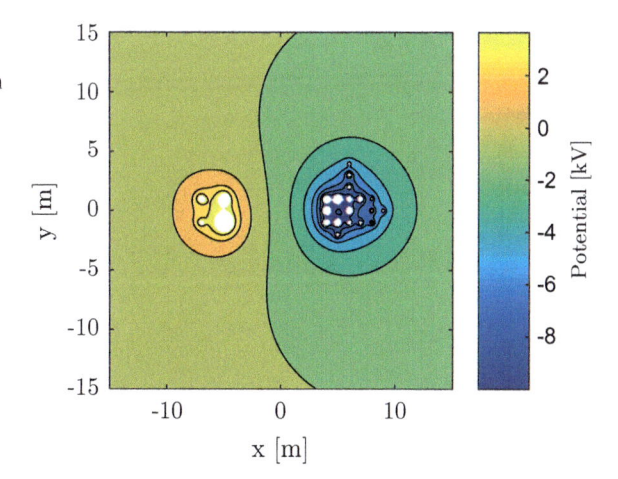

Fig. 5 Surface plot showing the 2D potential field of the solution model tug-debris system

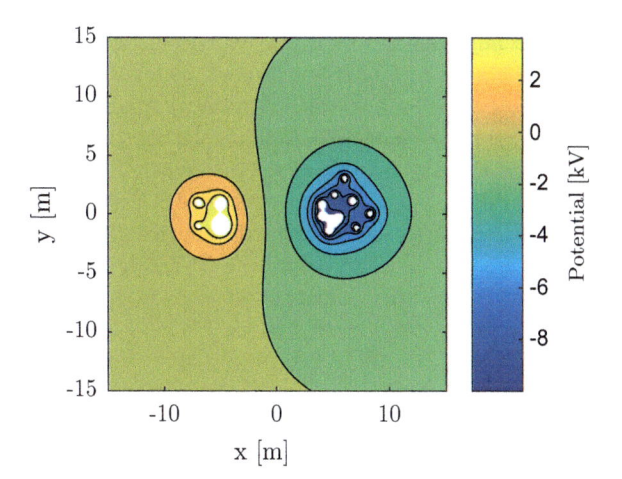

Fig. 6 Force of the solution model as well as its error with respect to the truth, over debris rotation angles

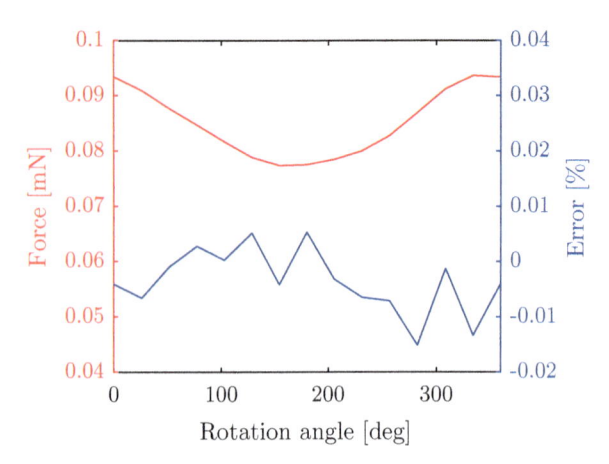

Fig. 7 Torque on tug and debris of the solution model as well as its error with respect to the truth, over debris rotation angles

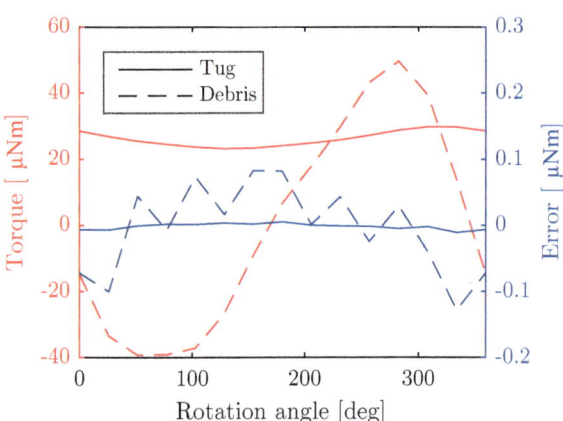

From these figures, it can be noticed that the potential field approximates a dipole field for larger separations. Moreover, while the field internal to the debris deviates significantly from the truth, the external field is accurately represented. Since the force and torque on both bodies is of interest and not the field structure internal of the debris, this deviation can be neglected. In order to determine the force and toque, Eqs. (5) and (6) are applied. Rotating the debris over time yields the distribution for force and torque over debris rotation angle, as give in Figs. 6 and 7 respectively.

From these figures, it can be seen that even though the spheres are placed at significantly different positions than in the truth model, the resulting force and torque are still accurate below 1%. The range of error in force is comparable to that of the measured potential and even though the torque on the debris has the largest error, it is still quite accurate.

4.2 Performance Sweep over Parameters

In order to show the influence of input parameters on the accuracy of the electrostatic characterization, a variation of parameters is performed. The same model is used as in the example set up in the previous section. Instead of applying a fixed number of 10 spheres to represent the debris, this number is varied. As can be seen in Fig. 8, a solution represented by too few spheres will introduce large errors in the solution. A cause for this error is symmetric sphere placement, which can be alleviated by adding another probe such that two potential curves have to be matched instead of just one. Furthermore it is deduced from Fig. 8 that even though the truth model consists of a large number of spheres, the solution will not be optimal with the same number of spheres. This may be because the optimizer has too many free parameters. A representation by as little as 4 spheres already generates a solution that models the system very well.

The fields and their angular dependance are smaller if the two spacecraft are farther apart. Their separation was also varied to see how performance was affected. Referring to Fig. 9, it can be concluded that the influence of this distance on the error in force and torque is negligible. The error appears to be a function of the geometry of the example set up and the ability of the solver to find an optimal state vector. However, the accuracy of LiDAR and probes may degrade due to the increased distance and smaller field strength.

Fig. 8 Error in force versus amount of spheres in the debris model

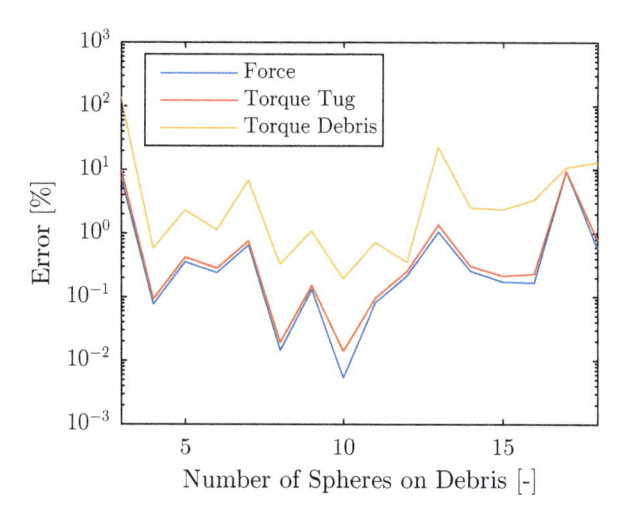

Fig. 9 Error in force versus separation distance between tug and debris center of mass

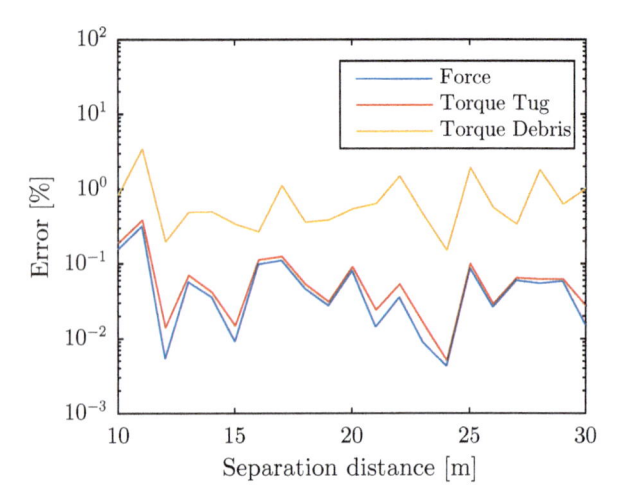

Fig. 10 Error in the solution in terms of force and torque due to an error in the measured separation distance

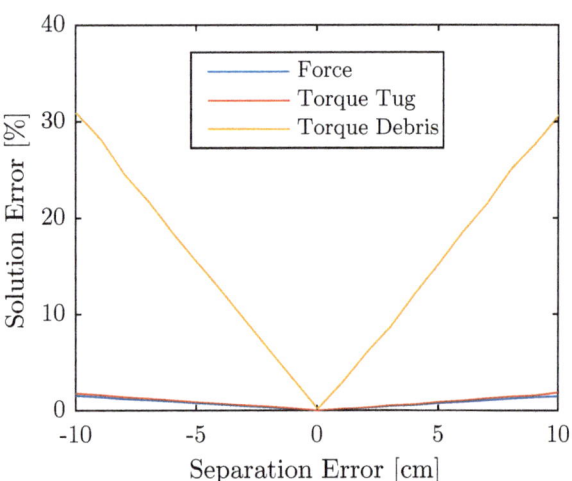

4.3 Sensitivity with Respect to Separation and Measurement Error

One of the assumptions in the described model is that the separation distance between tug and center of rotation of the debris is known. In order to see what an error in measurements of this distance does to the accuracy of the results, a sensitivity analysis is performed.

Figure 10 is acquired with the assumption that LiDAR will introduce measurement errors in the cm range. Although the force between tug and debris is still accurately represented by the solution, the torque on the debris will deviate rapidly from the truth model. In order to stay in the percent error range, millimetre accuracy is therefore required from LiDAR. In application of LiDAR, distance is measured

Fig. 11 Error in the solution in terms of force and torque due to an error in the potential measurements

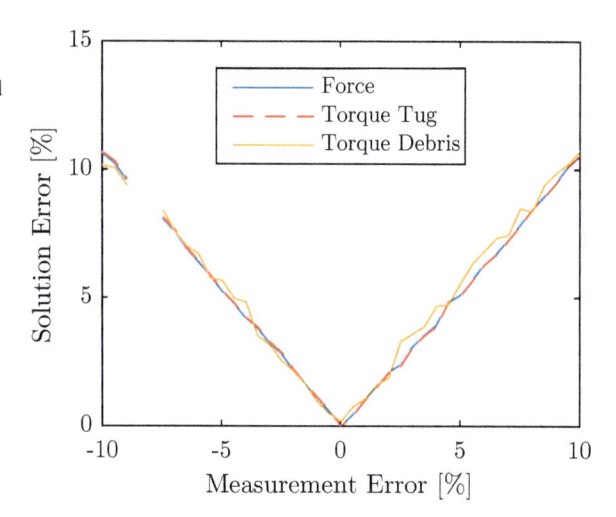

from the sensor towards a reflecting surface. Since the separation distance is defined from center of the tug to the center of rotation of the debris, this discrepancy has to be accounted for.

Assuming that the separation distance is perfectly determined, the error in the solution will still be non-zero due to errors in the probe measurements. In Sect. 2, it is determined that measurements near the electron beam will deviate by less than 10 V. Adding the interference of booms and space weather to this measurement error is expected to result in a percent error scale. By running the RSEC solver with a percent error in the measurements, Fig. 11 is obtained.

The omitted data in Fig. 11 corresponds to outliers due to the solver's inability to converge to a proper solution. In the figure, a clear trend can be distinguished, showing an almost 1-on-1 linear relation between measurement percent error and solution percent error for both force and torque.

5 Conclusion and Recommendations

Assuming knowledge of the tug potential, the center-to-center separation between tug and debris, and measurements of space potential or electric field in the vicinity of the craft, the voltage and an MSM model of the debris can be found using the method outlined here. This is done by constructing a predicted curve of potential vs. time for each probe and comparing it to the measured curve. Although special attention is paid to the ET, the method proposed could be used for electrostatic characterization of objects of interest such as asteroids. The potential field of this solution distribution matches the truth MSM model within percentage error. Even more so, the force and torque are also within percentage range. By performing a variation of parameters and sensitivity analysis, it can be found that only a small number of spheres is required to accurately characterize the electrostatic behaviour of the debris. This is under the conditions that the measured separation distance has errors in the millimetre range and external influences such as the photo-electron cloud and potential field due to the tractor beam are negligible.

Future validation of these results is envisioned by using a field-mill to measure the electric field between spherical probes in a test-bed. Furthermore, the planar model described in this paper does not account for 3D satellite dimensions and tumbling motion. The next step in developing the RSEC model therefore includes implication of general 3D shapes and dynamics. This model can be optimized by using appropriate estimators, such as a Kalman filter in place of Matlab's *fmincon* solver. Using an estimator is expected to improve the accuracy of the results and offer better feed-forward estimations.

References

1. Albuja, A.A.: Rotational dynamics of inactive satellites as a result of the YORP effect. Ph.D. Thesis, University of Colorado Boulder (2015)
2. Ariafar, S., Jehn, R.: Long-term evolution of retired geostationary satellites. In: 4th European Conference on Space Debris, vol. 587, p. 681 (2005)
3. Bennett, T., Schaub, H.: Touchless electrostatic three-dimensional detumbling of large GEO debris. In: AAS/AIAA Spaceflight Mechanics Meeting, Santa Fe, New Mexico (2014)
4. Bombardelli, C., Pelaez, J.: Ion beam shepherd for contactless space debris removal. J. Guid. Control. Dyn. **34**(3), 916–920 (2011)
5. Boyd, T., Sanderson, J.: The Physics of Plasmas. Cambridge University Press, Cambridge (2003)
6. Bristow, J., Folta, D., Hartman, K.: A formation flying technology vision. AIAA Paper **5194**, 19–21 (2000)
7. Chrystal, P., McKnight, D., Meredith, P.L., Schmidt, J., Fok, M., Wetton, C.: Space debris: on collision course for insurers? Technical Report, Swiss Reinsurance Company Ltd, Zürich, Switzerland (2011)
8. Chubb, J.: Two new designs of 'field mill' type fieldmeters not requiring earthing of rotating chopper. IEEE Trans. Ind. Appl. **26**(6), 1178–1181 (1990)
9. Denton, M., Thomsen, M., Korth, H., Lynch, S., Zhang, J., Liemohn, M.: Bulk plasma properties at geosynchronous orbit. J. Geophys. Res. Space Phys. **110**(A7) (2005). https://doi.org/10.1029/2004JA010861
10. Farrell, W., Stubbs, T., Vondrak, R., Delory, G., Halekas, J.: Complex electric fields near the lunar terminator: the near-surface wake and accelerated dust. Geophys. Res. Lett. **34**(14) (2007). https://doi.org/10.1029/2007GL029312
11. Gibson, W.: The Method of Moments in Electromagnetics. CRC, Boca Raton (2014)
12. Grard, R.: Properties of the satellite photoelectron sheath derived from photoemission laboratory measurements. J. Geophys. Res. **78**(16), 2885–2906 (1973)
13. Hartzell, C.M.: The dynamics of near-surface dust on airless bodies. Ph.D. Thesis, University of Colorado Boulder (2012)
14. Hogan, E., Schaub, H.: Relative motion control for two-spacecraft electrostatic orbit corrections. J. Guid. Control. Dyn. **36**(1), 240–249 (2012)
15. Hogan, E., Schaub, H.: Impacts of tug and debris sizes on electrostatic tractor charging performance. Adv. Space Res. **55**(2), 630–638 (2015)

16. Inter-Agency Space Debris Coordination Committee (IADC): IADC space debris mitigation guidelines. Technical Report, IADC-02-01, Inter-Agency Space Debris Coordination Committee (2007)
17. Jackson, T., Zimmerman, M., Farrell, W.: Concerning the charging of an exploration craft on and near a small asteroid. In: 45th Lunar and Planetary Science Conference (2014)
18. Jin, J.: The Finite Element Method in Electromagnetics. Wiley, Hoboken (2014)
19. Jo, S., Kim, S., Park, T.: Equally constrained affine projection algorithm. In: Conference Record of the Thirty-Eighth Asilomar Conference on Signals, Systems and Computers, vol. 1, pp. 955–959 (2004)
20. Johnson, N.: Protecting the GEO environment: policies and practices. Space Policy **15**(3), 127–135 (1999)
21. Kaplan, M., Boone, B., Brown, R., Criss, T., Tunstel, E.: Engineering issues for all major modes of in situ space debris capture. In: Proceedings of the AIAA SPACE 2010 Conference & Exposition, Anaheim (2010)
22. Kemp, R., Sellen, Jr. J.: Plasma potential measurements by electron emissive probes. Rev. Sci. Instrum. **37**(4), 455–461 (1966)
23. Mizera, P., Fennell, J., Croley, D., Gorney, D.: Charged particle distributions and electric field measurements from s3-3. J. Geophys. Res. Space Phys. **86**(A9), 7566–7576 (1981)
24. Moorer, D., Schaub, H.: Electrostatic spacecraft reorbiter. US Patent 8,205,838 B2 (2012)
25. Mullen, E., Gussenhoven, M.: Scatha survey of high-level spacecraft charging in sunlight. J. Geophys. Res. **91**(A2), 1474–1490 (1986)
26. National Aeronautics and Space Administration (NASA): Satellite box score. Orbital Debris Q. News **20**(1–2), 12–14 (2016)
27. Pfaff, R., Borovsky, J., Young, D.: Measurement Techniques in Space Plasmas: Particles. American Geophysical Union, Washington, DC (1998)
28. Schaub, H., Jasper, L.: Circular orbit radius control using electrostatic actuation for 2-craft configurations. Adv. Astronaut. Sci. **142**, 681 (2011)
29. Schaub, H., Moorer, D.F.: Geosynchronous large debris reorbiter: Challenges and prospects. J. Astronaut. Sci. **59**(1–2), 161–176 (2014). https://doi.org/10.1007/s40295-013-0011-8
30. Secker, P., Chubb, J.: Instrumentation for electrostatic measurements. J. Electrost. **16**(1), 1–19 (1984)
31. Seubert, C., Stiles, L., Schaub, H.: Effective coulomb force modeling for spacecraft in earth orbit plasmas. Adv. Space Res. **54**(2), 209–220 (2014)
32. Stevenson, D., Schaub, H.: Multi-sphere method for modeling electrostatic forces and torques. Adv. Space Res. **51**(1), 10–20 (2013). https://doi.org/10.1016/j.asr.2012.08.014
33. Stevenson, D., Schaub, H.: Optimization of sphere population for electrostatic multi-sphere method. IEEE Trans. Plasma Sci. **41**(12), 3526–3535 (2013)
34. Ulamec, S., Biele, J.: Surface elements and landing strategies for small bodies missions–philae and beyond. Adv. Space Res. **44**(7), 847–858 (2009)
35. Vasavada, H., Schaub, H.: Analytic solutions for equal mass four-craft static coulomb formation. J. Astronaut. Sci. **56**(1), 17–40 (2008)

Active Debris Removal and Space Debris Mitigation using Hybrid Propulsion Solutions

Stefania Tonetti, Stefania Cornara, Martina Faenza, Onno Verberne, Tobias Langener, and Gonzalo Vicario de Miguel

Abstract This paper presents the results of a study carried out in the frame of the ESA General Studies Programme (GSP), dealing with the feasibility of performing active debris removal by using a hybrid propulsion system embarked on the chaser spacecraft. While the study focuses mainly on the use of a hybrid rocket propulsion system on-board a chaser spacecraft that performs active debris removal, it also addresses the application of this innovative propulsion technology for debris mitigation purposes. Hybrid propulsion systems seem to be a promising alternative to conventional liquid propellant in-orbit propulsion systems, in terms of complexity, cost, operational advantages, whilst also offering the use of non-toxic propellants. The study has been carried out by Elecnor Deimos, expert in mission analysis, and Nammo Raufoss, expert in hybrid propulsion.

1 Introduction

Hundreds of satellites populate the Earth-bounded orbits, and the number of satellites orbiting around the Earth is rapidly increasing. Among the objects in orbit around our planet, about 95% are classified as space debris. These debris objects are a threat as they can collide with the active satellites, in turn creating more debris objects and possibly even restricting access to important orbits such as the Sun-synchronous orbit (SSO) region. Several collisions have already occurred and the

---

S. Tonetti (✉) · S. Cornara · G. Vicario de Miguel
Deimos Space S.L.U., Madrid, Spain
e-mail: stefania.tonetti@deimos-space.com; stefania.cornara@deimos-space.com; gonzalo.vicario@deimos-space.com

M. Faenza · O. Verberne
Nammo Raufoss AS, Raufoss, Norway
e-mail: martina.faenza@nammo.com; cj.verberne@nammo.com

T. Langener
ESA ESTEC, Noordwijk ZH, The Netherlands
e-mail: tobias.langener@esa.int

population of debris will keep growing if no measures are taken to mitigate the generation of space debris by implementing proper policies and mission design standards, as well as to remove space debris in the future. By removing existing objects from orbit, the risk of collisions can be greatly reduced and access to important orbits retained.

This paper presents the results of a 1-year study, funded by the ESA General Studies Programme (HYPSOS: Hybrid Propulsion Solutions for Space Debris Remediation Study), and dealing with investigating the feasibility of performing active debris removal (ADR) by means of hybrid propulsion solutions. While the study focuses mainly on the use of a hybrid rocket propulsion system on-board a dedicated chaser spacecraft that performs ADR, it also preliminarily addresses the application of this innovative propulsion technology for space debris mitigation (SDM) purposes.

The debris proliferation is posing challenges to missions in densely populated orbits (LEO in particular), thus promoting the need for providing propulsion capabilities corresponding with debris remediation measures, (e.g. controlled deorbit of large debris after capture over inhabited areas such as the South Pacific Ocean Uninhabited Area SPOUA or re-orbit above the LEO protected region). Several studies [1–14] have been performed in the last years to preliminary assess the feasibility of dedicated debris remediation missions, and their associated requirements, in order to remove several large objects per year from the highly populated orbits.

Active systems represent autonomous spacecraft able to reach the target debris, to capture it in some way and to de-orbit the space debris object or move it to special graveyard orbits. In the following is presented a survey of main ADR missions and studies conducted in the last years, with emphasis on European ones.

The e.Deorbit is an ESA study in the frame of the Clean Space Initiative. The e.Deorbit pre-assessment study was performed in ESA's Concurrent Design Facility (CDF) between 26 June and 13 September 2012 and some aspects of the mission were already defined [2]. "Phase A" was completed in 2014 and the mission is now in "Phase B1". The mission is targeted to be launched in 2023. The e.Deorbit mission is a single debris active removal mission aimed to be launched from Kourou by Vega as a single launch. e.Deorbit key challenges are: removing a noncooperative debris with Envisat mass and size and in tumbling mode, stabilizing the stack composed by chaser and Envisat, fitting in Vega launch capacity.

IBS-IOD (Ion Beam Shepherd IOD Mission) is an ESA study lead by Deimos in the frame of "Requirements and Concepts for IOD Missions for Breakthrough Concepts and Approaches". The IBS debris removal technique, proposed by the Technical University of Madrid (UPM) in March 2010 [15, 16], is contactless, i.e. the transmission of forces and torques to a target body without any mechanical interaction.

SOADR (Service Oriented Approach to the Procurement and Development of an Active Debris Removal Mission) is an ESA funded "Phase 0" study [17]. The study team was lead by SSTL, with Aviospace, Elecnor Deimos and the University of Surrey as sub-contractors. The aim of the study is to make a preliminary investigation of the technical feasibility of actively removing a single large debris object from SSO orbit, and to further investigate the possibility of providing this mission (and further missions) as commercial services rather than traditional procurements. The study 'user case' was an Active Debris Removal (ADR) mission targeting ESA-owned Envisat via controlled re-entry into atmosphere.

DEOS (Deutsche Orbitale Servicing Mission) [18] is a DLR project that will focus on Guidance and Navigation, capturing of non-cooperative as well as cooperative client satellites, performing orbital maneuvers with the coupled system and the controlled de-orbiting of the two coupled satellites. The project prime contractor is Airbus Defence and Space (Friedrichshafen).

RemoveDebris [19] is a European Union Framework 7 (FP7) research project to develop and fly a low cost in-orbit demonstrator mission that aims to de-risk and mature key technologies needed for future ADR missions. The mission, which launch is foreseen in early 2017 from ISS, is composed of one chaser which holds the payloads and two targets CubeSats which are ejected to test the technologies.

CNES has a wide interest in debris removal with the OTV (Orbital Transfer Vehicle) program which focuses on the removal of large pieces of space debris between now and 2020 [20].

SPADES (Solid Propellant Autonomous DE-Orbit System) is an ESA study [21] intended to be a system installed in the satellite prior to launch able to provide basic velocity increment for de-orbiting or autonomously de-orbit a spacecraft when control of the spacecraft has been lost. It could serve on large and small LEO satellites, MEO and GEO satellites, upper stages and jettisoned components, multiple ADR missions.

The Italian D-Orbit company (http://www.deorbitaldevices.com/) is producing decommissioning devices for re-orbiting GEO satellites (with high or low thrust) and de-orbiting LEO satellites from 700 km altitude. In June 2017 D-Orbit launched D-SAT mission, a satellite with a fully functional independent decommissioning system able to perform a precise controlled de-orbit in few hours.

Hybrid propulsion seems very promising to cope with the space debris problem, offering also a more environmentally friendly fuel. The hybrid propulsion system considered is based on 87.5% H_2O_2 as oxidizer. H_2O_2 in 87.5% concentration has a consolidated heritage as oxidizer in propulsion systems (TRL 9) in Europe and its handling and storing does not pose any hazard as such, if compatible materials are used and proper procedures and guidelines are followed. The combination of H_2O_2 and hydrocarbon-based fuels has the advantage of having very good performances in terms of theoretical specific impulse, competitive with many of the best candidates in hybrid propulsion. High concentration hydrogen peroxide has favorable characteristics because of: its high density, its storability at room temperature at ambient pressure and its possibility of being decomposed through a catalyst. The catalytic decomposition produces a gaseous oxidizing mixture of oxygen and water vapor at a temperature high enough to guarantee ignition. The motor is capable to self-sustain the combustion process at a wide range of operating conditions without the need of dedicated systems to control the combustion. This peculiarity is particularly beneficial when a synergy between different propulsive architectures is desirable: high concentration H_2O_2 can be used as propellant for monopropellant thrusters and in combination with a hybrid motor to provide a solution which covers a wider range of propulsive functions with a single propulsion architecture. The same oxidizer and pressurizer tanks and main fluid system components are used to drive all different kind of thrusters, both hybrid and mono-propellant.

The paper is organized as follows. In Sect. 2, a detailed survey is performed, in order to identify size-mass-altitude distribution of space objects belonging to ESA and EU from completed, on-going and future missions. Section 3 presents the active debris removal application. After a parametric sizing of orbital maneuvers required for accomplishing a dedicated debris removal mission, the modeling of the Propulsion System (PS) is described, with the purpose of creating reliable and accurate tools to size and model the propulsion system and the appraisal of the PS, targeted at assessing the hybrid-based propulsion system for selected representative mission scenarios. The assessment, apart from the detailed sizing of the main components, includes a comparison with the corresponding conventional propulsion system and an indication of the TRL at component and system level. Section 4 tackles an outlook for SDM, with the goal of considering the adoption of hybrid propulsion system both as main engine embarked on future satellites and as a kit to be integrated on launchers upper stages to be compliant with End-of-Life (EoL) disposal policies. Finally, Section 5 contains the main conclusions of the paper.

Two contributors, with the supervision of ESA-ESTEC, have taken part in the development of the study: Elecnor Deimos, expert in mission analysis and design, and Nammo Raufoss, expert in rocket propulsion.

2 Detailed Survey on European LEO Missions

In order to identify a set of interesting spacecraft and their corresponding orbits to perform ADR and SDM, a detailed survey of European Low Earth Orbit (LEO) in-orbit spacecraft, under current design and planned future missions, has been performed. Because of liability reasons, only objects belonging to ESA, ESA member states and ESA cooperating states are considered. The information has been retrieved mainly from ESA DISCOS database [22] for in-orbit missions and mainly from eoPortal Satellite Missions Database (https://directory.eoportal.org/web/eoportal/).

A detailed survey identified 130 missions with mass greater than 10 kg to be potential objects of study for ADR and SDM analyses. The extracted information revealed that: satellite masses range from few kg to 7821 kg of EnviSat; 80% of the missions are in SSO, near SSO, polar or near-polar orbits; altitude ranges from almost 400 to 1450 km and inclination between 20 and 100°.

Based on the main conclusions of this survey, *the region defined by the SSO orbits between 350 and 850 km seems to be the most interesting for debris remediation and mitigation*. The propulsion metrics retrieved considering SSO orbits allow designing a hybrid propulsion system suitable for the majority of LEO missions.

3 Active Debris Removal

In the active debris removal mission scenario, it is assumed that a dedicated satellite, the chaser, is sent to target/capture/de-orbit a single object in space. The chaser is then sized and instrumented to be fully independent once released by the carrier in its injection orbit. As stated by the scope of the study, the propulsion system is based on hybrid rocket technology.

It is considered that the carrier for such kind of missions is Vega [23] and it will impose the limitations on the maximum admissible chaser mass and envelope.

The target objects to be de-orbited considered in this study are in most cases satellites, which almost certainly have flexible appendages and deployed panels. These items represent points of weakness in the structure and thus impose strict requirements on debris remediation missions. A strict requirement in terms of maximum tolerated acceleration in order to prevent the risk of breaking down the target during re-entry is enforced to avoid producing additional debris. Several studies about debris mitigation and remediation mention this aspect but just a few give quantitative numbers; all the studies performed by ESA [2, 21] consider a maximum tolerated acceleration on the debris of 0.04g during the de-orbiting phase.

The HYPSOS study has considered therefore this value as its reference for maximum tolerated acceleration, but assessing at the same time the benefit/impact on the propulsion system and overall chaser design of having a less stringent limitation.

3.1 Mission Scenario

Parametric ΔV needed for the propulsive phases of an active debris removal mission are computed in order to provide inputs for sizing the hybrid propulsion system embarked on the chaser spacecraft. When possible, results have been compared with the outcome of the ESA e.Deorbit Phase-A study [2, 24] focused on EnviSat de-orbiting. For the ADR mission analysis, three different propulsive phases are studied. They are presented in the following sections.

3.1.1 Transfer to Target Orbit

Considering *Vega* [23] as baseline launcher, the best injection altitude that maximizes the chaser mass at the target orbit is selected by retrieving launcher performance, parameterizing the injection altitude and computing the ΔV necessary to acquire the target orbit. The transfer to the target orbit starts with the injection of the chaser spacecraft into an intermediate orbit called the injection orbit. This orbit has been chosen to be circular and coplanar with the target orbit in order to avoid highly consuming out-of-plane maneuvers. In order to assess which injection altitude maximizes the mass placed into the target orbit, a range of altitudes from 300 km to 850 km has been considered. Starting from the injection orbit, the chaser spacecraft performs a Hohmann transfer for target orbit acquisition. Maneuvers are assumed symmetrical with respect to the apsidal point, performed with tangential thrust and considering gravity losses.

The outcome of the parametric analysis, considering an average specific impulse (Isp) delivered by the hybrid propulsion of 300 s, showed that, in the case of the Vega launcher, the highest chaser mass at the target orbit is achieved when the injection altitude is the lowest possible (300 km) and orbital raising to target orbit is performed by the chaser itself. Total ΔV varies from about 40 m/s for 350 km target orbit altitude to 300 m/s for 850 km target orbit altitude.

3.1.2 Approach to Target

The proximity phase is necessary for proper approaching the target object. This phase typically starts from a distance of a few kilometers and ends at the vicinity of the target. The chaser is considered to perform a safe rendezvous and to mate with an uncooperative target. This phase analyses far and close rendezvous (refer to Fig. 1 for the rendezvous profile chosen).

Parametric analyses have been performed, varying both chaser mass between 1000 and 1550 kg and debris orbital altitude between 350 and 850 km. Two monopropellant thrusters aligned with each body axis were assumed for Reaction Control System (RCS); a total of 12 thrusters, plus 12 for redundancy, of 20 N each are considered, with a delivered specific impulse of 150 s (expected for H2O2 monopropellant thrusters in this class) and a control frequency of 1 Hz.

The results showed that a high number of burns are required for station keeping and that ΔV mainly depends on the orbit altitude: lower orbits have faster relative dynamics and require more ΔV. Total ΔV for both far and close rendezvous can vary from 14 to 18 m/s. In the study the maximum value has always been adopted.

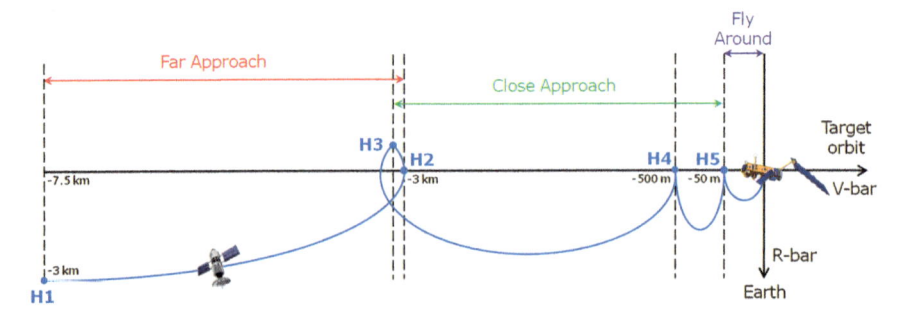

Fig. 1 Graphical representation of the selected rendezvous profile

3.1.3 Controlled De-Orbit

The controlled re-entry consists in *lowering the perigee to a given re-entry altitude* (60 km in this study) such that the maneuver guarantees the impact of the stack formed by the debris and the chaser spacecraft over an unpopulated area in the South Pacific Ocean, not exceeding the casualty risk threshold imposed by the debris mitigation guidelines [25–28]. One of the most relevant drivers in the controlled re-entry is the gravity loss: the higher the gravity losses, the bigger the ΔV, propellant mass and burning time required. For low thrust-to-mass ratios the number of maneuvers drives the ΔV expenditure for the disposal phase: the higher the number of maneuvers, the lower the gravity losses. A survey of feasible thrust and system mass combinations for hybrid propulsion has been carried out, together with multi-maneuver perigee lowering strategies, leading to the selection of solutions guaranteeing high-enough thrust-to-mass ratio to keep gravity losses negligible, while not exceeding maximum acceleration values. Initial orbit altitudes range from 350 km to 850 km. The perigee lowering strategies studied consider: one burn to lower the perigee at 60 km for direct re-entry; two burns to lower the perigee altitude to an intermediate value (200 km in this study); three burns to gradually lower the perigee down to 60 km. Thrust-to-mass ratio values considered, typical for hybrid propulsion system, range from 0.04 N/kg to 0.8 N/kg and propulsion system specific impulse is fixed to 300 s. The outcome of the analysis showed that, for the magnitude of the ΔV considered in the ADR scenario, thrust-to-mass ratios above 0.375 N/kg allow keeping the gravity losses below 1%. The thrust-to-mass ratios considered for the hybrid propulsion system in the present study are all above the limit where the gravity losses become negligible. In this case, the total ΔV does not depend on the perigee lowering strategy chosen and it varies from a minimum of 80 m/s starting from 350 km to a maximum of about 220 m/s starting from 850 km.

3.1.4 Delta-V Summary

A summary of the ΔV computed for some reference mission cases is presented in Table 1.

Table 1 ΔV summary for some reference cases

Target name	Target orbit (km)	Target mass (kg)	ΔV Transfer to target (m/s)	ΔV Approach to target (m/s)	ΔV Controlled de-orbit (m/s)
EarthCare	393	1860	53.2	18.1	95.2
Deimos-2	620	310	178.7	17.4	157.8
EnviSat	760	7821	253.0	17.0	193.1
MetOp-SG-A	817	3000	283.7	16.8	207.5

3.2 Hybrid Propulsion System Sizing

The hybrid-monopropellant propulsion system considered is based on 87.5% H2O2 as oxidizer and HTPB or HDPE as fuel and its architecture is presented in Fig. 2. The hybrid motor is responsible for phases 1, targeting, and phase 3, de-orbiting. The monopropellant part is responsible for phase 2, rendezvous, and for RCS.

Parametric ΔV collected from the three propulsive phases with adequate margins has been considered to initially perform a simplified mass budget estimation, based on Tsiolkovsky rocket equation, in order to get an overview on the thrust class versus burning time required by each of the 130 mission scenarios retrieved in the mission survey. Starting from this information, a few interesting scenarios, spanning

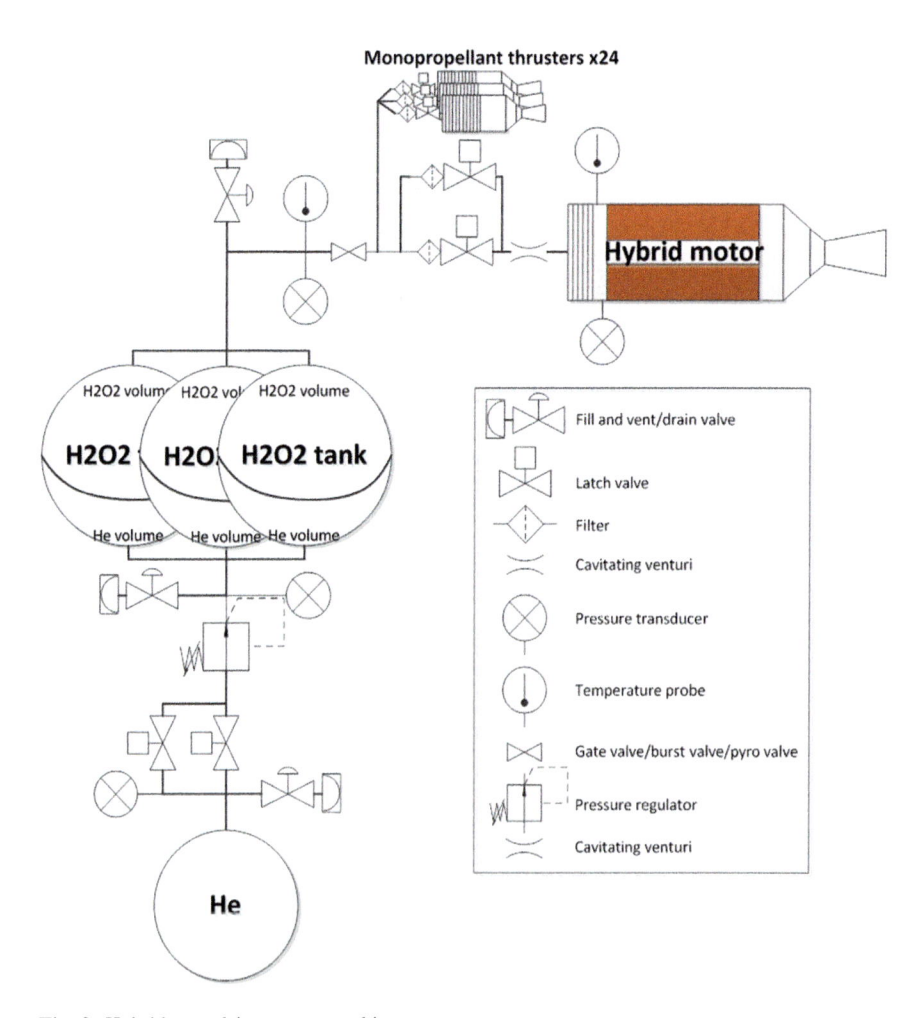

Fig. 2 Hybrid propulsion system architecture

Active Debris Removal and Space Debris Mitigation using Hybrid Propulsion Solutions 171

different ranges of thrust class and burning time, have been selected to perform a more detailed investigation. For this purpose, a *complete and detailed performance model* has been implemented with the aim of sizing in detail the hybrid propulsion system and predicting the temporal behavior of the motor based on the requirements and assumptions.

All the main components of the propulsion system have been sized (mass and envelope of: oxidizer and pressurizing tanks, combustion chamber, nozzle) and the hybrid motor performance and behavior have been assessed (propellants consumption and thrust profile).

In order to assess the hybrid propulsion system in a more thorough way, an additional goal of the HYPSOS study was *to compare the hybrid propulsion system configurations considered for ADR with their corresponding conventional systems based on bi-propellant technology*. For this purpose, a reliable sizing methodology of such bi-propellant system, when applied to the considered scenario, has been implemented as well.

3.2.1 Hybrid Propulsion System Detailed Sizing for Envisat

This scenario has been investigated because of its known interest in the European space community. ESA is already developing a dedicated mission to de-orbit this object through the study called "e.Deorbit" [2]. Envisat mass is estimated around 7900 kg and it flies at an altitude of about 760 km. The ΔV required for each phase of the chaser mission with corresponding margins and propulsion system involved are summarized in Table 2.

For this specific scenario, besides the hybrid propulsion system architecture with one single hybrid motor, an additional configuration has been investigated which makes use of multiple (smaller) hybrid motors working in parallel to deliver the thrust required to perform the orbital maneuvers. The reason behind this choice has been to investigate a configuration exploiting the higher volumetric compactness of the smaller hybrid motors.

Table 3 summarizes the resulting mass of the propulsion system for the two different architectures respectively, in comparison with the corresponding bi-propellant system. For the single motor configuration, a mass saving of about 6.9% is achieved considering a hybrid PS. With respect to the single-motor configuration, the propulsion system with 3-motors is heavier but more compact and it will allow for a more flexible architecture and ease the integration with the launch vehicle.

Table 2 ΔV budget for Envisat

Mission phase	ΔV (m/s)	Margin (%)	Propulsion system
Targeting	253	5	Hybrid
Rendezvous and RCS	18	100	H2O2 monopropellant
De-orbit	193	5	Hybrid

Table 3 Comparison of the hybrid propulsion system with the corresponding bi-propellant one for Envisat scenarios

	Hybrid single motor	Hybrid 3 motors	Bi-propellant
Propulsion system wet mass (kg)	844	899	907
Propulsion system dry mass (kg)	113	148	153
Propellant mass fraction	0.87	0.83	0.83

4 Outlook for Debris Mitigation

Adherence to the post-mission disposal guidelines is the absolute key driver for the environmental impact reduction and the new missions have to be designed in order to be able to autonomously and in a reliable way perform post-mission disposal at EoL, by means of:

- direct *control re-entry*, as already described;
- if the risk on ground is lower than 10^{-4}, *un-controlled re-entry* in less than 25 years [25, 27].

From a mission design point of view, the second option ensures the compliance with the space debris mitigation requirements for Earth observation missions, while minimizing the required propellant.

An exhaustive study has been carried out in order to determine whether a perigee-lowering maneuver is needed or not to comply with the 25 years rule for all the 130 missions selected in the first part of the study. In case it is needed, it is determined the highest (i.e. less costly) altitude onto which the spacecraft shall be maneuvered in order to guarantee a safe uncontrolled decay within 25 years and the corresponding ΔV. Resulting ΔV needed have then been considered to size a dedicated hybrid propulsion de-orbiting kit for EoL disposal in future ESA missions.

The parametric uncontrolled re-entry analysis has been performed considering a range of starting orbit altitudes between 350 and 850 km and ballistic coefficient between 10 and 180 kg/m^2.

The results showed that for low target orbits (below 550 km), the re-entry is always performed without any maneuver. For those cases where a maneuver is needed, the perigee altitude decreases as the ballistic coefficient increases, so the amount of ΔV required increases. Besides, for a fixed ballistic coefficient, the perigee altitude decreases as the reference altitude of the orbit increases due to the fact that the higher the re-entry orbit, the less drag undergoes the spacecraft around the apogee.

If either the altitude or the ballistic coefficient increases, the ΔV required increases too, reaching a maximum value of 119 m/s for a ballistic coefficient of 180 kg/m^2 and an initial orbit altitude of 850 km. Compared with the controlled re-entry, this strategy is cheaper in terms of ΔV, but it requires more time.

Feasibility of performing post-mission disposal of future ESA missions in LEO and European upper stage rocket bodies in GTO by means of hybrid propulsion has been addressed.

A representative LEO mission scenario represented by the FLEX Earth Explorer has been chosen for a detailed sizing of the hybrid propulsion system for debris remediation by implementing an un-controlled re-entry strategy. While controlled re-entry is proposed for Ariane 5 upper stage de-orbiting from GTO.

4.1 Future ESA LEO Mission: FLEX

FLEX (FLuorescence EXplorer) has been chosen as the eighth Earth Explorer mission within ESA's Earth Observation Programme [30]. FLEX, slated for launch after 2020, will fly in formation with Sentinel-3 in a SSO orbit at 815 km altitude. The FLEX propulsion subsystem will provide the necessary thrust for correction of launcher injection errors, formation flying acquisition with Sentinel-3, orbit maintenance for ground-track control at all latitudes (including a small ΔV allocation to cope with formation control in possible contingency situations), collision avoidance to avoid collision with space debris objects, End-of-Life disposal to comply with EoL guidelines. It will use a hydrazine system, with an assembly of four 1N thrusters, pressurized with helium and operated in blow-down mode. At the end of Phase B1, the estimated mass budget of hydrazine system and propellant sum up at about 80 kg. After 5 years (nominal mission phase + mission extension), the mission foresees as baseline scenario an in-plane maneuver to lower the orbit perigee to an altitude that guarantees safe uncontrolled decay within 25 years.

Deimos was involved in the FLEX Phase A/B1 study and the corresponding ΔV budget at the end of Phase B1 is summarized in Table 2. The ΔV for injection errors correction, collision avoidance and orbit maintenance are taken from Deimos FLEX Mission Analysis Report [31]. The ΔV for End-of-Life disposal to lower the perigee in order to guarantee re-entry into the atmosphere in less than 25 years was computed for FLEX in the frame of the HYPSOS study. The total ΔV to be delivered during FLEX lifetime is about 138 m/s and the minimum ΔV to be provided during orbit control is of 0.04 m/s.

The propulsion system should be the only one embarked on-board; this represents an appealing advantage if combined with the suitability of H2O2-based hybrids to operate synergistically with a H2O2 monopropellant system, the former being responsible of the EoL disposal maneuver while the latter provides the attitude control. Margins applied on top of the ideal ΔV are reported in Table 4.

The resulting architecture of the propulsion system considered in this case is the one reported in Fig. 3. Following the approach already adopted by ESA in preliminary sizing the propulsion system, [30], four monopropellant thrusters with nominal thrust of 1N each have been considered for the ACS.

Table 5 summarizes the overall characteristics of the hybrid PS for FLEX.

Table 4 ΔV budget for FLEX mission

Mission phase	ΔV (m/s)	Margin (%)	Propulsion system
Injection errors and formation acquisition ΔV (in-plane and out-of-plane manoeuvers)	29.9	20	H2O2 monopropellant
Collision avoidance and orbit control ΔV (in-plane and out-of-plane manoeuvers)	34.8	20	H2O2 monopropellant
EoL disposal ΔV	73.0	5	Hybrid
Total ΔV	137.7		

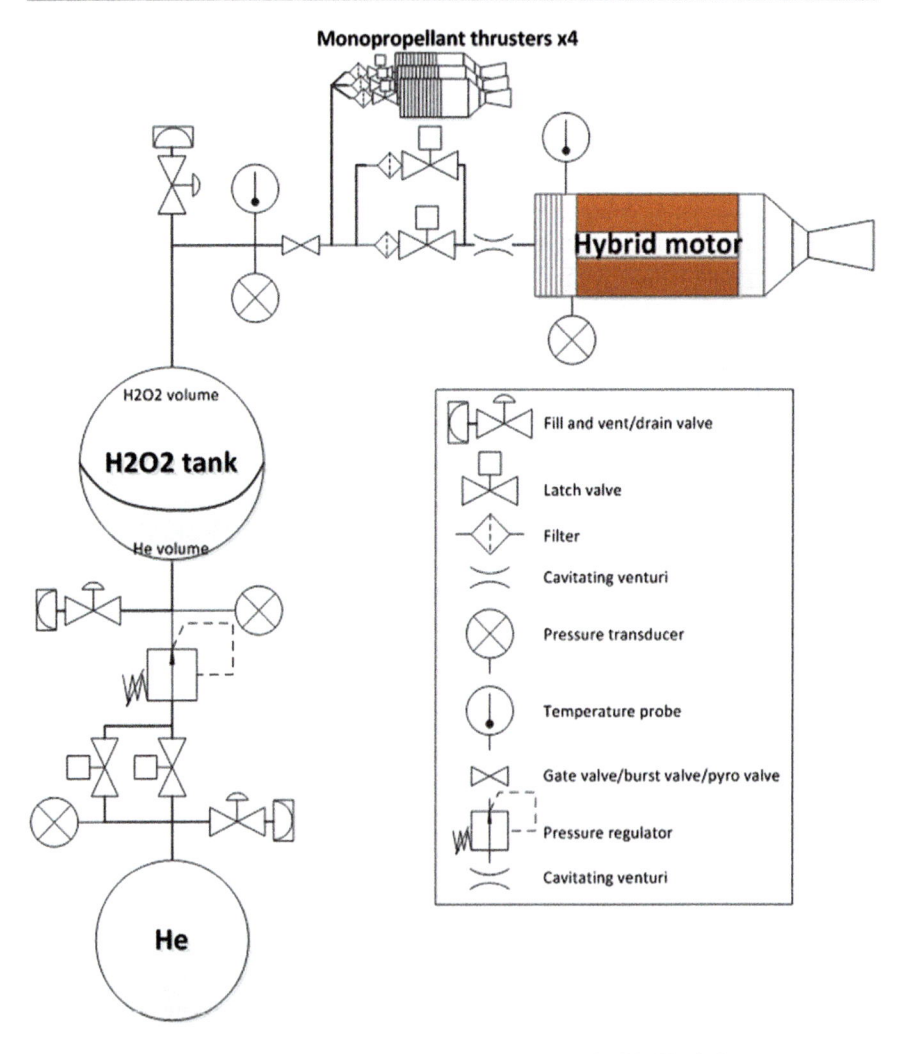

Fig. 3 Fully independent hybrid propulsion system architecture for FLEX mission

Table 5 Main results for the hybrid propulsion system of FLEX

Consumed propellant, monopropellant (kg)	48	Propulsion system wet mass (kg)	86
Consumed propellant, hybrid (kg)	24	Propulsion system dry mass (kg)	15
Burning time (s)	200	Oxidizer tank capacity (l)	52
Peak thrust (N)	340	Pressurizing gas tank capacity (l)	8
Peak acceleration (m/s^2)	0.4		

4.2 Upper Stage: Ariane-5 ECA

Among all the scenarios considered for space debris mitigation, upper stages disposal provides a very interesting scenario for space debris mitigation, where H2O2-based hybrid propulsion system could represent the best compromise between performance and complexity/costs. Upper stages are geometrically simple bodies, sized to withstand severe thermomechanical loads at launch and separation; as such, they do not impose any requirement on maximum tolerated acceleration during de-orbiting as stringent as the one applicable to satellite. This in turns gives more flexibility and freedom for designing the propulsion system since it allows for a quite sharp and short maneuver, without demanding requirements associated to slow and long duration actuations. The maneuver is performed at beginning of life that is a few hours after launch and this removes all the issues and complexity associated to guaranteeing long reliability of the propulsion system in space, in particular in storing on board the propellants without affecting the performance. A single burn is needed, meaning that no restart capability is required for the propulsion system, thus preventing it from being subjected to thermal cycling and lowering risk of failures due to multiple actuation of the valves.

The current European launchers are: Vega, Soyuz and Ariane 5. Vega [23] and Soyuz [32] already comply with the ESA space debris mitigation policies, with the upper stage performing a last burn to re-enter into the atmosphere after releasing the payload. Ariane 5 upper stage, instead, lacks of fuel to perform re-entry. Not even the next generation European launcher, Ariane 6, seems to implement the policy, at least based on the information publically available. This section focuses on proposing strategies to de-orbit Ariane 5 upper stage from Geostationary Transfer Orbit (GTO) exploiting hybrid propulsion technologies.

The initial orbital parameters of the GTO orbit reached by the upper stage of Ariane 5 ECA (Evolution Cryotechnique type A) are 35,943 km apogee altitude and 250 km perigee altitude [33]. The optimised controlled re-entry performed lowering the perigee of the spacecraft from 250 to 60 km will require a ΔV of 20 m/s, including gravity losses. The de-orbiting system will have to carry a maximum host mass of 6335 kg.

The architecture of the propulsion system considered in this case is shown in Fig. 4. Respect to the architectures conceived for ADR, the present one results quite simplified because only the de-orbiting phase has to be taken into account.

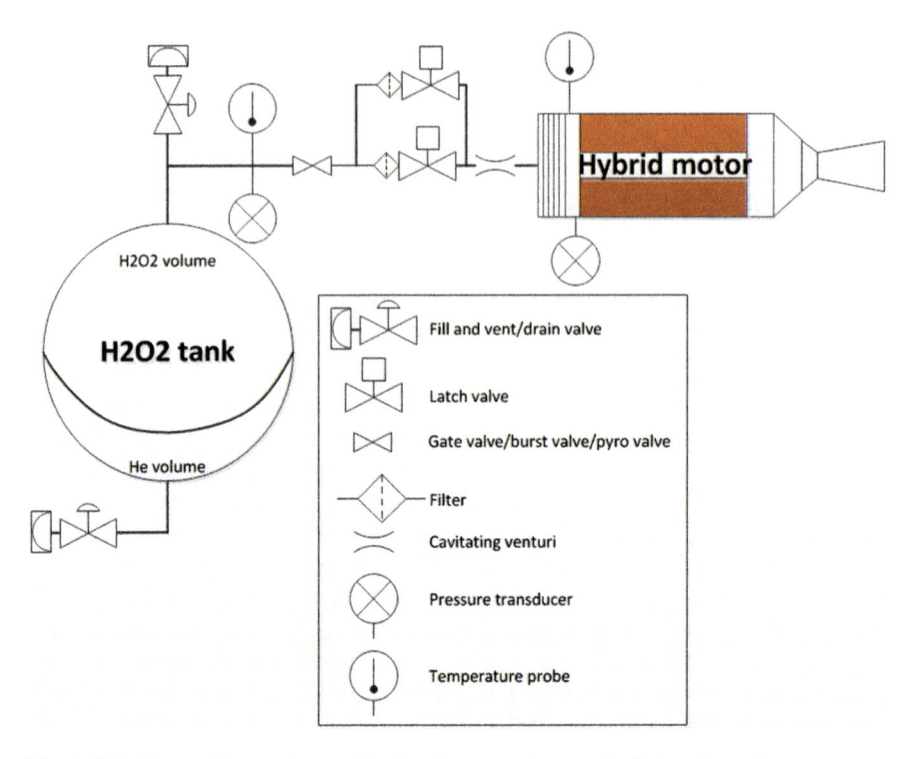

Fig. 4 Hybrid propulsion system architecture for upper stage post-mission disposal

Table 6 Main results for the orbit lowering maneuver applied to Ariane-5 ECA

Consumed propellant (kg)	49	Propulsion system wet mass (kg)	81
Burning time (s)	24	Propulsion system dry mass (kg)	33
Peak thrust (N)	3189 × 2	Oxidizer tank capacity (l)	117
Peak acceleration (m/s^2)	0.98		

Table 6 summarizes the overall characteristics of the maneuver performed by the hybrid propulsion system for the considered scenario.

Because of integration benefits, a configuration with two hybrid motor has been selected in the end as the most favorable, allowing for axis-symmetric mounting and providing an easier vectoring of the thrust through the center of gravity of the upper stage.

Taking into account the geometry and components distribution on the upper stage, the configuration of the propulsion system with two hybrid motors and two tanks is the most favorable. The motors can be mounted diametrically opposed with each one its tank in the vicinity, either at the bottom or at the top end of the upper stage fairing, being fixed at its inner surface. It is recommended that the motors are integrated in the upper stage with the nozzle divergent hung outward of the envelope for a most effective and safe maneuver.

Active Debris Removal and Space Debris Mitigation using Hybrid Propulsion Solutions   177

Fig. 5 3D view of hybrid propulsion system main components accommodation

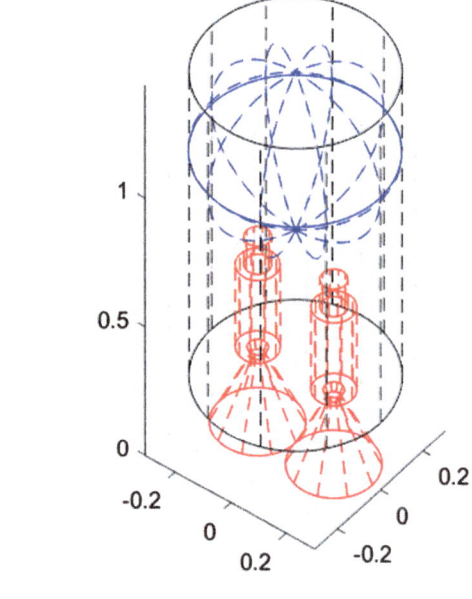

Fig. 6 Ariane 5 upper stage, section view schematic with possible locations of the hybrid PS for de-orbiting (sketch on scale)

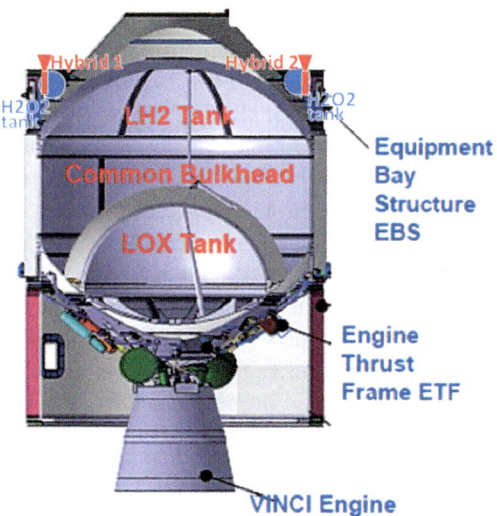

Figure 5 is a 3D view of the system with two hybrid motors and one tank and its overall envelope. As an example, Fig. 6, shows a possible location of the components at the top of the fairing of the upper stage.

5 Conclusions

In this paper have been presented the results of a 1-year project funded under ESA-GSP and developed by Nammo and Deimos. The goal of the study was investigating the implementation of a propulsion system based on hybrid rocket technology for active debris remediation missions. Promising results have been obtained thanks to the simplicity and intrinsic safety of hybrids, whilst offering competitive performances.

The paper provides an extensive mission survey on ESA/EU missions to retrieve debris dispersion and masses. ΔV analysis for the main propulsive phases has been carried out and contributions for targeting, de-orbiting and rendezvous have been calculated for each scenario.

Modeling tools have been implemented to assess, first preliminary on all the scenarios and then in detail on selected scenarios, the hybrid propulsion system. Size and mass of the main components as well as time evolution of the propulsion system have been computed. A modeling tool has been implemented to assess the corresponding bi-propellant propulsion system for each scenario, in order to compare the two systems. In general, it has been observed that a hybrid-propulsion-based system benefits of a simpler architecture with a lighter impact in terms of wet mass.

TRL and delta-development of the main technologies included in the hybrid propulsion system have been evaluated. Projects are already on-going in Europe for developing and qualifying the key technologies of the system and the results achieved so far allow assigning a TRL6 at component level to all of them.

The paper shows that the methodologies and the tools developed along the study are flexible and can be employed to tackle different scenarios, from active debris removal to space debris mitigation.

Hybrid propulsion systems have been proven to be a promising alternative to classical propellant systems to implement in a safe and more environmentally friendly way the space debris mitigation policies and in particular the post mission disposal from both LEO and GTO regions.

Future ESA and European launchers and LEO missions should start considering this technology already in the early design phases as baseline for the propulsion system.

Acknowledgments This work is part of a study funded by the ESA General Study Programme. The authors would like to thank the ESA General Studies Programme (GSP) office and the Clean Space Initiative.

References

1. Biesbroek, R.: The e.DeorbitCDF Study. A design study for the safe removal of a large space debris. In: 6th IAASS Conference, Montreal, Canada (2013)
2. Biesbroek, R., Innocenti, L.: CDF Study Report e.Deorbit. CDF Study Report: CDF-135(C), ESA, September 2012

3. Soares, T., Innocenti, L., Carnelli, I.: CDF Study Report SPADES – assessment of solid propellant deorbit module. CDF Study Report: CDF-137(C), ESA, April 2013
4. Liou, J.-C.: Engineering and technology challenge for active debris removal. Prog Propuls Phys. **4**, 735–748 (2013)
5. Liou, J.-C.: Orbital Debris Modeling and Future Orbital Debris Environment. Orbital debris lecture (ASEN 6519), NASA Orbital Debris Program Office, Johnson Space Center, Huston, Texas, University of Colorado Boulder, September 18, 2012
6. DeLuca, L.T., Bernelli, F., Maggi, F., Tadini, P., Pardini, C., Anselmo, L., Grassi, M., Pavarin, D., Francesconi, A., Branz, F., Chiesa, S., Viola, N., Bonnal, C., Trushlyakov, V., Belokonov, I.: Active space debris removal by a hybrid propulsion module. Acta Astronaut. **91**, 20–33 (2013)
7. Liou, J.-C.: A parametric study on using active debris removal for LEO environment remediation. In: 61st International Astronautical Congress (IAC), Prague, Czech Republic, September 27–October 1 2010
8. Martin, T., Pérot, E., Desjean, M.-C., Bitetti, L.: Active debris removal mission design in low earth orbit. Prog Propuls Phys. **4**, 763–788 (2013)
9. Castronuovo, M.M.: Active space debris removal – a preliminary mission analysis and design. Acta Astronaut. **69**, 848–859 (2011)
10. DeLuca, L.T., Lavagna, M., Maggi, F., Tadini, P., Pardini, C., Anselmo, L., Grassi, M., Tancredi, U., Francesconi, A., Branz, F., Chiesa, S., Viola, N., Trushlyakov, V.: Active removal of large massive objects by hybrid propulsion module. In: 5th European Conference for Aeronautics and Space Sciences (EUCASS), Munich, Germany, July 1–5 2013
11. Wormnes, K., et al.: ESA Technologies for Space Debris Remediation. Technical report. European Space Agency (ESA) (2013)
12. Schonemborg, R.A.C., Schyer, H.F.R.: Solid propellant de-orbiting and reorbiting. In 5th European Conference on Space Debris, March 2009
13. Burkhardt, H., et al.: Evaluation of propulsion systems for satellite end-of-life de-orbiting. In: 38th AIAA/ASME/SAE/ASEE Joint Propulsion Conference and Exhibit, AIAA No. 2002-4208, Indianapolis, IN, July 7–10 2002
14. Gibbon, D.M., Haag, G.S.: Investigation of an Alternative Geometry Hybrid Rocket for Small Spacecraft Orbit Transfer. Technical report, Surrey Satellite Technology LTD, Final Report 0704-0188, Guilford, Surrey, UK, July 27 2001
15. Bombardelli, C., Peláez, J. Sistema de modificación de la posición y actitud de cuerpos en órbita por medio de satélites guía. Patent number P20103035411, Presented at the Spanish Patent Office on March 11, 2010. PCT Patent Application PCT/ES2011/00001
16. Bombardelli, C., Peláez, J.: Ion beam shepherd for contactless space debris removal. J. Guid. Control. Dyn. **34**(4), 1270–1272 (2011)
17. SSTL: Service oriented approach to the procurement/development of an active debris removal mission. Exec. Summ. Rev. **2**, 4 (2013)
18. Wolf, T.: Deutsche Orbitale Servicing Mission, presented at Astra (2011)
19. Forshaw, J.L., Aglietti, G.S., Navarathinam, N., Kadhem, H., Salmon, T., Pisseloup, A., Joffre, E., Chabot, T., Retat, I., Axthelm, R., Barraclough, S., Ratcliffe, A., Bernal, C., Chaumette, F., Pollini, A., Steyn, W.H.: RemoveDEBRIS: an in-orbit active debris removal demonstration mission. Acta Astronaut. **127**, 448–463 (2016)
20. Bonnal, C.: Active debris removal: current status of activities in CNES, P2ROTECT workshop, Ankara, Turkey (2012)
21. Soares, T., Innocenti, L., Carnelli, I.: CDF Study Report SPADES – Assessment of Solid Propellant Deorbit Module. CDF Study Report CDF-137(C), ESA (2013)
22. DISCOS web page https://discosweb.esoc.esa.int, data retrieved in June and July 2013
23. Vega User's Manual 4.0, 2014
24. EDEORBIT Mission Analysis Report, DEIMOS Space, v1.1 (2014)
25. Requirements on Space Debris Mitigations for Earth Observation Satellites, ESA, v1.0 (2009)
26. European Space Debris Safety and Mitigation Standard, Final draft (2002)
27. European Code of Conduct for Space Debris Mitigation, v1.0 (2004)

28. ESA Space Debris Mitigation Handbook, v1.0 (2003)
29. IADC Space Debris Mitigation Guidelines, v1.0 (2007)
30. Carbonsat FLEX: report for the mission selection, ESA SP-1330/2: FLEX, June 2015
31. FLEXAB1-DMS-TEC-TNO02, Deimos FLEX Mission Analysis Report - Phase A2/B1, issue 3.2, 9 November 2015
32. Soyuz User's Manual, 2.0, March 2012
33. Ariane 5 User's Manual, issue 5, revision 1

The Puzzling Case of the Deep-Space Debris WT1190F: A Test Bed for Advanced SSA Techniques

Alberto Buzzoni, Siwei Fan, Carolin Frueh, Giuseppe Altavilla, Italo Foppiani, Marco Micheli, Jaime Nomen, and Noelia Sánchez-Ortíz

Abstract We report on somewhat unique photometric and spectroscopic observations of the deep-space debris WT1190F, which entered Earth atmosphere off the Sri Lanka coast, last 2015 November 13. This striking case has been imposing to the worldwide SSA community as an outstanding opportunity to effectively assess origin and physical nature of such extemporary impactors and appraise their potential threat for Earth. Our observations indicate for WT1190F an absolute magnitude R = 32.45 ± 0.31, with a flat dependence on the phase angle, and slope 0.007 ± 0.002 mag deg^{-1}. The detected short-timescale variability suggests a "four-facet" geometry, with the body likely spinning with a period P = 2.9114 ± 0.0009 s. In the BVRI color domain, WT1190F closely resembled the Planck deep-space probe, a feature that points to an anthropic origin of the object. This match, together with a depressed reflectance around 4000 and 8500 Å may be suggestive of a "grey" (aluminized) surface texture. An analysis is in progress to assess the two prevailing candidates to WT1190F's identity, namely the Athena II upper stage of the Lunar

---

A. Buzzoni (✉) · I. Foppiani
INAF – Osservatorio Astronomico di Bologna, Bologna, Italy
e-mail: alberto.buzzoni@oabo.inaf.it; italo.foppiani@oabo.inaf.it

S. Fan · C. Frueh
School of Aeronautics and Astronautics, Purdue University, West Lafayette, IN, USA
e-mail: fan11@purdue.edu; cfrueh@purdue.edu

G. Altavilla
INAF – Osservatorio Astronomico di Bologna, Bologna, Italy

INAF - Osservatorio Astronomico di Roma, Monte Porzio Catone (RM), Rome, Italy
e-mail: giuseppe.altavilla@oabo.inaf.it

M. Micheli
INAF - Osservatorio Astronomico di Roma, Monte Porzio Catone (RM), Rome, Italy

ESA SSA-NEO Coordination Centre, ESRIN, Frascati, Italy
e-mail: marco.micheli@esa.int

J. Nomen · N. Sánchez-Ortíz
DEIMOS Space S.L.U., Madrid, Spain
e-mail: jaime.nomen@deimos-space.com; noelia.sanchez@deimos-space.com

Prospector mission, and the ascent stage of the Apollo 10 lunar module (LEM LM-4) "Snoopy", by comparing observations with the synthetic photometry from accurate mock-up modeling and reflectance rendering.

1 Introduction

A common drawback when dealing with deep-space impactors (intending all those natural and artificial objects heading Earth at roughly escape velocity), is that they are usually discovered hours or just minutes before reaching our planet. The small asteroids 2008 TC_3, 2014 AA, or even the disrupting Chelyabinsk event are remarkable examples in this sense [1–4]. In this framework, the case of WT1190F offered a somewhat unique opportunity of deeper investigation as object's recognition, and in particular its fatal orbital evolution was successfully assessed [5] weeks in advance of its final fate over the sky of Sri Lanka [6]. This left room, therefore, for a wider and much deeper study of the inherent properties of this quite unusual and still mysterious body.

Discovered by the Catalina Sky Survey [7], on 2015 October 3 (and then pre-covered in different sky surveys back to year 2009), WT1190F was first recognized as a possible small (metric-sized) NEO asteroid captured in a prograde chaotic motion around the Moon-Earth system. The object moved on orbital timescales between 19 and 40 days, along a very eccentric ($0.33 \le e \le 0.98$) translunar ($490{,}000 \le a \le 655{,}000$ km) trajectory with strongly variable inclination ($3° \le i \le 78°$).

A forward-integrated orbit [8] led eventually to predict for WT1190F an Earth impact on 2015 November 13, at 06:18 UT, entering atmosphere with a steep incident angle about $20°$, at a speed of 10.6 km s^{-1}. This would be the first predicted impact of space debris on such an eccentric asteroid-like orbit. The strongly perturbed orbit led to infer [5] a large Area-to-Mass ratio (AMR) in the range $0.006 \le \text{AMR} \le 0.011$ m^2 kg^{-1}, a feature that better pointed to an anthropic origin of the object, possibly a relic of some lunar mission, although of fully unknown origin.

2 The Observing Dataset

As a part of the WT1190F worldwide observing campaign, we have been tracking this so puzzling object using the 1.52 m "Cassini" telescope of the Loiano Observatory (Italy, MPC code 598) [9, 10] (see Fig. 1) and the two 0.40 m and 0.28 m DESS telescopes of the DEIMOS Observatory of Mt. Niefla (Spain, code Z66) [11]. Our observations covered the returning leg of WT1190F's last orbit up to very late moments before Earth impact, with the aim to physically characterize the body and shed light on its real nature.

Fig. 1 An illustrative frame of WT1190F from the Loiano observatory in the night of 2015 November 7–8. The image is taken in the Johnson-Cousins R band with a 420 s exposure time. The object was 518,000 km away from us. Thanks to the "on-target" telescope tracking, WT1190F is clearly detected as a point source near the field center, with an apparent magnitude R = 20.45

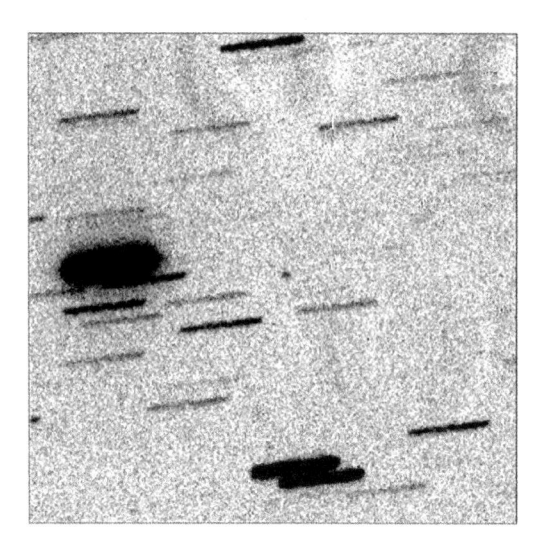

In addition to the Mt. Niefla V photometry and the Loiano BVRI observations, in the night of 2015 November 12–13 we also took advantage of the "on-target" tracking capabilities of the "Cassini" telescope to acquire a low-resolution (R = $\lambda/\Delta\lambda$ 250) spectrum of the object, about three hours before the final atmosphere entry. The spectrum covered the full optical and NIR wavelength range, nominally between 3500 and 9500 Å at 35 Å px-1 dispersion, and allowed us to obtain a quite accurate measure of WT1190F's relative reflectance.

3 Absolute Magnitude and Spinning Properties

A first striking result of our analysis, when matching our observations with the full magnitude database, as from the DASO Circulars, is that WT1190F's brightness displayed a quite flat dependence on phase angle, φ (some 0.007 ± 0.002 mag deg^{-1}), leading to an absolute magnitude R = 32.45 ± 0.31. The flat magnitude trend with the phase angle, and the lack of any "surge effect" of object brightness when approaching the $\varphi \to 0$ configuration, can be regarded as an important signature of the artificial origin of WT1190F. Man-made space artifacts tend, in fact, to level out their reflectance properties with changing the illumination conditions [12, 13], partly due to a smoother surface texture and to the averaging action of quick body spinning.

The enhanced apparent luminosity of WT1190F during its final Earth approach made possible a unique investigation of its short-timescale variability. A direct evidence for a flashing behavior just appeared in the final images of the November 13 observations, both from Loiano and Mt. Niefla (see Fig. 2). These results were also confirmed by independent observations of the two amateur telescopes of Campo dei Fiori (Italy, code 120) [14] and Great Shefford (UK, code J95) [15] observatories.

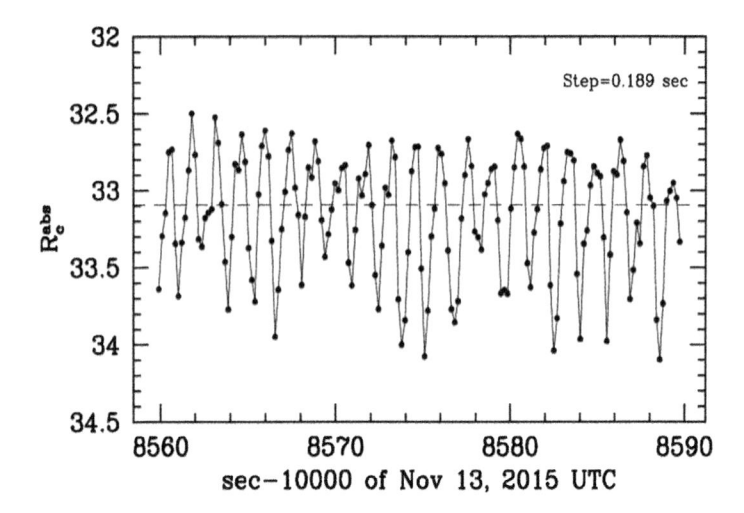

Fig. 2 The derived photometry (in absolute magnitude scale) from a 30 s R trailing image of WT1190F taken with the "Cassini" telescope of Loiano observatory along the 2015 November 13 observations. The object appears to "flash" with an apparent period shorter than 1.5 s, likely as a consequence of a quick spinning motion. The photometric behavior is sampled in time with a step of 0.189 s, as reported in the plot. The dashed line marks the mean absolute luminosity along this set of observations

The combined analysis of the data points to a flashing period $P_{flash} = 1.4557 \pm 0.0013$ s with the photometric signature consistent with the presence of four orthogonal mirroring facets in object's geometry. Simple symmetry arguments lead eventually to conclude that WT1190F was in fact spinning with a period $P_{spin} = 2P_{flash} = 2.9114 \pm 0.0009$ s, that is twice the flashing period.

4 Colors and Reflectance

According to our extended BVRI photometry, no appreciable color variation for WT1190F was detected along the full orbit, from translunar distances down to Earth, in spite of any change of the phase angle. As shown in Fig. 3, in the B-V vs. V-R color diagram, the object appeared slightly "redder" than the Sun and fully consistent in color with a star of spectral type K3.

When compared, with other relevant deep-space spacecraft, like the L2 probes Planck and Gaia [17], the target displayed a substantially similar color as for the Planck spacecraft, characterized by a polished aluminized surface (see Fig. 4). These conclusions are corroborated also by the the reflectance curve of WT1190F, as obtained from the Loiano low-resolution spectroscopy (see Fig. 5). To a closer analysis, the curve shows, in fact, two significant "dips" around the 4000 and 8500 Å spectral regions, possibly a signature of "grey" (aluminized) material onboard [18].

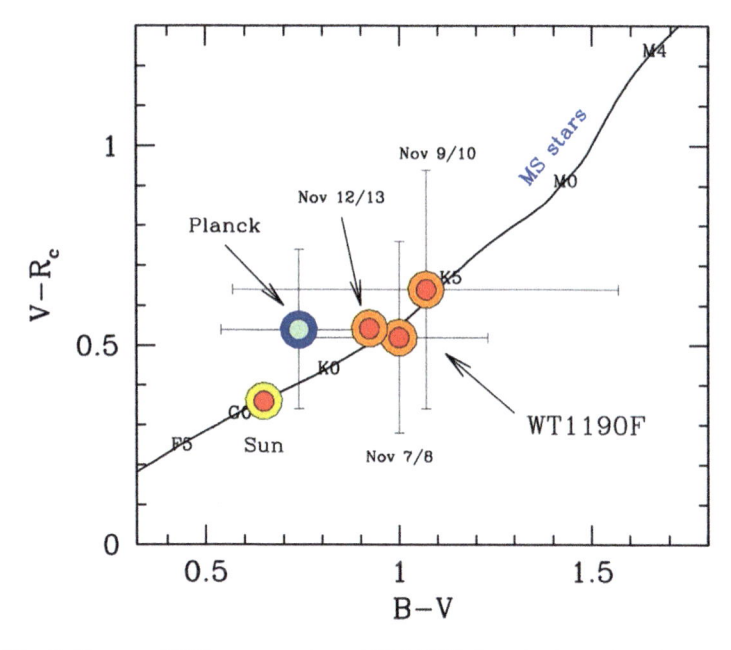

Fig. 3 The (B-V) versus (V-R) average colors of WT1190F along the three observing runs of 2015 November, from Loiano (red/orange dots). The target colors are compared with other reference objects, namely the Planck spacecraft, the Sun, and the locus of Main Sequence stars [16] (solid curve, labeled with the stellar spectral type). Note that WT1190F appears to be slightly "redder" than the Sun, but very close in color to Planck. It would also quite well match the colors of a star of spectral type K3

Fig. 4 The deep-space astronomical observatory Planck. The probe, in operation between 2009 and 2013, was placed in an Earth-corotating orbit around the Sun-Earth Lagrangian point L2, some 1.5 million km away. Note the prevailing grey color of the surface texture

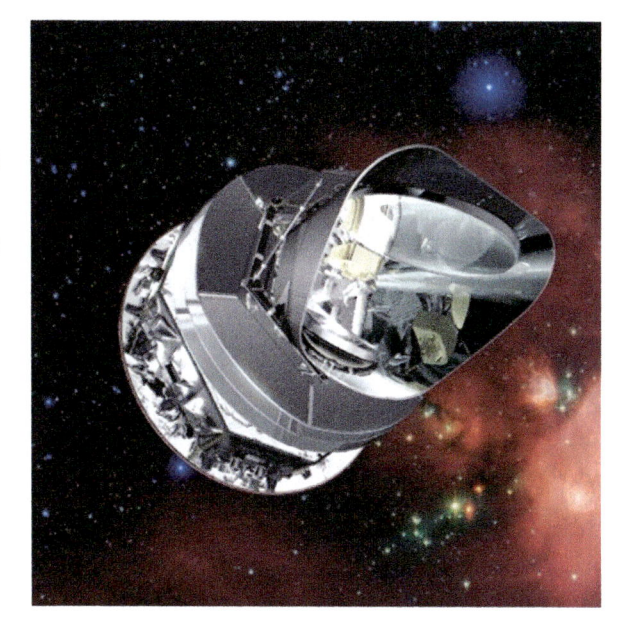

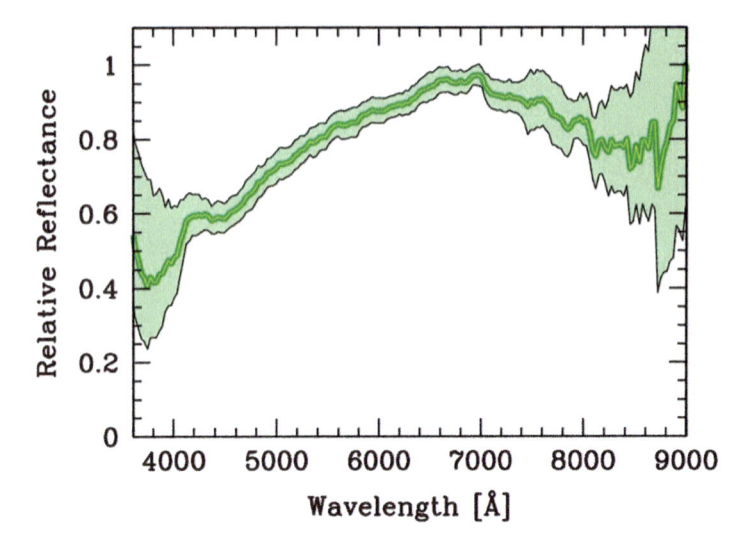

Fig. 5 The WT1190F reflectance curve, together with its ±1σ statistical uncertainty band, as obtained from the Loiano spectroscopy. The curve is normalized at the 7000 Å value. Note two clear "dips" in the curve about 4000 and 8500 Å, a possible signature of some "aluminized" material onboard

5 Toward Assessing WT1190F's Identity

The wide observing evidence hereby collected clearly points for WT1190F to be a man-made artifact, most probably a relic of a past lunar mission. In addition, a still remarkable residual spin of the body may call for a relatively young age of WT1190F. The chaotic orbital motion may even support this scenario, as longer in-orbit lifetime would have greatly increased the chance of a Moon impact.

On the other hand, the same weak gravitational boundaries with the Earth-Moon system could also provide opposite evidence, by setting the case in a more historical context. It could be, in fact, that in its 2009 discovery WT1190F was in a returning path to Earth, after being recaptured from a heliocentric orbit. If the latter is the case, then the object could be much older and its origin should be moved back to the pioneering lunar missions of the 1960s.

According to these two possible scenarios, the current debate on WT1190F's ultimate identity has been focusing on two prevailing candidates. In particular, the Athena II Trans-Lunar Injection Stage (TLIS), that carried the Lunar Prospector probe to the Moon in year 1998, could be the most viable "young" contender. On the other hand, a 47 year old best candidate could be identified in the ascent stage of the lunar module (LEM LM-4) "Snoopy", released in heliocentric orbit on 1969 May 23, after completion of the Apollo 10 mission.

To consistently assess the two different scenarios, accurate mock-up modeling (relying on the 3D CAD design software SolidWorks) and reflectance rendering (through the open-source suite MeshLab for mesh rendering) is in progress in order to reproduce synthetic photometry to be compared with the observed dataset. Each

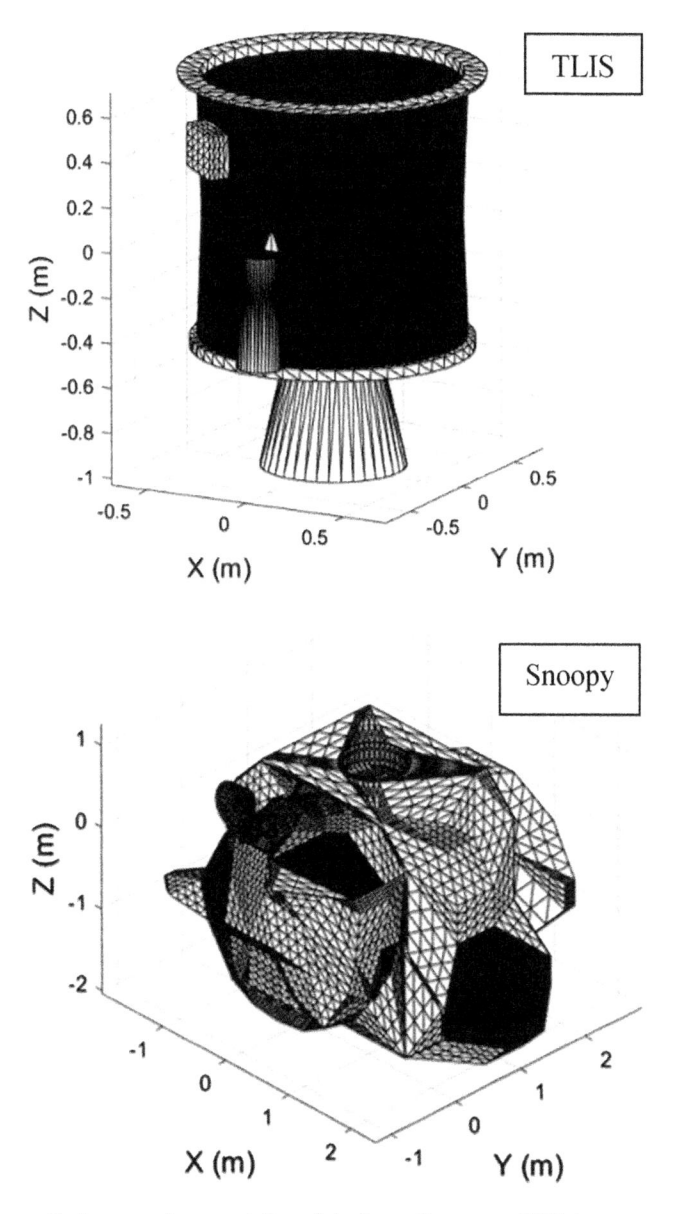

Fig. 6 Our preliminary mock-up modeling of the Lunar Prospector TLIS (upper panel) and the Apollo 10 LEM "Snoopy" (lower panel). The derived synthetic photometry may likely provide an effective tool to assess the ultimate nature of WT1190F

target is rendered in our models as a composite structure of 10,000 planar facets. An illustrative example of our preliminary meshing assembly of the TLIS and "Snoopy" mock-up [19] is shown in Fig. 6. A detailed analysis of the resulting synthetic photometry is deferred to a forthcoming exhaustive paper [20].

6 Conclusions

The case of WT1190F, the first Earth impactor discovered with more than a day of advance notice, provided an ideal real life test case for how to quickly organize an observing campaign with multiple instruments and techniques. To shed light on WT1190F's ultimate nature we relied on a combined observational strategy that used the 1.52 m "Cassini" telescope of the Loiano Observatory, in Italy, to track the most distant orbital arc of the object, eventually accompanied by the DEIMOS telescopes at Mt. Niefla (Spain) for the very last approaching phase to Earth. We aimed at characterizing the body both from a dynamical and physical point of view, via astrometry and multicolor BVRI photometry and low-resolution spectroscopy.

According to our observations, WT1190F displayed a quite flat luminosity dependence on the phase angle, leading to an absolute magnitude of $R = 32.45 \pm 0.31$ mag at "opposition" geometry (i.e. $\varphi \to 0$). Both the photometric trend with phase angle, together with a somewhat chaotic dynamical regime, as from the available astrometry, appear to be clear signatures of WT1190F's artificial nature as a man-made space artifact, possibly a metric-sized device related to some lunar mission. Our diagnostics is also corroborated by the evident spinning properties of the body, as caught by our observations along the final approaching leg to Earth, with a flashing period (possibly half the physical spinning period) of $P = 1.4557 \pm 0.0013$ s.

The study of object's colors and reflectance revealed that WT1190F looked like a star of spectral type K3, although with two significant "dips" in its reflectance spectrum around 4000 and 8500 Å, likely a signature of "grey" (aluminized) material onboard.

These elements may provide an important piece of evidence to the current debate on WT1190F's identity. Two prevailing candidates are in fact under scrutiny. While the 1998 Lunar Prospector TLIS upper stage could be the most viable contender, a much older scenario may however be invoked dealing with the ascent stage of the Apollo 10 lunar module "Snoopy", released in heliocentric orbit in 1969, and possibly re-captured by Earth in the recent years. In-progress accurate mock-up modelling and reflectance rendering of both candidates, carried on by our group, will soon provide supplementary data allowing us to better constrain the distinctive photometric signatures of both the TLIS and "Snoopy" spacecraft, to be compared with the observations and eventually lead, we hope, to conclusive arguments about WT1190F's nature.

References

1. Borovička, J., Charvát, Z.: Meteosat observation of the atmospheric entry of 2008 TC_3 over Sudan and the associated dust cloud. Astron. Astrophys. **507**(2), 1015–1022 (2009)
2. Jenniskens, P., et al.: The impact and recovery of asteroid 2008 TC_3. Nature. **458**(7237), 485–488 (2009)

3. Farnocchia, D., Chesley, S.R., Brown, P.G., Chodas, P.W.: The trajectory and atmospheric impact of asteroid 2014 AA. Icarus. **274**, 327–333 (2016)
4. Borovička, J., Spurný, P., Brown, P., Kalenda, P., Shrbený, L.: Atmospheric behavior of the Chelyabinsk impactor. In: Miunonen, K., et al. (eds.) Proceedings of Asteroids, Comets Meteors 2014, ACM14, Univ. of Helsinky, in electronic form, at http://www.helsinki.fi/acm2014/pdf-material/ACM2014.pdf
5. Gray, W.J.: Project Pluto web site available at http://projectpluto.com/temp/wt1190f.htm (2015)
6. Jenniskens, P., et al.: Airborne observations of an asteroid entry for high fidelty modeling: space debris object WT1190F. In: Proceedings of the 54-th AIAA Science and Technology forum and exposition (SciTech 2016), AIAA, San Diego CA, USA (2016)
7. Matheny, R.G., et al.: DASO Circ. No. 520, IAU Minor Planet Center, USA, ed. G.V. Williams (2015)
8. Farnocchia, D.: s46 orbit solution, JPL Horizons SSD Interface at http://ssd.jpl.nasa.gov/horizons.cgi (2015)
9. Buzzoni, A., Altavilla, G., Micheli, M., Bruni, I., Gualandi, R.: DASO Circ. No. 532, IAU Minor Planet Center, USA, ed. G. V. Williams (2015)
10. Altavilla, G., et al.: DASO Circ. No. 539, IAU Minor Planet Center, USA, ed. G.V. Williams (2015)
11. Sánchez-Ortiz, N., Nomen, J., Hurtado, M.: DASO Circ. No. 539, IAU Minor Planet Center, USA, ed. G.V. Williams (2015)
12. Miles, R.: The unusual case of 'asteroid' 2010 KQ: a newly discovered artificial object orbiting the Sun. J Br Astron Assoc. **121**(6), 350–354 (2011)
13. Murtazov, A.K.: Physical simulation of space objects' spectral characteristics for solving the reverse problem of their photometry. Am. J. Mod. Phys. **2**(6), 282–286 (2013)
14. Buzzi, L., Colombo, G.: The "G. Schiaparelli" Obs. web site at http://www.astrogeo.va.it (2015)
15. Birtwhistle, P.: The Great Shefford Obs. web site at http://peter-j95.blogspot.it (2015)
16. Pecaut, M.J., Mamajek, E.E.: Intrinsic colors, temperatures, and bolometric corrections of pre-main-sequence stars. Astrophys. J. Suppl. Ser. **208**(1), 9–30 (2013)
17. Buzzoni, A., Altavilla, G., Galleti, S.: Optical tracking of deep-space spacecraft in Halo L2 orbits and beyond: the Gaia mission as a pilot case. Adv. Space Res. **57**(7), 1515–1527 (2016)
18. Henninger, J.H.: Solar absorptance and thermal emittance of some common spacecraft thermal-control coatings, NASA Reference Publication, no. 121, NASA, Greenbelt, USA (1984)
19. Fan, S., Frueh, C., Buzzoni, A.: A light curve simulation of the Apollo Lunar Ascent module. In: Proceedings of the AIAA/AAS Astrodynamics Specialist Conference, AIAA SPACE Forum, (AIAA 2016-5504) (2016)
20. Buzzoni, A., Altavilla, G., Fan, S., Foppiani, I., Frueh, C., Micheli, M., Nomen, J., Sanchez-Ortiz, N.: Physical characterization of the deep-space debris WT1190F: a testbed for advanced SSA techniques, in preparation (2018)

Development of a Debris Index

Francesca Letizia, Camilla Colombo, Hugh G. Lewis, and Holger Krag

Abstract Environmental indices for space objects have been proposed to identify good candidates for active debris removal missions and to deal with the licensing process of space objects before their launch. A way to rank the environmental impact of spacecraft may be based on the assessment of how their fragmentations would affect operational satellites. In particular, the effect of a breakup can be measured by the resulting collision probability for a set of target spacecraft. A grid in semi-major axis, inclination, and mass is used to define possible initial conditions of the breakup. Once the value of the index is known for any point in the grid, a simple interpolation can be used to compute the value of the index for any object. The current work aims to extend a previous formulation, which focussed only on the effect of collisions, by including also the effect of explosions and considering the likelihood of these fragmentations.

1 Introduction

The space around the Earth is populated by an increasing number of objects and most of them are not operational ones. According to the European Space Agency (ESA),[1] out of the 23,000 catalogued objects, only around 1000 are operational

---

[1] http://www.esa.int/Our_Activities/Operations/Space_Debris/About_space_debris, last access 15/06/2016.

F. Letizia (✉) · H. G. Lewis
University of Southampton, University Road, Southampton SO17 1BJ, UK
e-mail: f.letizia@soton.ac.uk; h.g.lewis@soton.ac.uk

C. Colombo
Politecnico di Milano, Via La Masa 34, 20156 Milan, Italy
e-mail: camilla.colombo@polimi.it

H. Krag
European Space Agency, ESA-ESOC, Darmstadt 64293, Germany
e-mail: Holger.Krag@esa.int

Fig. 1 Example of energy label

satellites whereas the rest is composed by spent satellites, mission related objects and, mainly, fragments produced by explosions and collisions.

The analysis of the evolution of these numbers over time has suggested the adoption of measures to limit the growth of the debris population such as the passivation of rocket bodies (to limit the risk of explosions) and the definition of protected orbital regions that should be left clear at the end of a mission. However, the efficacy of these measures is still under discussion. In the recent years, a greater awareness of the threat posed by space debris to the future access to space is emerging and initiatives such as ESA Clean Space actively promote the idea of a sustainable use of space. From this point of view, the guidelines for space debris mitigation may take inspiration from the ones developed to create a more sustainable use of resources on Earth to limit global warming.

Among different indicators that have been developed to measure the sustainability of our way of life (e.g. CO_2 footprint), the labelling of large household appliances appears to be a successful example, able to shift the market towards more efficient and more environmental friendly products. The European energy label (Fig. 1) was introduced in 1994 for cold appliances (e.g. freezers, refrigerators), and then extended in the following years to washing machines and dishwasher [11]. In the years since its adoption, the seven-level coloured scale has become a well known indicator of energy efficiency, applied (unofficially) also to cars, buildings, and planes.

The labelling of appliances was introduced to fill the so-called *energy-efficiency* gap [2], i.e. the fact that consumers were not aware of the consumption of their appliances. This had a direct impact both on the *private* level in terms of the cost of bills, and on the *society* level in terms of the energy demand and the environmental consequences. The eco-labelling contributed to orient the market towards more efficient products, with an increase of the market share of A-level appliances [2, 13]. It also helped to define a required minimum level of efficiency, for example with the ban of new refrigerators with classes D to G [11]. Finally, it contributed to create awareness in consumers and producers, so that now energy efficiency is among the drivers in the choice of a product [13]. Similarly to [10], this work analyses how the labelling approach could be applied to tackle the space debris issue.

The task to define a *debris label* for spacecraft should start from the analysis of the main differences with respect to the case of household appliances. The first important difference is that the labels for appliances are targeted to the final users to orient their decision while buying. For satellites this approach is not feasible as currently missions are developed ad-hoc to provide specific data and services. For this reason, labelling a spacecraft should address mostly the spacecraft operators, for example with respect to their interface with space agencies and external organisations.

Connected to this point, it should be observed that the *private* cost of operating a spacecraft with a high *debris index* is less direct than the case of bills for a household. For example, putting spacecraft in a congested orbit could increase the operational cost due to the need of performing more collision avoidance manoeuvres. On the other hand, the decision to dispose a spacecraft at the end of its mission may not bring a direct economic benefit to its operator. This observation suggests that a *debris label* would make sense only if implemented within processes such as licensing of the spacecraft before the launch, insurance, or provision of collision avoidance services by external providers.

Another important decision to make for such an index is what should be measured. It was observed that the long term evolution of the space debris environment is highly affected by the fragmentation of large intact objects. Figure 2 shows the evolution of the number of objects in orbit with time and one can observe the effect of the fragmentation of Fengyun-1C and of the Iridium-Cosmos collision. A fragmentation can be caused by explosion (for example due to a failure on-board) or by a collision with another object. In both cases, a cloud of fragments is generated: the cloud, initially dense and localised, spreads under the effect of different forces, so that a fragmentation is able to affect objects in different orbital regimes.

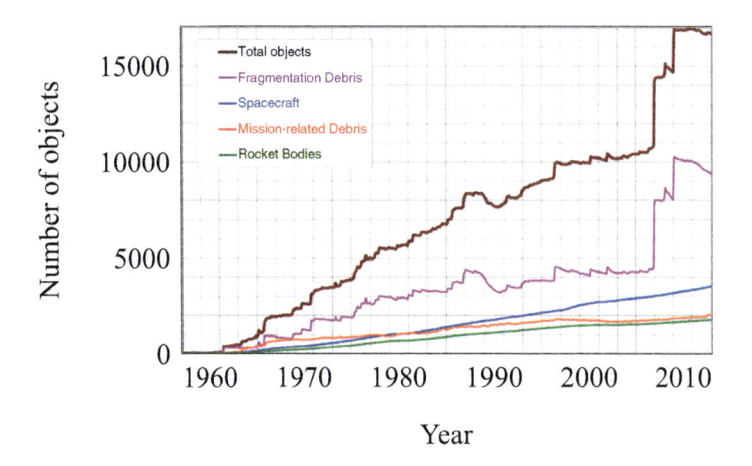

Fig. 2 Growth of the catalogued population of objects in Earth orbit [3]

A way to measure the severity of the consequences of these fragmentation is to look at the increase in the collision risk for operational satellites. It is very important to underline that this is only one possible option; alternative approaches may be based on the analysis of the fragments still in orbit after a certain time period or the increase in the collision risk for the whole population (so considering not only operational satellites, but also spent satellites and rocket bodies) [5, 12]. The reason why this work suggests to look at the effects on operational satellites is because this can be more easily connected to the cost to operators due to fragmentations (*private* cost). For example, a recent analysis by ESA Space Debris Office [6] has shown how the fragmentation of Fengyun-1C and the Iridium-Cosmos collision have affected the number of conjunctions and collision avoidance manoeuvres for some ESA missions. In addition, the collision risk for operational satellites may also be seen as an indicator of the availability of future access to space (*shared* cost) because the orbital regions with most operational satellites are the ones that offers a privileged point of view for Earth observation. For example, this is the case of sun-synchronous orbits, which allows the Earth to be observed with constant illumination conditions. Therefore, they are expected to be an important asset also in the future.

For these reasons, the proposed index is based on the evaluation of the consequences of fragmentations on operational satellites. In addition, the likelihood of these fragmentations to happen should be considered. For explosions, the probability can be estimated starting from historical data on fragmentations in orbit, whereas the probability of collisions depends on the orbital region where the spacecraft operates. In summary, the index will have the following structure

$$\text{Index} = p_e \cdot e_e + p_c \cdot e_c \tag{1}$$

where p_e is the probability of an explosion happening, e_e measures the effects of the explosion on operational satellites, p_c is the probability of a collision happening, and e_c measures the effects of the collision on operational satellites. The term e_c was already developed [9], so this work will focus on the explanation of the other three terms.

2 Method

To assess the effect on operational satellites, a set of representative targets is defined. This is done to avoid having to propagate the trajectories of all operational spacecraft and to build a reference set that is robust to the variation of some elements in the population. In this way, there is no need to regenerate the results after each new launch. A way to define this representative set is to look at the distribution the cross-sectional area of operational satellites in semi-major axis and inclination. A grid in these two dimensions is introduced and, for the cells where most targets are concentrated, a representative target is defined, with mass and area equal to the

Fig. 3 *Reference map*: variation of the term e_c with the orbital parameters [9]

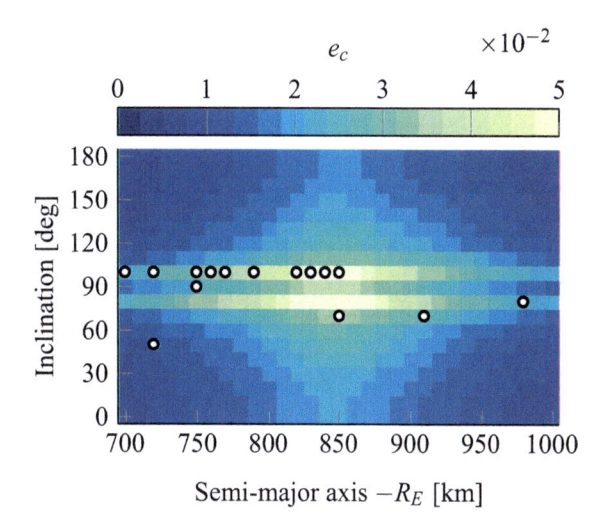

average values among the object in the cell, and orbital parameters equal to centre of the cell.

Once the target set is defined, the effect of fragmentations can be evaluated. A key point of the suggested approach is not to compute the index only for specific objects, but rather to study its dependence on parameters such as orbit altitude, inclination, and the spacecraft mass. The same grid in semi-major axis and inclination is now used to defined possible orbits where the fragmentations occurs. Figure 3 shows the variation of the component e_c of the index obtained in [9], computed as

$$e_c = \sum_{j=1}^{N_T} w_j p_{c,j}, \qquad (2)$$

where $p_{c,j}$ is the cumulative collision probability for each representative target and w_j a weighting factor to consider that each representative target represents a different share of the total area distribution. The grey markers refer to the representative targets identified with the approach based on the cross-sectional area.

One of the advantages of studying the index dependence on these parameters (rather than only evaluating single spacecraft) is that maps such as Fig. 3 clearly show which are the most critical orbits. Observe also that Fig. 3 was obtained simulating always the same fragmentation and changing only its location; in particular, the mass involved in the fragmentation is fixed. It can be shown that if the fragmenting mass is changed, the value of the index changes accordingly following a power law [9]. This follows directly from the equations of the breakup model used to generate the fragments. This behaviour is particularly convenient because it means that no additional simulations are required if one wants to obtain the same map as in Fig. 3 for a different value of the fragmenting mass; it is sufficient to rescale the result already obtained.

In this way, only the reference map in Fig. 3 is needed to compute e_c for any specific spacecraft. This requires to rescale the reference map to the value of the mass of the studied object that we want to evaluate and to interpolate the reference map to find the value of the index for its specific orbital parameters. This means that the process of computation of the index is split into two parts: the generation of the reference map and the actual computation of the index. The generation of the map requires operations that are computationally expensive and that rely heavily on the availability of efficient methods for debris cloud propagation [8] and computational resources. Once the reference map is generated, this can be saved and stored. When the index needs to be computed for some specific objects, this can be done by simply rescaling and interpolating, as explained in the previous paragraph. These operations are fast and can be easily implemented in different programming languages.

Following this approach, the index can be computed in a matter of seconds for all the objects in a database. This is important because it could be expected that the index may be computed also outside research organisations, for example in companies and institutions with no access to the propagation methods and the computational resources required by the generation of the reference map. This rationale is kept also in the development of the new terms.

2.1 Effect of Explosions

For the case of the effect of explosions (e_e), a similar approach to the one previously described can be adopted. Explosions tend to produce larger fragments with lower speed compared to collisions, so different equations are used for the generation of the fragments. In addition, in the NASA breakup model [4], the mass of the exploding spacecraft does not appear explicitly (differently from the case of collisions). In particular, the number of fragments generated by an explosion with size equal or larger than L_c is given by

$$N(L_c) = 6SL_c^{-1.6}, \qquad (3)$$

where L_c is in m and S is type-dependent unitless number that acts as a scaling factor for the explosions. The initial version of the breakup model uses only $S = 1$, but a later update to the model suggests that its value can change between 0.1 and 1, depending on the explosion type [7]. The parameter S can be used to introduce a dependence on the mass in Eq. (3). First, one can observe that the mass of generated fragments depends linearly on S; keeping the constraint $S \leq 1$, the maximum fragment mass is around 160 kg. Following the observation that explosions usually involved only specific components and not the whole space objects, the following relationship for S was found

Development of a Debris Index 197

Fig. 4 *Reference map*: variation of the term e_e with the orbital parameters

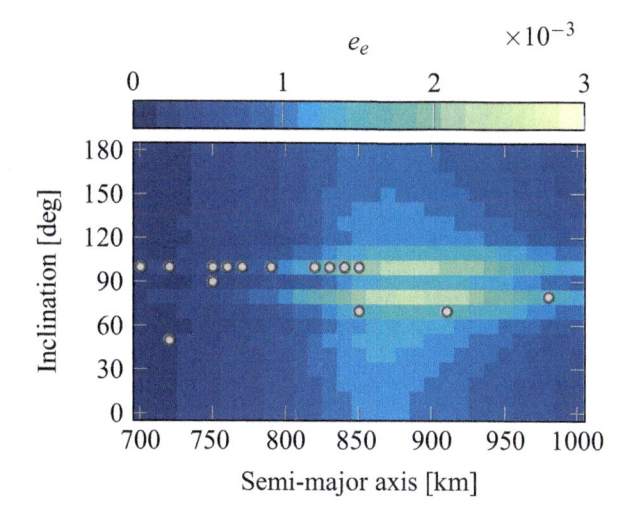

$$S = \begin{cases} k \frac{m_{obj}[kg]}{10,000[kg]}, & \text{for } k\, m_{obj} < 10,000\,\text{kg} \\ 1 & \text{for } k\, m_{obj} \geq 10,000\,\text{kg} \end{cases}, \quad (4)$$

with $k = 1$ for payloads and $k = 9$ for rocket bodies. The different value of k for payloads and rocket bodies derives from the analysis on the number of fragments produced per kg for the two classes of objects according to the data available in DISCOS[2] (Database and Information System Characterising Objects in Space).

Following the same approach used for e_c, also in the case of explosions different fragmentations were simulated changing the orbital parameters of the orbit where the breakup occurs, while keeping constant the value of the mass of the fragmenting objects. The effect was measured on the same representative targets to assess their sensitivity to the breakup conditions. Figure 4 shows the result of this analysis by visualising the value of e_e obtained, similarly to the case of collisions, as

$$e_e = \sum_{j=1}^{N_T} w_j p_{c,j}, \quad (5)$$

where $p_{c,j}$ is the cumulative collision probability for each representative target due to the fragments generated by the explosions and w_j the weighting factors defined in Eq. (2). Similarly to the results in Fig. 3, one can observe the role of the inclination with two horizontal bands that correspond to fragmentations where the targets will cross the resulting fragment cloud in the latitude regions with the maximum fragment density. Compared to the case of the collisions in Fig. 3, the dependence on the altitude is more localised for the case of explosions.

---

[2] https://discosweb.esoc.esa.int.

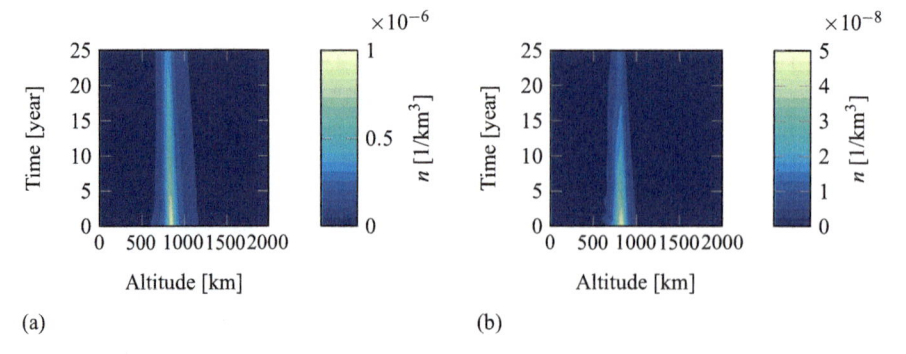

Fig. 5 Evolution of the density profile with time for two fragmentations starting at 850 km of altitude. (**a**) Collision; (**b**) Explosion

This is due to the fact that, as already mentioned, explosions and collisions produce fragments with different characteristics. In particular, the fragments generated by explosions have a lower variation in velocity, so they remain more concentrated around the orbit where the fragmentation occurred.

The behaviour can be visualised by studying the persistence of the fragments in orbit, for the two cases, as shown in Fig. 5 that represents the evolution of the cloud density profile with time for a collision and an explosion starting from an orbit with altitude equal to 850 km. In the case of the collision, the complete fragmentation of a satellite of 10,000 kg is simulated, whereas for the explosion the maximum fragment mass in equal to 160 kg and this explains difference in the order of magnitude for the density. Also from this representation, one can notice how the explosion appears to affect a smaller range of altitudes and the corresponding cloud to decay quicker than the one generated by the collision.

2.2 Probability of Collision

The probability of a collision happening (p_c) can be estimated by using the analogy with the kinetic theory of gas, so that the cumulative collision probability is written as

$$p_c = 1 - \exp(-\rho \Delta v A_c \Delta t), \tag{6}$$

where ρ is the debris density at the spacecraft altitude, Δv is the collision velocity, A_c the cross-sectional area, and Δt is the time.

As only catastrophic collisions are considered, the value of ρ depends on the mass of the studied object. For this reason, ρ should express the density of objects able to trigger a catastrophic collision. This means that, given the mass of the object, the energy threshold for catastrophic collisions (40 kJ/g) should be applied to find the

Fig. 6 Impact velocity distribution obtained with MASTER for an orbit at 814 km of altitude and 98° of inclination

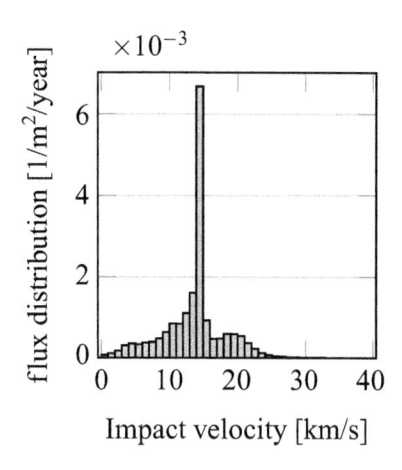

limit on the impactor size. Then, the corresponding density is derived from the ESA tool MASTER (Meteoroid and Space Debris Terrestrial Environment Reference).

For the relative velocity, MASTER is used again to identify the most likely impact velocity given the orbital parameters of the studied object. An example of this analysis is shown in Fig. 6 for an orbit at 814 km of altitude and 98° of inclination. The distribution can be used to associate to each orbital configuration, the most likely impact velocity. For example, in Fig. 6 the peak is at $\Delta v = 14.5$ km/s; alternatively, the reference impact velocity can be found by computing the integral mean of the distribution and in this case $\Delta v = 13.6$ km/s is obtained. This second option is the one used in the following. This process was repeated on a set of reference orbits, using the same grid as in Fig. 3, so with a spacing of 10 km in semi-major axis and 10° in inclination. The eccentricity is put equal to zero for all the cases; in addition, no variation in the other parameters (i.e. the longitude of the ascending node Ω and the argument of periapsis ω) is considered as it is assumed that the background population is uniformly distributed with respect to these parameters.

Also the terms used to build p_c can be precomputed and interpolated, so that the analysis of the collision risk can be quickly performed on all the objects in a database. For example, Fig. 7 shows the ten payloads with the highest value of the collision risk ($I_c = p_c \cdot e_c$) among the objects in the ESA DISCOS database and in orbit between 700 and 1000 km. One can observe how with I_c is considered instead of only e_c as in [9], all the top objects belong to the peak areas with at altitudes between 760 and 870 km. In particular, Cosmos 2502 and 2455, both in an orbit at 905 km of altitude, were in the top ten objects when only e_c is considered, but their relative criticality is reduced when also p_c is considered because that orbital region is less critical in terms of background debris population. Similarly for Cosmos 2486 and 2441, in orbit at 720 km of altitude, that are in the top ten when only the effects of a fragmentation are considered, but ranks over the 120th place when also the

Fig. 7 Top ten payloads with the highest value of $I_c = p_c \cdot e_c$

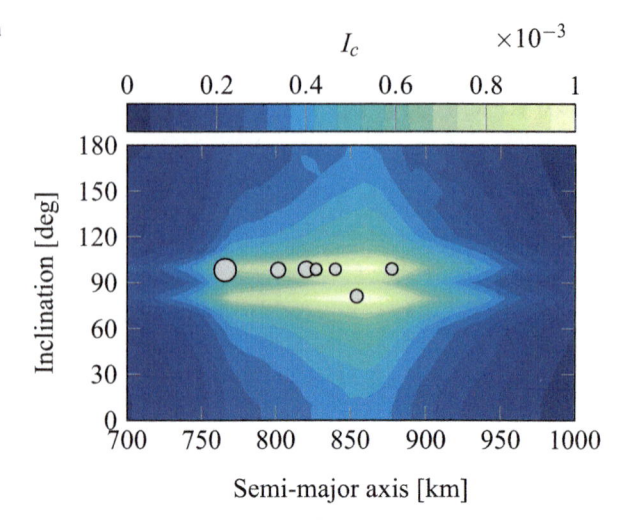

collision probability is taken into account. The first object is the same (Envisat) for both classifications (i.e. only e_c or I_c).

2.3 Probability of Explosion

The estimation of the probability of explosion is performed starting the data on historical fragmentations available on ESA DISCOS. First, the data in DISCOS was analysed by looking at the classification of all events based on the cause of fragmentation. Figure 8 shows the frequency of fragmentation causes for payload and rocket bodies: as expected, the distribution is different for two classes and this should be reflected in the estimation of the p_e term should take into consideration this classification. In addition, the probability of explosion should take into account only events due to propulsion failures, battery failures or unknown fragmentation cause. The events due to collisions or deliberate destruction are not relevant to the development of a model of the explosion probability, whereas fragmentations due to atmospheric forces or attitude failures cannot be modelled with the breakup model used for the term e_e. In addition, only fragmentations occurred in Low Earth Orbit (LEO) will be considered in the following.

The statistical modelling of the term p_e was built by looking at fragmentations occurred in LEO, involving objects launched after 1985 and distinguishing between payloads and rocket bodies. The adopted approach is the following: for each class of objects (i.e. payloads and rocket bodies), the number of fragmentations in a year is registered and it is assumed that it can be described with a Poisson distribution. The chi-squared test is used to verify if this hypothesis is acceptable. If so, the parameter of the Poisson distribution gives the estimated average number of fragmentation in

Development of a Debris Index

Fig. 8 Classification of past fragmentations involving payloads (PL) and rocket bodies (RB)

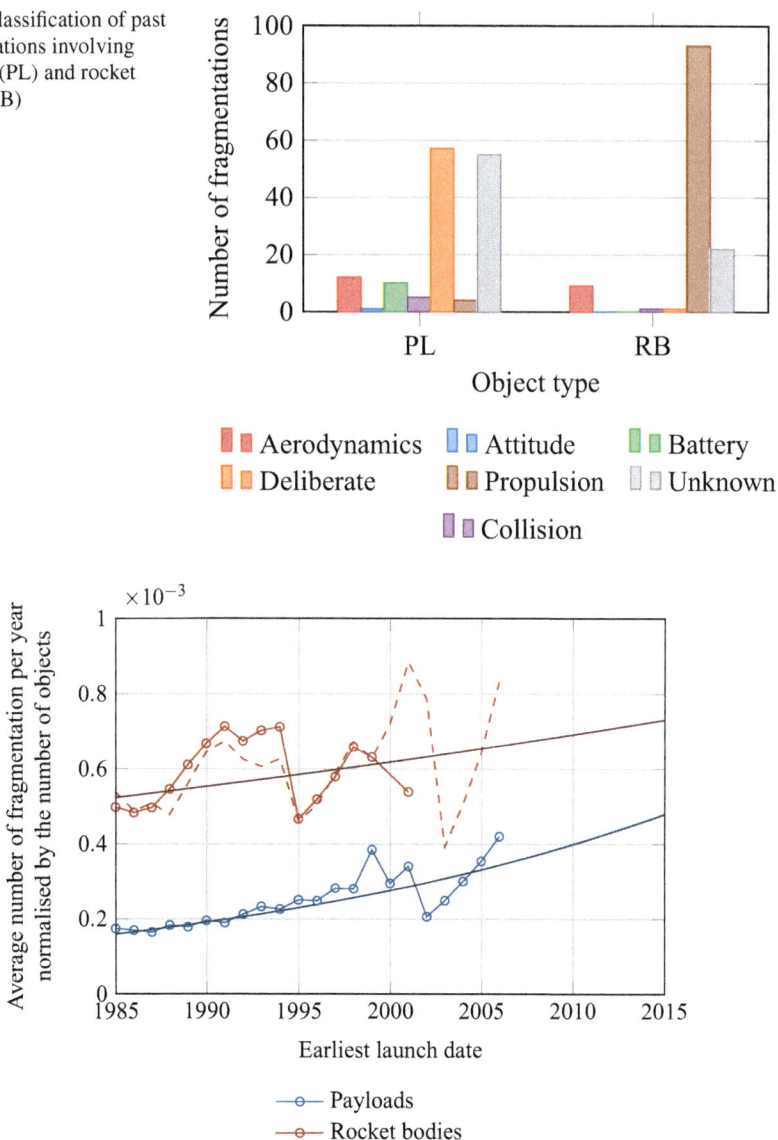

Fig. 9 Average number of fragmentations per year normalised by the number of objects

a year. In order to obtain a value that can be inserted in Eq. (1), the average number of fragmentations per year needs to be normalised with the number of launched objects.

Figure 9 shows the value of the average number of fragmentations normalised with the number of objects considering the fragmentations and the launched objects

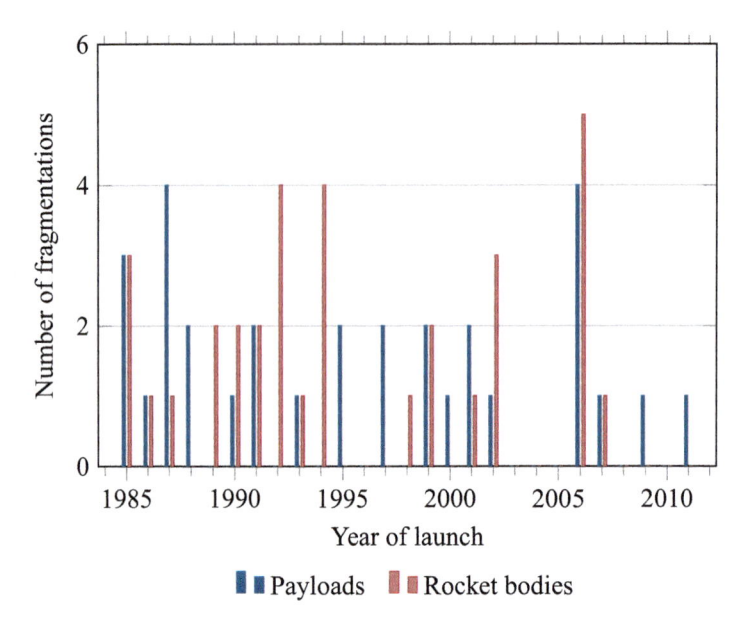

Fig. 10 Distribution of fragmentation events in LEO as a function of the year of launch. The numbers refer only to events due to battery or propulsion failures and with unknown fragmentation cause

in LEO after a certain epoch. The choice to define a value of explosion probability for all the objects launched after a certain date instead of defining a value for each year or each decade is due to the fact that the latter approach leads to subsets with few samples where the chi-squared test is not conclusive and, more in general, the number of events is too low to obtain robust statistics. Already in the years after 2006, the average number of fragmentations in a year cannot be found because the total number of events is too low to apply the chi-squared test. In addition, in the years 2000 and 2002–2006 the estimation for the rocket bodies fail. For this reason, the value is estimated using the Poisson approximation for all the events (rocket bodies and payloads) and subtracting the value obtained from the payloads. In this way, the dashed curve in Fig. 9 is obtained.

One can observe how rocket bodies have a larger probability of explosion than spacecraft. In addition, the increasing trend in the curve means that when considering more and more recent objects, the number of objects is reduced, but the number of fragmentations decreases at a lower rate. This is consistent with the observed trend of fragmentations with the launch year of the objects (Fig. 10). In addition, one can observe how the peaks of fragmentations for objects launched in 2002 and 2006 is reflected in the change of the slope of the curve in Fig. 9. These curves can be, in a first step, approximated with an exponential function as shown in Fig. 9. To limit the risk associated to extrapolation, the exponential function is used only up to the present epoch and then this value is applied for all the future launches.

Table 1 Fitting parameters for the probability of explosion

Parameter	Payload	Rocket bodies
α	1.59193E−04	5.22821E−04
β	3.66550E−02	1.11388E−02

In summary,

$$p_e = \begin{cases} 0 & \text{for } y_L < 1985 \\ \alpha \exp[\beta(y_L - 1985)] & \text{for } 1985 \leq y_L \leq 2016 \\ \alpha \exp[\beta(2016 - 1985)] & \text{for } y_L > 2016 \end{cases}, \quad (7)$$

where α and β are the fitting parameters reported in Table 1. It is important to observe how the expression for p_e in Eq. (7) introduces a temporal dependence, so that, in accordance to historical data, objects launched more recently present a *higher* probability of explosion. Another way to explain the ascending trend of p_e with time is that, as shown in Fig. 10, the number of explosions per year of launch does not improve in recent years. So while the average number of fragmentations per year is quite stable, moving along the *x*-axis in Fig. 9 a smaller and smaller subset of space object is considered, hence the ascending trend.

3 Eco-Labelling of Space Missions

Now that all the four terms have been defined, the debris index can be computed for any space objects in the altitude range of 700–1000 km. The expected output of the proposed debris index is to have a metric able to distinguish space missions on the basis of two main aspects: the mass of the spacecraft and the orbital regime where the spacecraft will operate. Current work is undergoing to take into consideration also the implementation of end-of-life disposal strategies.

The numerical value of this metric would depend on the distribution of operational satellites used to evaluate the effect of a fragmentation, so a process of normalisation is suggested. Only in this way metrics computed in different times (with a different underlying population of operational satellites) are comparable. This would also facilitate the interpretation of the numerical value of this debris index.

An attempt in this direction was already performed for the classification of the effect of a collision (term e_c). In that case, some severity levels (Table 2) were derived from the FMECA (Failure Modes, Effects, and Criticality Analysis) applied during the quality assessment of space missions. The transition between two levels was marked by reference fragmentations.

Figure 11 shows the classification applied to several missions and, as expected, large missions in sun-synchronous orbits (e.g. Sentinel 3) have larger value of e_c than small missions (e.g. Exactview). A similar approach could be adopted also for the whole index as defined in Eq. (1). The most challenging aspect of the process

Table 2 Definition of severity categories [1] and possible meaning for the description of the consequences of a breakup

Severity	Dependability effects	Safety effects	Breakup consequences	Symbol
Catastrophic	Failure propagation	Severe detrimental environmental effects	Subsequent collisions	■
Critical	Loss of mission	Major detrimental environmental effects	Major increase in collision risk	■
Major	Major mission degradation		Increase in collision avoidance manoeuvres	■
Minor	Minor mission degradation		Negligible	■

Fig. 11 Example of fragmentation severity classification for some representative missions

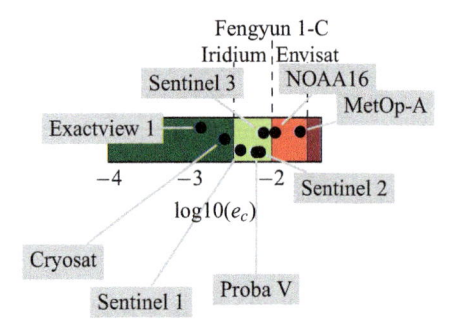

would be to define levels and reference scenarios to build a scale that enables an immediate understanding as in the case of the labelling of household appliances.

A classification of the effects on the space debris environment of a mission is going to be accepted only if all the relevant stake holders are involved in its definition (especially in the formulation of the reference scenarios and the corresponding criticality levels). This means that agencies, operators, manufacturers, and users should be involved in the process. Only in this way it can be avoided that such a classification appears to *blame* specific players.

In addition, such classification should be associated also to a *positive* message. For example, agencies may consider to implement a lean licensing process for *A-level* spacecraft. This would be interesting in particular for small satellites missions that would see a benefit in be more compliant with the guidelines, while now some operators may be tempted to launch their small satellites in a crowded orbital region just because a cheap launch opportunity is available. Similar advantages may be envisioned also in terms of insurance of cost of collision avoidance services provided by external companies. All these measures would enhance the *private* interest of satellite operators to adopt the proposed classification and avoid that it exists only for communication purposes.

… Development of a Debris Index

4 Conclusions

This work described a possible formulation for a debris index, that is a metric for the impact of a space object (i.e. spacecraft or rocket body) on the space debris environment. The index is a development of a previous formulation that looked at how the potential fragmentation of the studied object would affect the collision probability for operational satellites. In the current work, this idea is expanded by considering the probability of these fragmentation happening, distinguishing between collisions and explosions. When assessing the effects of collisions and explosions it was observed how the latter tend to have a more localised effect. It was shown how this is related to the fact that, according to employed breakup model, explosions produce larger fragments with a lower velocity variation compared to collisions. For what concerns the probability of collision, it was estimated using the analogy with the kinetic theory of gases. ESA MASTER was used to retrieve the density of debris objects at different altitudes and the most likely impact velocity for different orbital regimes (defined in terms of semi-major axis and inclination). Finally, the probability of explosion was estimated starting from the data available in Discos on past fragmentations. Only events occurred in LEO after 1985 were considered; in addition, only fragmentations due to propulsion failures, battery failures or unknown cause were analysed. Applying the distinction between spacecraft and rocket bodies, a Poisson distribution is used to approximate the distribution of the number of fragmentation in a year and the resulting average value. This number is then divided by the number of objects launched in the considered time period to estimate a value of probability. By repeating this procedure for different epochs of launch, it is possible to obtain an estimation of the probability term that depends not only on the type, but also on how long the object has been in orbit. By putting together the four terms one can obtain a more complete representation of the fragmentation risk associated to a space object and, therefore, to its exposure and potential contribution to the space debris population. Future work will further enhance this representation by addressing the distinction among objects with different disposal strategies, the application to constellations and the extension of the applicability region of the index to the whole LEO. Finally, a possible normalisation of the index was proposed based on the definition of four levels of severity and the definition of some reference fragmentations. For the moment the definition of the levels considers only the term related to the effects of the potential fragmentation of the studied object, but the same procedure can be applied also to the complete formulation of the index. Future work will analyse the definition of meaningful reference threshold when all the terms of the index are considered.

Acknowledgements Francesca Letizia acknowledges the support from the EPSRC Doctoral Training Partnership at the University of Southampton, grant EP/M508147/1. Francesca Letizia thanks ESA Space Debris Office for the support received during her visiting period in ESOC where part of this work was developed. The authors acknowledge the use of the IRIDIS High Performance Computing Facility, and associated support services at the University of Southampton, in the completion of this work. Data supporting this study are openly available from the University of Southampton repository at http://doi.org/10.5258/SOTON/405121

References

1. European Cooperation for Space Standardisation: Space product assurance: failure modes, effects (and criticality) analysis (FMEA/FMECA). Technical Report, ECSS-Q-ST-30-02C, ESA Requirements and Standards Division (2009)
2. Heinzle, S.L., Wüstenhagen, R.: Dynamic adjustment of eco-labeling schemes and consumer choice – the revision of the eu energy label as a missed opportunity? Bus. Strateg. Environ. **21**(1), 60–70 (2012). https://doi.org/10.1002/bse.722
3. IADC Steering Group: Space Debris. IADC Assessment Report for 2011 (2013)
4. Johnson, N.L., Krisko, P.H.: NASA's new breakup model of EVOLVE 4.0. Adv. Space Res. **28**(9), 1377–1384 (2001). https://doi.org/10.1016/S0273-1177(01)00423-9
5. Kebschull, C., Radtke, J., Krag, H.: Deriving a priority list based on the environmental criticality. In: 65th International Astronautical Congress, International Astronautical Federation, iAC-14.A6.P48 (2014)
6. Krag, H., Merz, K., Flohrer, T., Lemmens, S., Bastida Virgili, B., Funke, Q., Braun, V.: ESA's Modernised Collision Avoidance Service. In: 14th International Conference on Space Operations (2016)
7. Krisko, P.H.: Proper implementation of the 1998 NASA breakup model. Orbital Debris Q. News **15**(4), 1–10 (2011)
8. Letizia, F., Colombo, C., Lewis, H.G.: Multidimensional extension of the continuity equation method for debris clouds evolution. Adv. Space Res. **57**(8), 624–1640 (2015). https://doi.org/10.1016/j.asr.2015.11.035
9. Letizia, F., Colombo, C., Lewis, H.G., Krag, H.: Assessment of breakup severity on operational satellites. Adv. Space Res. **58**(7), 1255–1274 (2016). https://doi.org/10.1016/j.asr.2016.05.036
10. Lewis, H.G., George, S.G., Schwarz, B.S., Stokes, H.: Space debris environment impact rating system. In: Ouwehand, L. (ed.) Sixth European Conference on Space Debris, ESA Communications (2013)
11. Mills, B., Schleich, J.: What's driving energy efficient appliance label awareness and purchase propensity? Energy Policy **38**(2), 814–825 (2010)
12. Rossi, A., Valsecchi, G.B., Alessi, E.M.: The criticality of spacecraft index. Adv. Space Res. **56**(3) (2015). https://doi.org/10.1016/j.asr.2015.02.027
13. Sammer, K., Wüstenhagen, R.: The influence of eco-labelling on consumer behaviour – results of a discrete choice analysis for washing machines. Bus. Strateg. Environ. **15**(3), 185–199 (2006). https://doi.org/10.1002/bse.522

Part IV
Re-entry Analysis and Design for Demise

A Multi-disciplinary Approach of Demisable Tanks' Re-entry

C. Bertorello, C. Finzi, P. Perrot-Minnot, G. Pinaud, J. M. Bouilly, and L. Chevalier

Abstract During an atmospheric re-entry, several spacecraft's components, in particular high pressure tanks are known to survive. Some have been found on different locations worldwide. The current design baseline of such tanks consists of Titanium liners with carbon composite overwrapped. This paper aims at presenting simulations of the thermal behavior of new baselines that would be destroyed during their atmospheric reentry. New tools and processes have been developed to quickly and accurately give technical feedbacks to engineers about the survivability of particular designs. Design for Demise, as new discipline, is playing a bigger role in new spacecraft developments, like the Ariane 6 launcher.

1 Introduction

The current baseline for high pressure tanks—operating pressure from around 120–400 bars—consists of titanium liner wrapped with carbon reinforced composite layers. This state of the art technology has been shown to enhance the chances of survivability during an atmospheric re-entry [1] i.e. for a satellite end of life. ESA, with the CleanSat initiative, has been focused on designing new tanks to ensure that they will most certainly be destroyed during their re-entry.

Airbus Safran Launchers (ASL) is now responsible to demonstrate the compliance of new launchers to the safety requirements specified by the French Space Operations Act. In this context, new methods and tools have been developed to analyze and predict the risk associated with an atmospheric reentry of a rocket launcher stage or a satellite at the end of the mission or, in the worst case after a failure. Together with the experience in re-entry studies of rocket and military elements, ASL is aiming at a better prediction of the survivability of spacecraft components.

C. Bertorello (✉) · C. Finzi · P. Perrot-Minnot · G. Pinaud · J. M. Bouilly · L. Chevalier
ArianeGroup (AG), Les Mureaux Cedex, France
e-mail: charles.bertorello@ariane.group

Table 1 Cases selected for simulation

Reference case	Material composition and element thickness of the liner and the overwrapped composite		
T1	Liner	Aluminum 1	~1 mm
	Composite	Carbon	~4 mm
T2	Liner	Aluminum 2	~2 mm
	Composite	Aramid	~12 mm

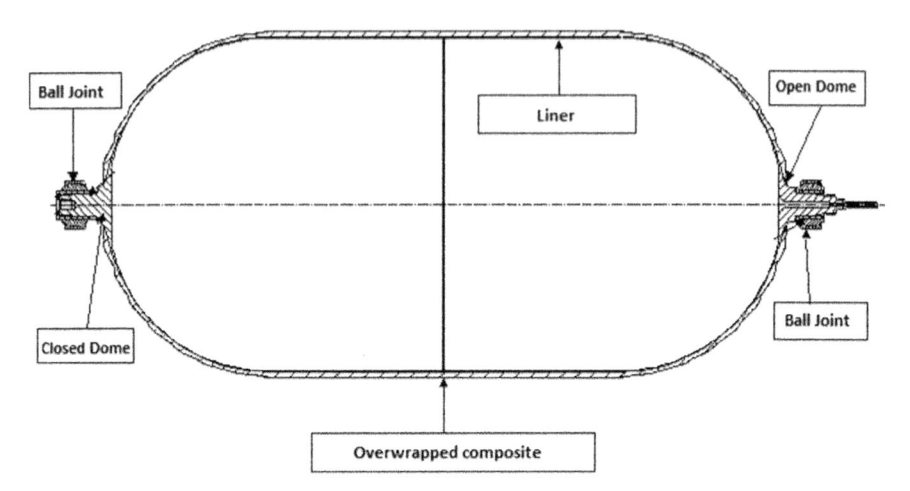

Fig. 1 High pressure tank general design

The focus of the present work is on the thermal and ablation responses of two pressure tanks; respectively T1 and T2; in a destructive re-entry trajectory, see Table 1.

A more general 2D/3D approach using AMARYLLIS software is used and compared to the fast 1D approach using object oriented software ADRYANS in its new version 5.0. Material properties are based on Airbus Safran Launchers own database. Future software and material database developments will be briefly discussed.

2 General Tank Design Presentation

High pressure tanks for space applications are made of a liner overwrapped with a multi-planar composite filament. Liner mostly ensures leak tightness. Composite Overwrap is designed to sustain specified pressure. Figure 1 shows the general design concept of a high pressure tank.

Fig. 2 Polar mount design

The current baseline for liner is Titanium. It has a high melting temperature which is not compliant with demisability requirements. In this study, liner material is changed to aluminum, with two grades being considered.

Composite current baseline is CFRP (Carbon Fiber Reinforced Product (carbon fiber with epoxy resin): it will be compared with another composite: Aramid with thermoplastic resin. Compared to the baseline, these materials have lower strength, leading to thicker designs.

2.1 Design Concept Interfaces

Airbus Safran Launchers experience leads to polar mount design. This concept of interface with platform structure is well known and already validated for larger/heavier tanks; such as the one on Ariane 5.

Tank mount consists of polar mounting (Fig. 2). This mounting concept is based on two ball-joints. Currently, the ball joint at open dome is clamped and the opposite ball joints concept is sliding mount. This sliding ball-joint allows axial displacement mainly due to internal pressure (no additional stress due to hyperstatic structural design).

Ball-joints are compatible with misalignment of the two ball-joints supports. For instance an autofrettage manufacturing pressure is applied on the tank and creates residual deformations via plastic deformations of the liner.

2.2 Case Definition and Materials

Table 1 describes the two different test tanks considered in this study.

They will be named respectively T1 and T2. The chosen liner and composite thicknesses are both common values, calculated by the dimensioning of pressure tanks considering nominal loading and bulking phenomena.

2.2.1 Liner

The melting temperature and melting enthalpy of aluminum is taken into account in the simulation as a fully reversible reaction. Nevertheless, liquid metal flow inside the tank is not simulated. The same value is considered identical for all the different aluminum since no data is currently available for the two different grades. ASL is currently looking at experiments to better estimate the properties at high enthalpy.

2.2.2 Composite

The remaining difficulties lie in the availability of high temperature properties of the material such as Aramid -Thermoplastic matrix composite especially since this material is not often used in such products. Even if carbon-Thermoplastic resin properties are more common, the same "ab-nihillo" procedure has been used to build the numerical material model of Aramid and Aramid thermoplastic. The same approach being used for both material (carbon and Aramid composite) will allow a fair comparison of the performance or survivability of the material with same level of confidence and fidelity between each family. Nevertheless, this methodology relies on strong assumptions and simplifications which all tend to put the predictions in pessimistic trends for demisability.

Elementary or material system characterization will still be necessary to enhance the model and demise analysis.

For a charring and ablative composite material the FE analysis required:

- The material density, conductivity, enthalpy
- A pyrolysis model, an ablation scheme

Elementary composition as well as combustion heat will be necessary to feed the model. TGA (thermo-gravimetrical analysis) gives an insight on thermal stability and pyrolysis product enabling the establishment of kinetics law for the degradation. Figure 3 shows typical TGA plots for Aramid.

In order to keep consistency with state of the art pyrolysis model approach, TGA shall be performed in inert gas. This allows the separation of the oxidation mechanism from pure thermal decomposition and to avoid an overestimation of the mass loss due to non-realistic pressure and oxidizing environment. Thermal conductivity is a one of the main source of uncertainties especially in the charred state due to the entanglement of the fibers.

Finally, the ablation is supposed to occur in a thin volume surrounding the gas-material interface which is still valid for dense material (before pyrolysis). For highly porous or a stack of unrevealed fibers, volume ablation might also occurs. This effect (volume ablation) would tend to decrease the radius of fiber and consequently increase mechanical flexural constraints and could finally lead to a mechanical rupture of the fibers. This mechanism is not modeled in the current approach.

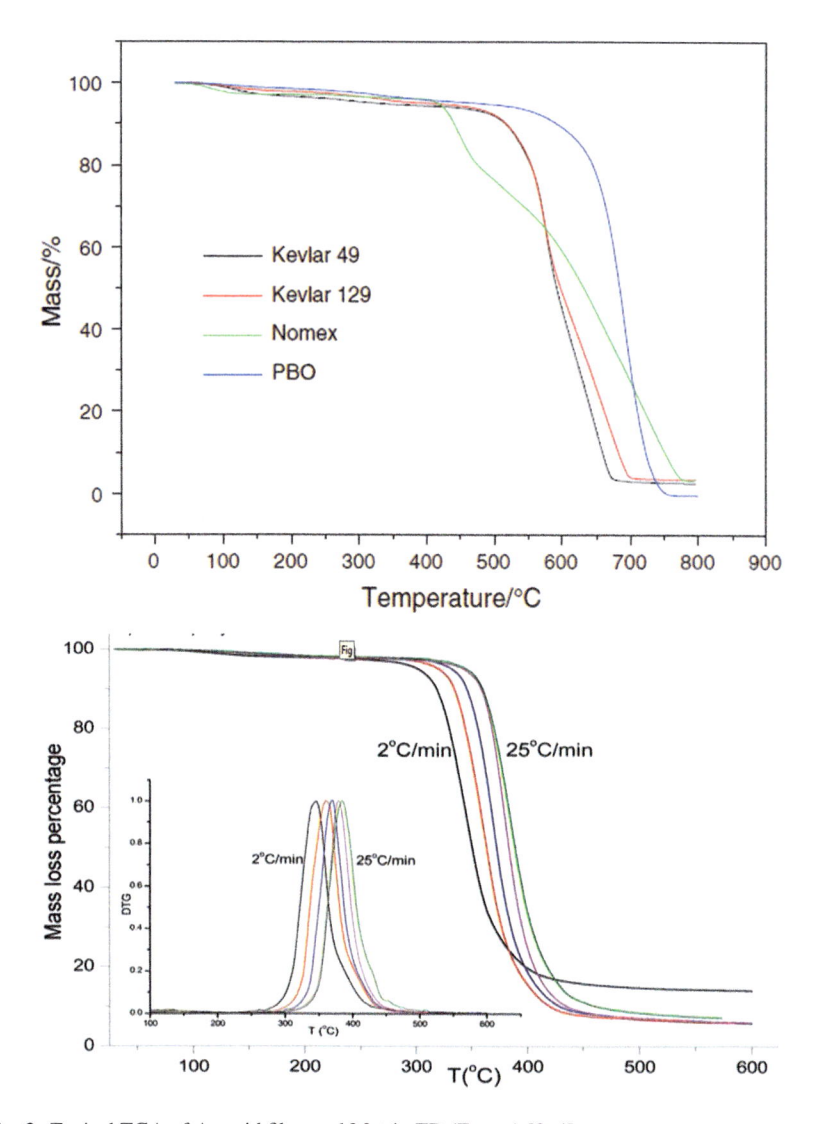

Fig. 3 Typical TGA of Aramid fiber and Matrix TD (Epoxy) [2–4]

The hybrid (equilibrium diffusion limited and finite rate chemistry) ablation scheme is taken from pure carbonaceous material but takes into account the effect of:

- The wall temperature
- The wall pressure
- The mass flow rate of pyrolysis and ablation products
- The heat exchange coefficient

This scheme is taken from previous internal studies on carbon ablator and is commonly accepted and validated for this domain of application.

Nevertheless, this scheme is driving the surface recession and shall be validated for Aramid fiber (and different thermoset or thermoplastic matrix). It is used in this study for all material since no specific scheme is currently available. ASL is working on developing schemes for those materials.

3 Trajectories and Attitude Assumptions

The first part of the re-entry trajectory concerns the spacecraft; satellite or launcher's stage. High pressure tanks are supposed attached to the spacecraft during the first phase of the atmospheric re-entry. Tanks are therefore protected from external aerothermal fluxes until the final break-up.

The spacecraft will be supposed to be in the 1000–1200 kg class with a span of approx. 7 m initially before break-up in the atmosphere with an aerodynamic reference surface of 17 m^2. Solar arrays are supposed to break around 95 km and aerodynamic coefficients are updated to continue the trajectory (Reference surface is set to 8 m^2 and the length to 4 m). Initial re-entry velocity is set to 7.6 km/s with a flight path angle of −0.5° at 120 km.

A final break-up altitude of 78 km is considered as a reference and two additional altitudes with an arbitrary offset of ±15.6 km (±20%) are added to the study for sensibility analysis. At these altitudes, tanks are instantaneously exposed to the aerothermal environment. Trajectories for the spacecraft and the remaining tanks are plotted on Fig. 4 concerning the T1 tank.

When T1 and T2 are ejected from the spacecraft, new trajectories are computed and presented on Fig. 5.

Note that trajectories are computed with three degree of freedom.

3.1 Aerothermal Fluxes

Aerothermal loads are computed using the stagnation point formulation of ADRYANS. Then the aerothermal fluxes have to be transformed for AMARYLLIS. Since we are using the random-tumbling motion hypothesis, mean aerodynamic fluxes and forces are taken in the case of a cylindrical or spherical tank. They are plotted on Fig. 6.

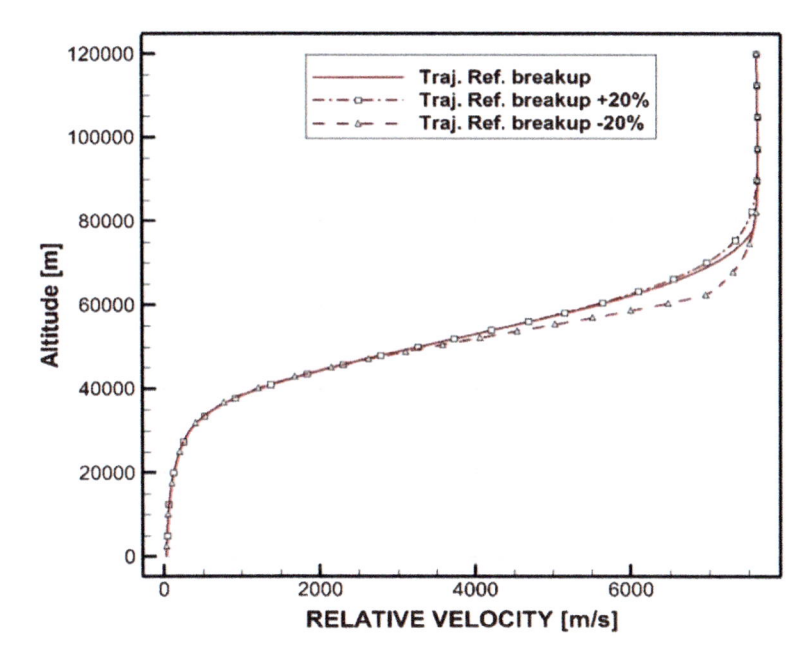

A Multi-disciplinary Approach of Demisable Tanks' Re-entry 215

Fig. 4 Trajectories, ref break-up is set to 78 km. T1 tank

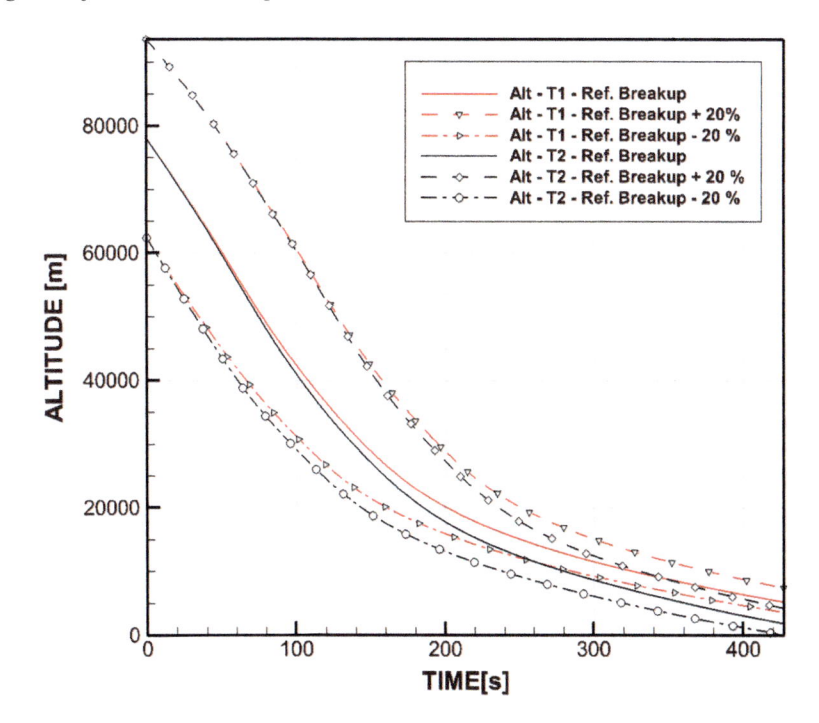

Fig. 5 Trajectories of T1 and T2 tanks after break-up. Ref break-up is set to 78 km

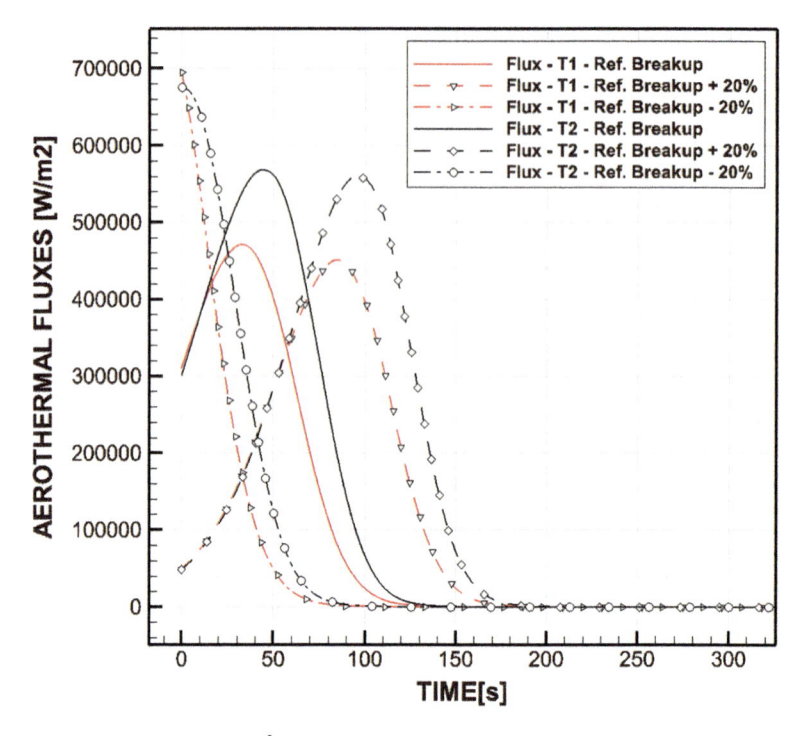

Fig. 6 Aerothermal fluxes (W/m^2). Ref break-up is set to 78 km

4 Simulation Tools

4.1 Tools Description

In the first approach, the industrial software AMARYLLIS [5, 6] has been used. AMARYLLIS is a numerical tool developed and distributed by the company SIEMENS-SAMTECH (Belgium), which is technically supported by the University of Liège. This code is one of the various modules of SAMCEF. This code is dedicated to the simulation (1D, 2D, 3D) of non-linear, transient thermal response of material undergoing thermochemical degradation (pyrolysis) and ablation (both chemical and mechanical). Such as all other module of SAMCEF, AMARYLLIS relies on the finite elements discretization technique. Amaryllis results showed good agreement with US standard ablation code CMA/FIAT simulations of a PICA-like theoretical material [5, 6].

The second approach has been done with the new version 5.0 of ADRYANS. It is a fast computer software that computes the survivability of space debris during its atmospheric reentry. It has been developed at Airbus Safran Launchers in close partnership with the French Space Agency CNES. The new version now supports:

1-dimensional thermal model, composite materials and stack of different materials (i.e. thermal protections), chemical simulation such as oxidation, ablation, pyrolysis of composite materials, etc.

As of now, no other object oriented code has the ability to model composite materials [7–9] that are present on launchers and satellites.

4.2 Output

Outputs of the simulations are visualized through the following graphs:

- Temperatures (K) with respect to time (s)
  - TC1: outer side of the composite
  - TC3: inner side of the liner
- Density change with respect to time (s)
  - at outer thermocouple TC3
  - at inner thermocouple TC2
- Recession/displacement (m) with respect to time (s)
- Pyrolysis gases mass rate ($kg/m^2/s$) with respect to time (s)

  Other properties can also be plotted if necessary.

4.3 Input Geometry, Meshing and Assumptions for SAMCEF AMARYLLIS

Input geometry in AMARYLLIS is modeled in this study as a 2D finite small zone of the tanks common section (see Fig. 4). Aerothermal loads are applied to face the outer element and hypothesis of adiabatic condition on other faces is made. The number of elements and the distribution of these elements were thought as to cope up with boundary conditions (Fig. 7).

4.4 Input Geometry and Assumptions for ADRYANS

ADRYANS models a 1D zone along the thickness of the tank common section. Properties requirements are almost identical to AMARYLLIS software and are extracted from the ASL's database.

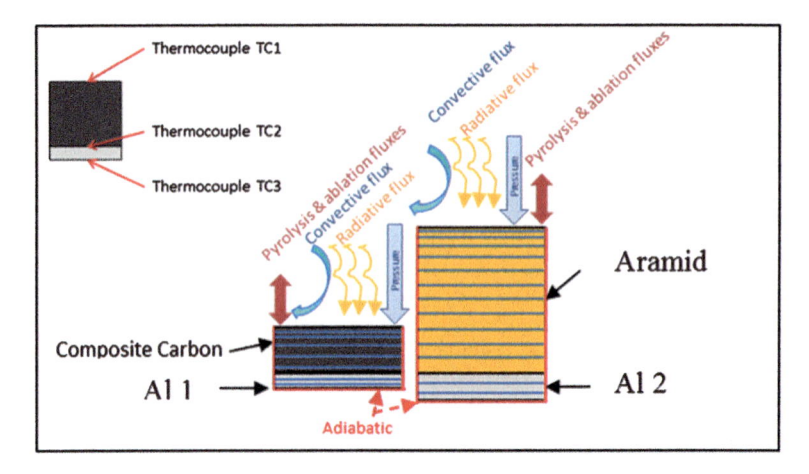

Fig. 7 FEM simulation set-up view and thermocouple location (TC1, TC2 and TC3)

5 Results Presentation and Cross-checking

5.1 Temperature

5.1.1 Tank T1

Trajectories Ref. Break-up and Ref. Break-up+20% will be analyzed simultaneously since the events are quite similar. Ref. Break-up -20% will be analyzed in a second time.

Figure 8 for the Ref. Break-up trajectory and for the Ref. Break-up+20% trajectory that the wall temperature of the composite exceeds respectively *1079 K* and *1250 K* at the pic heat flux for T1 with AMARYLLIS. The comparison shows a fairly good agreement between the two codes with a difference on the maximum temperature of +2 K (when comparing ADRYANS to SAMCEF). Note that the AMARYLLIS temperature is hotter during the temperature decrease phase (after the peak and also during the final cooling phase).

In both code we can observe at the beginning of the trajectory the outgassing of the composite due to pyrolysis of the thermoset matrix is blowing the boundary layer which results in a slight decrease of the wall temperature (higher decrease in AMARYLLIS will be discussed later) as well as a blockage of the ablation. This higher outgassing in the AMARYLLIS software is due to a different modelling of the gas transport between the composite layers and the outside atmosphere.

When the composite material is fully charred at *90 s* (both codes) for the Ref. Break-up trajectory and *120 s* (both codes) for the Ref. Break-up+20% trajectory, this protective effect disappears entirely and the entire aerothermal heat flux is applied at the wall.

Recession is then ignited again under an oxidation of the carbon fiber and resin residues reaction controlled by finite rate chemistry. After *150 s* for Ref. Break-up

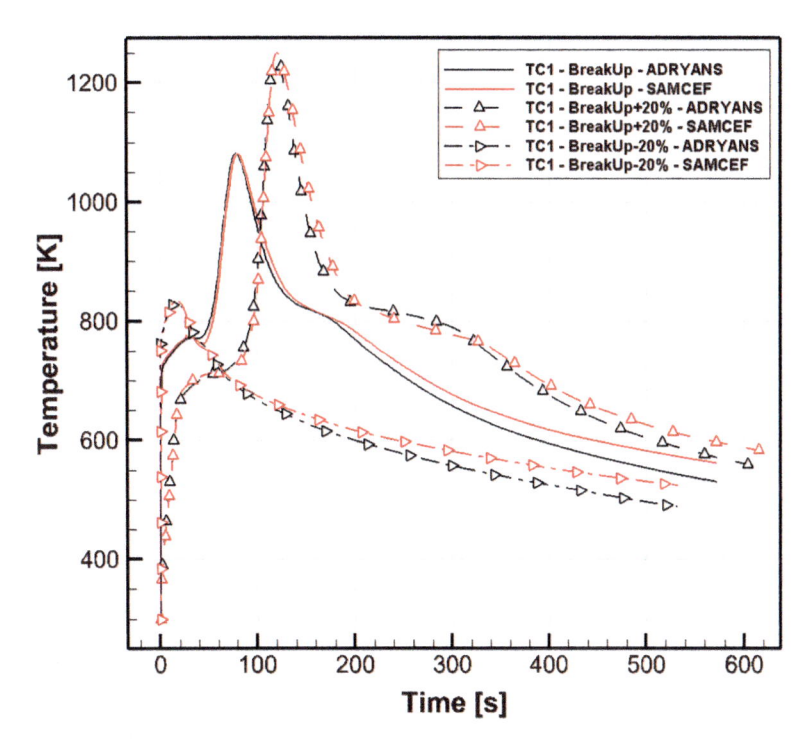

Fig. 8 Simulated composite wrapping temperatures of the T1 tank for the three trajectories—AMARYLLIS & ADRYANS computation

trajectory and *210 s* for Ref. Break-up+20%, wall temperature is not high enough and rates of reaction are negligible (see Sect. 5.2). No major differences on the time have been found between ADRYANS and AMARYLLIS.

The temperature of the inner side of the liner is increasing with a delay due to the thermal diffusivity of the two materials; see Fig. 9. As the liner temperature reaches the melting temperature, the conductive heat is then transferred in the latent heat of melting, creating a temperature plateau *for 65 s* for Ref. Break-up trajectory (for both code despite the delay observe) and $\approx 150\ s$ for Ref. Break-up+20%. It is assumed that the wall melts in a layer growing from the outer surface inside, and no flow of the liner is considered. As the liner inner face temperature has reached this plateau for a very long duration, we suppose the liner to be fully melted at the end of the temperature plateau. Therefore, we consider the aluminum to be melted above an altitude of 20 km for Ref. Break-up trajectory and 15 km for Ref. Break-up+20% for the AMARYLLIS results. ADRYANS's values are similar with a slight delay since the plateau is reached 10 s before AMARYLLIS.

For Ref. Break-up-20%, maximum heat flux occurs directly on the spacecraft and not on the tank: therefore the integrated heat load is lower and not sufficient to perform a complete pyrolysis reaction of the composite layer. As a consequence, the liner will not reach its melting temperature. Aluminum is softened and reaches a temperature of *641 K* in AMARYLLIS simulation, *642 K* in ADRYANS.

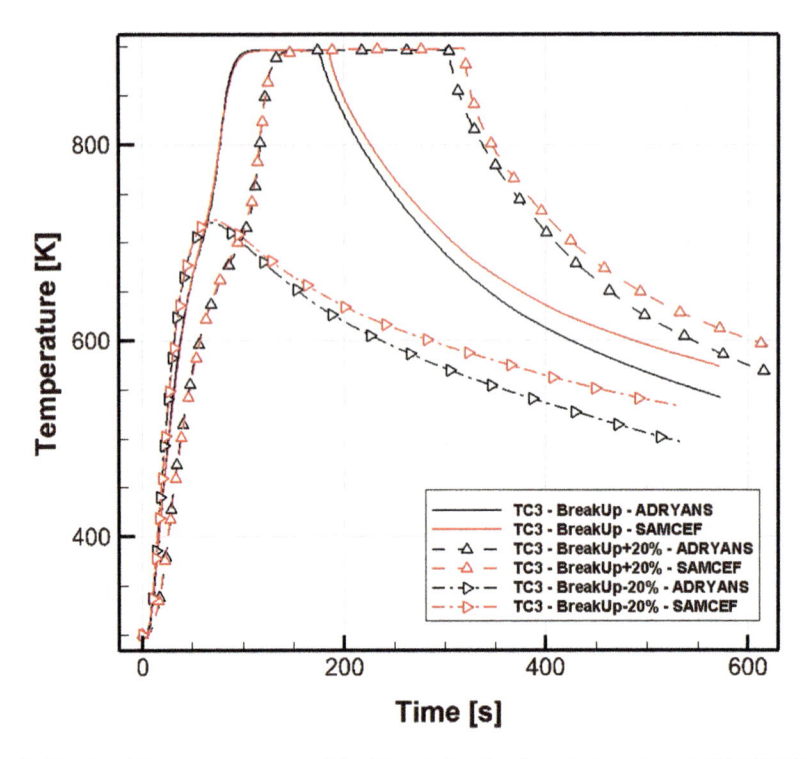

Fig. 9 Simulated liner temperatures of the T1 tank for the three trajectories—AMARYLLIS & ADRYANS computation

The tank should keep its shape due to the remaining virgin composite layer with sufficient mechanical strength (see Sect. 5.3).

The final velocity being around 30 m/s, the kinetic energy is in any case above the casualty risk criteria.

5.1.2 Tank T2

Trajectories Ref. Break-up, Ref. Break-up+20% will be analyzed simultaneously since the events are quite similar. Ref. Break-up-20% results being very close to the others, this case is not plotted hereafter.

The wall temperature of the composite exceeds *1170 K* at the pic heat flux for T2 tank in AMARYLLIS (difference of −15 K in ADRYANS) for Ref. Break-up trajectory and *1185 K* for Ref. Break-up+20% (difference of +19 K in ADRYANS). Due to the continuous outgassing of the composite by pyrolysis of the thermoplastic matrix and Aramid fiber, surface heat flux and ablation are blocked all along the trajectory. Indeed, the high thickness of the composite provides sufficient protection to the pyrolysis and heat front to penetrate slowly in depth and never reach the liner interface.

After *180 s* for Ref. Break-up trajectory and *240 s* for Ref. Break-up+20%, wall temperature is not high enough the ablation process stops. The total mass loss of the outer composite (due to ablation and pyrolysis) is approximatively 8% of its initial mass for Ref. Break-up and 9% for Ref. Break-up+20% with AMARYLLIS.

Figure 11 shows that the liner temperature does not exceed *500 K* with ADRYANS and *430 K* with AMARYLLIS, meaning that the liner is far from melting temperature and keeps enough mechanical strength to maintain the tank in its initial shape.

The final velocity being around 40 m/s, the kinetic energy is in any case above the casualty risk criteria.

5.1.3 Sum-Up

Figures 8 and 10 show the outer wall temperature of the tank computed using the AMARYLLIS software and the ADRYANS's tool. Differences can be observed. One of the main reasons come from the difference of modelling between the two software regarding the pyrolysis reaction. In fact, SAMCEF AMARYLLIS gas mass flux equation depends on pressure variations while ADRYANS equation depends on density variations inside the composite material.

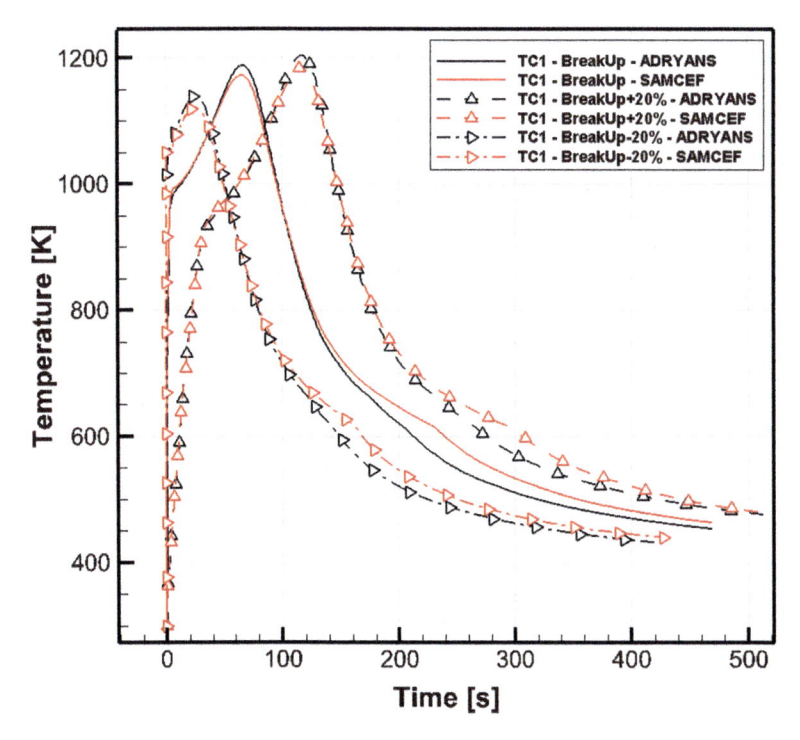

Fig. 10 Simulated composite wrapping temperatures of the T2 tank for the three trajectories- AMARYLLIS & ADRYANS computation

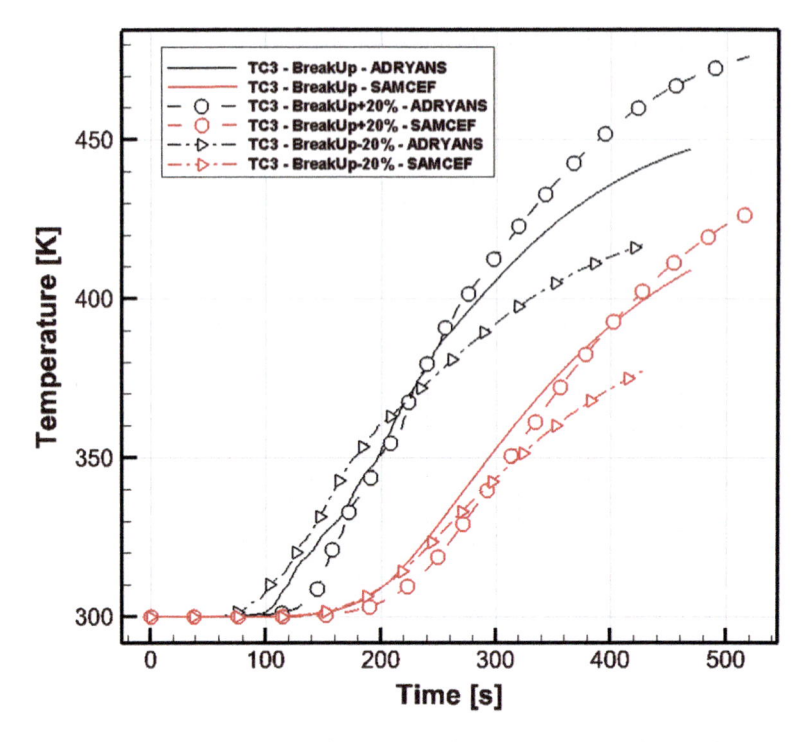

Fig. 11 Simulated liner temperatures of the T2 tank for the three trajectories—AMARYLLIS & ADRYANS computation

5.2 Ablation, Mass Loss and Recession Graphs

5.2.1 Tank T1

Figure 12 shows the simulated ablation rate occurring on T1 for the three trajectories. Few peaks can be observed at the beginning of the computation when the temperature is increasing rapidly for the trajectory with a break-up 20% lower than the reference and the reference trajectory as well. Concerning the trajectory with 20% increase of the altitude of break-up, ablation is occurring less rapidly at the beginning during the first 20 s.

This is due to the extreme heat fluxes occurring at the beginning when the material is not protected by the outgassing occurring with the pyrolysis chemical reaction.

Then the same behavior between the two codes is observed on the three trajectories: a sudden increase of ablation rate then a slower decrease when the maximum heat flux has been crossed.

Figure 13 shows the recession of the surface due to the ablation of the composite material. The two codes tend to simulate the same behavior. Differences are due to the temperature that tends to be hotter in the AMARYLLIS case especially in the decrease phase. Hotter temperatures will result in more ablation.

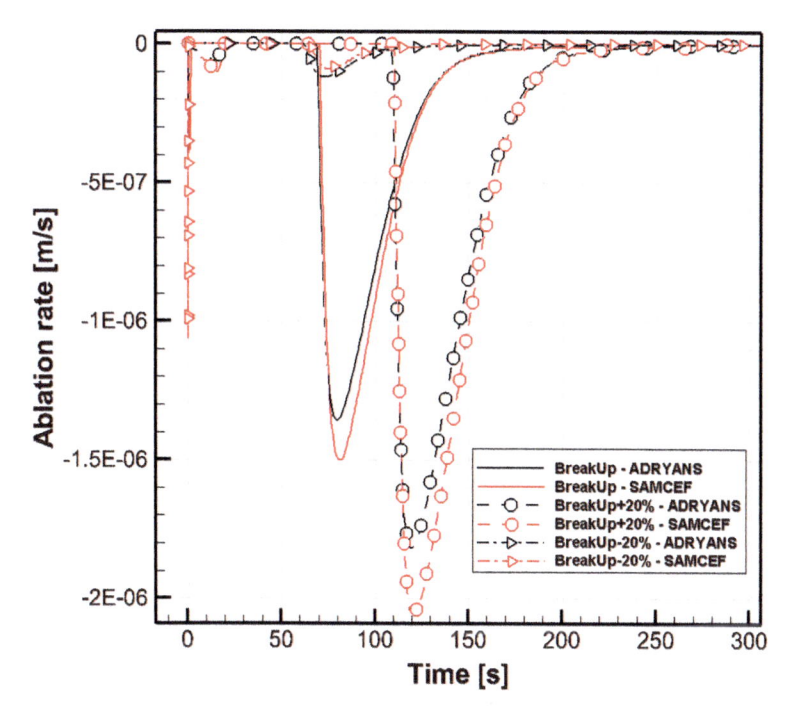

Fig. 12 Simulated ablation rate of the composite wrapping of the T1 tank for the three trajectories AMARYLLIS & ADRYANS computation

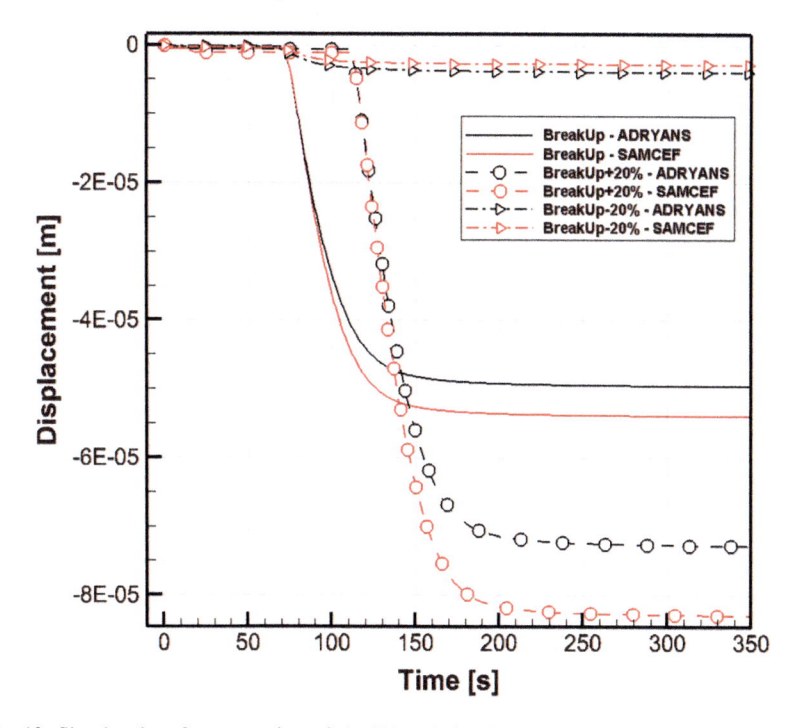

Fig. 13 Simulated surface recession of the T1 tank for the three trajectories AMARYLLIS & ADRYANS computation

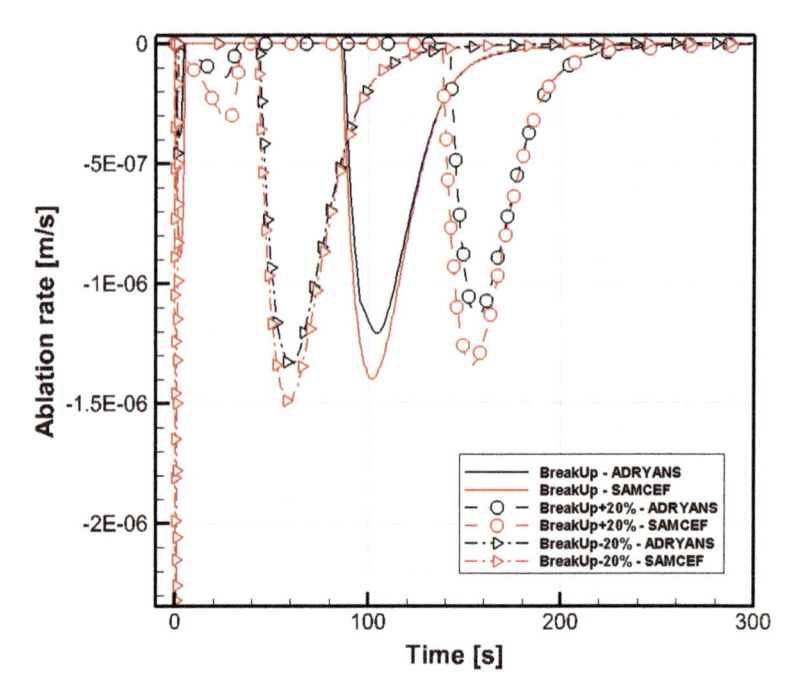

Fig. 14 Simulated ablation rate of the composite wrapping of the T2 tank for the three trajectories AMARYLLIS & ADRYANS computation

5.2.2 Tank T2

Figure 14 shows the simulated ablation rate occurring on T2 for the 3 trajectories. Few peaks can be observed at the beginning of the computation when the temperature is increasing rapidly for the reference trajectory and the pyrolysis reaction is delayed compared to the heat wave. Since the outgassing of the pyrolysis takes time to occur, the ablation of the surface is allowed. Figure 15 shows a small recession at the beginning of the trajectory and then a bigger one when the pyrolysis reaction is over.

5.2.3 Sump-up

Discrepancies in ablation results can be explained by the lower temperature on the outer surface of the sphere due to the outgassing of the pyrolysis reaction. The gas mass flow rate is protecting the sphere from the convective heat flux.

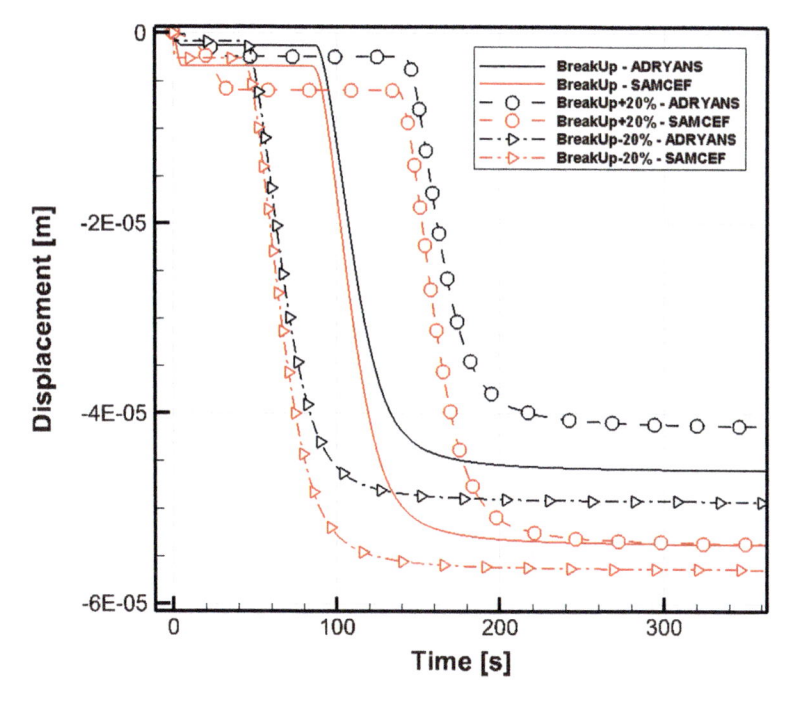

Fig. 15 Simulated surface recession of the T2 tank for the three trajectories AMARYLLIS & ADRYANS computation

5.3 Pyrolysis Results: Pyrolysis Gas Mass Flux and Evolution of Density in the Composite Layer

5.3.1 Tank T1

Pyrolysis gases evolution can be observed on Fig. 16. Major differences exist between the two computer codes: SAMCEF AMARYLLIS gas mass flux equation depends on pressure variations while ADRYANS equation depends on density variations inside the composite material. This influences the results of ablation. It can be noted that the gas flow rate follows the same trend between the 3 trajectories. It is happening when the pyrolysis reaction is activated and reaching a maximum value around the maximum temperature peak.

Figures 17 and 18 illustrate the density change at the external side and inner side of the composite layer. Both codes tend to simulate the same behavior depending mainly on the temperature of the outer surface that imposes the thermal gradient.

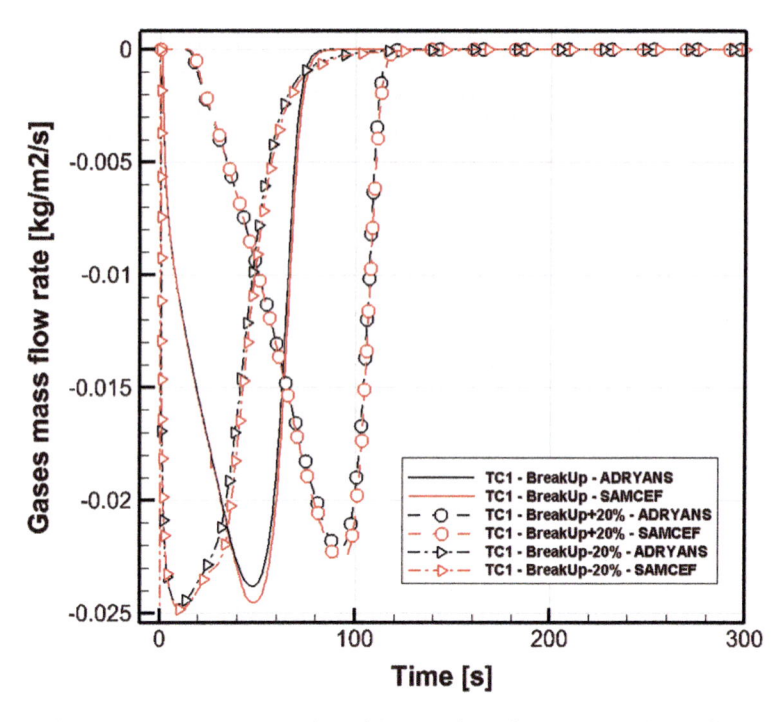

Fig. 16 Simulated pyrolysis gas mass flux of the outer face of the composite layer of the T1 tank for the three trajectories

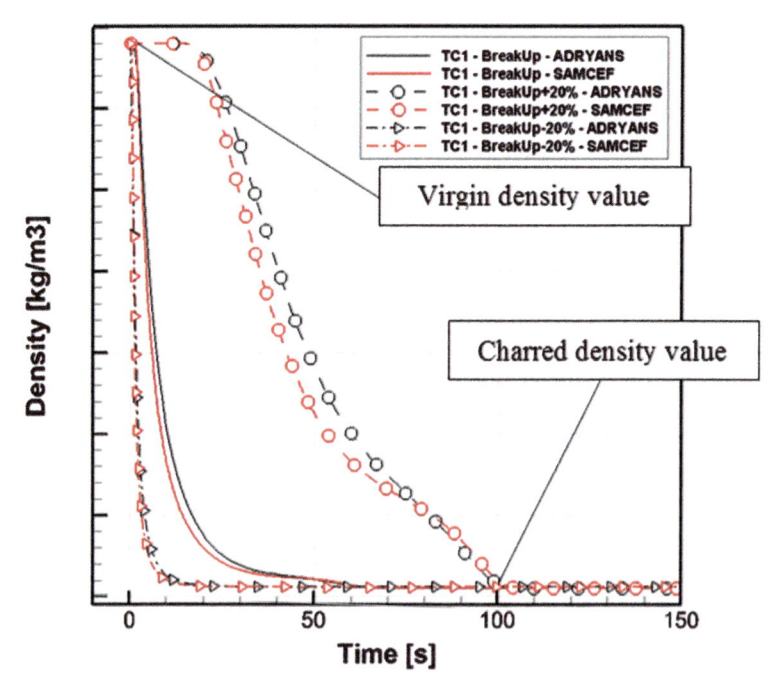

Fig. 17 Simulated density variation of the outer face of the composite layer of the T1 tank for the three trajectories

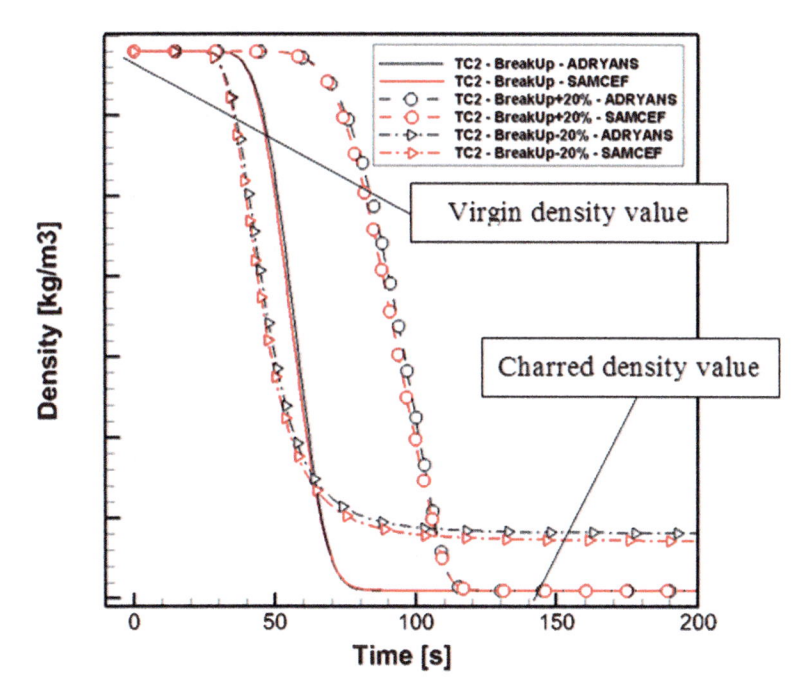

Fig. 18 Simulated density variation of the inner face of the composite layer of the T1 tank for the three trajectories

5.3.2 Tank T2

Same comments apply on the T2 case. Nevertheless, the pyrolysis reaction is not reaching the inner part of the Amarid layer. See mass flow rate of the gases on Fig. 19 and the density change on Fig. 20.

6 Global Conclusion on Designs

Actual Baseline for high pressure tanks is state of the art composite overwrap pressure vessel, developed to store pressurant gas (Helium, Nitrogen). Those tanks are made of a titanium liner with a carbon reinforced composite overwrap. As mentioned, it is not the preferred design for demise.

Main gap between this baseline and the proposed demisable tank relies on changing materials (liner and/or composite) to improve demisability. Main impact at system level is linked to the additional mass resulting of this material change. In our case materials that improve demisability (aluminum to replace titanium and Aramid to replace carbon fibers) have lower strength. Aramid positive impact has still to be shown (by testing e.g.)

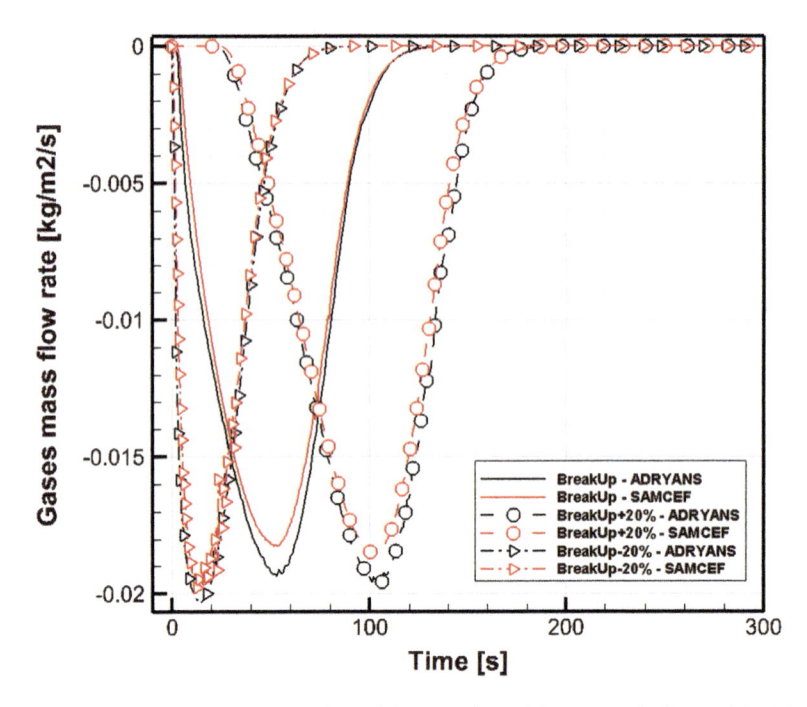

Fig. 19 Simulated pyrolysis gas mass flux of the outer face of the composite layer of the T2 tank for the three trajectories

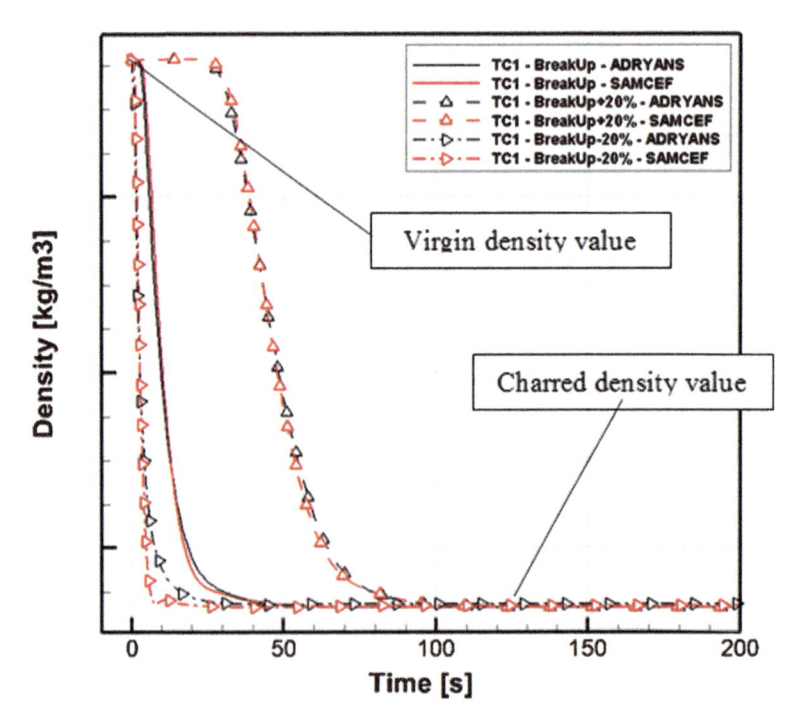

Fig. 20 Simulated density variation of the outer face of the composite layer of the T2 tank for the three trajectories

Current ball joints are steel ball joint. This type of massive part is prohibited due to demisability aspects. Two ways could be considered in future studies:

- Screen ball joint: replace with aluminum ball joint.
- Study feasibility on new polar mount design: bolted polar mount, other solutions such as equatorial mounts...

Several improvements will be conducted in the following years and are detailed in Sect. 7.

7 Foreseen Developments

7.1 Material Database Enhancement

Airbus Safran Launcher in cooperation with CNES is characterizing at high temperature both composite (CFRP laminate and overwrap composites...) and metallic material samples (Titanium, Aluminum, Inconel...), representatives of common structure materials presents on launchers of the Ariane family.

Test environment will be representative of reentry environment simulated by ASL internal plasma torch test facilities:

- Testing at temperatures close to the melting point for metallic materials while common literature on metallic focuses on functional temperature ranges rather than extreme temperatures.
- Representativeness of thermos-ablative degradation mode for composites will be enhanced. This topic is not yet being well covered in civil application.

These new data will help reduce errors due to material properties uncertainties for reentry simulations.

7.2 Design Choice, Development and Tests

Several tests shall be carried in the future for the tank design final choice.

- Composite coupon testing with plasma torch to study demisability at elementary scale and to evaluate the thermal protection potential of the matrix. These tests shall be carried under representative pressure, heat fluxes and fiber lay-up.
- Liner coupon tests: mechanical strength, low cycle fatigue, crack propagation testing
- Fabrication tests: compatibility of filament winding composite material and process with liner material.
- Material development

These tests shall confirm the final choice between T1 and T2 design concepts.

Looking at the demisability perspective, we can note that the longer the material is protected in the spacecraft the better it will survive. Satellite or spacecraft design that integrates controlled break-up devices once the trajectory is well known can be imagined.

8 Conclusion

The simulation of tanks re-entry coupled with a multidisciplinary approach allows engineers to design high pressure tanks and test their survivability to aerothermal fluxes. Tables 2 and 3 summarize the loss of mass and mechanical properties on the T1 and T2 tanks following three different re-entry trajectories. It appears that Aramid mechanical properties induce composite layer thickness that ends up acting as a thermal shield for the inner aluminum layer during reentry. This design seems then less favorable regarding demisability than the carbon fiber composite-aluminum design. Limitations of the current material models are well identified: thermochemical hypotheses shall be completed by characterization tests at high temperature. Moreover, further investigations on the mechanical behavior of fully charred composites are needed in order to better understand the behavior of charred composite layers and partially molded liner of tanks during reentry. This analysis simplifications and assumptions tend to put the predictions in a pessimistic trend for demisability. A brief description of experiments for the material characterization has been described. Tools associated with the presented methods have been also described and aim at better and quickly assist the engineer in automated and optimized schemes.

Differences in modelling between Amaryllis® and Adryans® are the reason for differences observed; actions are currently conducted to continue their development. These differences have little impact on demisability conclusions for preliminary design simulations so both can be used for this purpose. Mass loss mainly comes from resin being charred.

Table 2 Thermal and mechanical summary on the T1 case

Reference case T1	Break-up altitudes		
	62.4 km	78 km	93.6 km
Liner	Not melted but soft loss of mechanical properties	100% melted	100% melted
Composite SAMCEF	~22% total mass loss	~24% total mass loss	~25% total mass loss
Composite ADRYANS	~7.8% total mass loss	~8.0% total mass loss	~8.2% total mass loss

Table 3 Thermal and mechanical summary on the T2 case

Reference case T2	Break-up altitudes—results from SAMCEF and ADRYANS		
	62.4 km	78 km	93.6 km
Liner	Not melted	Not melted	Not melted
Composite SAMCEF	~5% total mass loss	~8% total mass loss	~9% total mass loss
Composite ADRYANS	~3% total mass loss	~4.8% total mass loss	~5.6% total mass loss

Acknowledgments Tools currently in development to be included in a future global reentry aerodynamic-thermo-mechanical simulation platform have been funded under a research and development agreement between Airbus Safran launchers and the French Space Agency (CNES). The authors would like to thank their management and ASL experts for their support to this contribution for the Stardust conference.

References

1. Summary of Recovered reentry debris, The aerospace corporation, www.aerospace.org
2. Azeem Arshad, M., et al.: Kinetics of the thermal decomposition mechanisms of conducting and non-conducting epoxy/Al composites. J. Mater. Environ. Sci. **5**(5), 1342–1354 (2014)
3. Cai, G.M., Yu, W.D.: Study on the thermal degradation of high performance fibers by TG/FTIR and Py-GC/MS. J. Therm. Anal. Calorim. **104**, 757–763 (2011)
4. Patel, P.: Investigation of the fire behaviour of PEEK based polymers and compounds. Doctoral Thesis University of Central Lancashire (2011)
5. Bouilly, J.M., Eekelen, T.: Inter-code calibration exercise series, Amaryllis results. In: 4th AF/SNL/NASA Ablation Workshop, March 2nd 2011, Albuquerque, New Mexico
6. Bouilly, J.M., Eekelen, T.: Inter-code calibration exercise series #2, Amaryllis results. In: 5th Nasa Ablation Workshop, Feb 28th-March 1st 2012, Lexington, Kentucky
7. Martin, C., et al.: Debris risk and mitigation analysis (DRAMA) tool final report. Qinetiq/KI/Space/CR050073 (2005)
8. Dobarco-Otero, J., et al.: The object re-entry survival analysis tool (ORSAT) version 6.0 and its application to satellite entry. In: 56th International Astronautical Congress (2005)
9. Omaly, P., Spel, M.: DEBRISK, a tool for re-entry risk analysis. In: 5th IAASS Conference, Versailles (2012)

Design-for-Demise Analysis using the SAM Destructive Re-Entry Model

James C. Beck, Ian E. Holbrough, James A. Merrifield, and Nicolas Leveque

Abstract In order to assess a number of design-for-demise techniques, a toy spacecraft has been constructed and is modelled with an approach which bridges the gap between the standard object-oriented and spacecraft-oriented approaches. Using the SAM destructive re-entry code, simulations have been performed on the complete spacecraft to the point where the major fragmentation events have occurred in six-degrees of freedom. The spacecraft is modelled as a set of components connected by joints, allowing the benefits of spacecraft oriented modelling to be achieved at a fraction of the computational cost. Using this analysis with the fragmentation altitudes as the indicator of the effectiveness of the design-for-demise techniques, the potential improvement from the use of demisable joints and inserts can be clearly identified. Assessment of the individual components which result from the fragmentation can then be carried out using a three-degrees of freedom approach as it is demonstrated that this represents the heating on basic shapes more reliably that spacecraft-oriented approaches. The flexibility of the SAM tool is demonstrated in the assessment of three critical items of spacecraft equipment.

1 Introduction

Two standard approaches have historically been used for the assessment of the casualty risk posed by re-entering spacecraft. The first is the object-oriented approach which in most instances considers a fixed spacecraft fragmentation altitude

---

J. C. Beck (✉) · I. E. Holbrough
Belstead Research Limited, Ashford, Kent, UK
e-mail: james.beck@belstead.com; ian.holbrough@belstead.com

J. A. Merrifield
Fluid Gravity Engineering Limited, Emsworth, UK
e-mail: jim.merrifield@fluidgravity.co.uk

N. Leveque
Airbus Defence and Space, Stevenage, UK
e-mail: nicolas.leveque@airbus.com

and resultant fragments modelled as three-degree-of-freedom (3dof) equivalent spheres representing simple shapes such as cylinders and cuboids. The second is the spacecraft-oriented approach where the complete spacecraft geometry is represented as a set of triangulated surfaces. The trajectory of the spacecraft is modelled in six-degrees-of-freedom (6dof) and the fragmentation of the spacecraft is modelled as occurring when only molten panels separate sections which are still solid. Similar fidelity aerothermodynamic heating and material sub-models are used in both cases. Strictly, the complexity of these sub-models is suitable for the object-oriented approach, and can be considered a considerable source of error for spacecraft-oriented modelling.

Significant differences are predicted between the different models, primarily due to the fragmentation representation [1] as similar results are obtained on objects of aspect ratio close to unity [2]. The unique approach used in the SAM tool is to model the aerodynamics and aerothermodynamics of the spacecraft using the real geometry in six-degrees-of-freedom as in a spacecraft-oriented model, but to consider the spacecraft as a collection of components connected by joints. This limits the number of geometric arrangements possible, improving the efficiency and allowing large numbers of simulations to become possible using a substantially representative geometric model. Furthermore, this allows the modelling of the spacecraft fragmentation to be performed predictably, and to be based on the failure of joints, which has been shown in basic tests to be a more realistic mechanism driving the fragmentation than the assumption of melt [3].

Once the fragmentation of the spacecraft has reached the component level, the simulation is propagated using an object-oriented approach, as the simple heating methodologies used are substantially more accurate on standard shapes, such as cuboids, cylinders and cones, than they are on arbitrary geometries, especially as the aspect ratio of the object increases. Therefore, there is a significant advantage in terms of both accuracy and efficiency to switch the simulations from 6dof to 3dof at this stage of the destructive re-entry simulation. Alternative demise phenomenologies and potential fragmentation pathways can be assessed powerfully using a 3dof code at component level, and a demonstration of the equivalence to mean 6dof calculations is provided for a number of important spacecraft components.

2 Toy Spacecraft

The toy spacecraft used in this work is based on the Sentinel-2 spacecraft. It is worth noting that an uncontrolled re-entry has been demonstrated as acceptable for this spacecraft such that no further design-for-demise techniques are necessary. Even so, it is a useful exercise to assess the effectiveness of concepts which could further improve the demise such that larger platforms can be considered for uncontrolled re-entries.

Design-for-Demise Analysis using the SAM Destructive Re-Entry Model

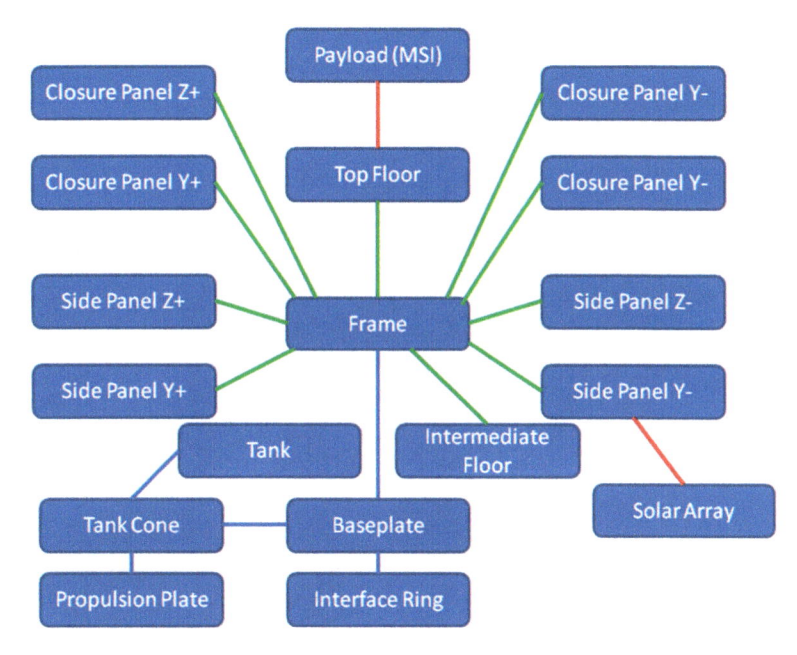

Fig. 1 SAM toy spacecraft schematic

In order to provide efficient solutions, the spacecraft has been modelled using 18 components. These components are simple primitive shapes, which have masses, inertias, orientations and locations such that the mass properties of the spacecraft can be calculated and the geometry of the spacecraft can be used to generate the aerodynamics and aerothermodynamics. The component/joint structure of the toy spacecraft is shown in Fig. 1. The frame structure is considered as a single component in the analysis.

The joints are shown with the components that they connect. The blue joints can only fail by melting, but the green components can be assigned a lower failure temperature to simulate the effect of insert weakening or temperature activated failure. The red joints can also be failed using a force based criterion to simulate the effects of launch locks and solar array break-off. The allocation of the green joints is based on the understanding that these joints use screws through epoxy inserts, and the blue joints use titanium screws through aluminium brackets (Fig. 1).

Each component can have multiple heating points. In this model, as the structural panel failure is key to the fragmentation of the spacecraft and the release of the subsystems, the mass of each component is split into a structure part and an equipment part. This provides results consistent with spacecraft oriented models which produce results consistent with low heat conduction to internal components.

The aerodynamics and aerothermodynamics are calculated on each configuration during the fragmentation process using Modified Newtonian theory and the Modified Lees correlation respectively. Examples of spacecraft configurations and the quality of gridding used for these calculations is shown in Fig. 2. This

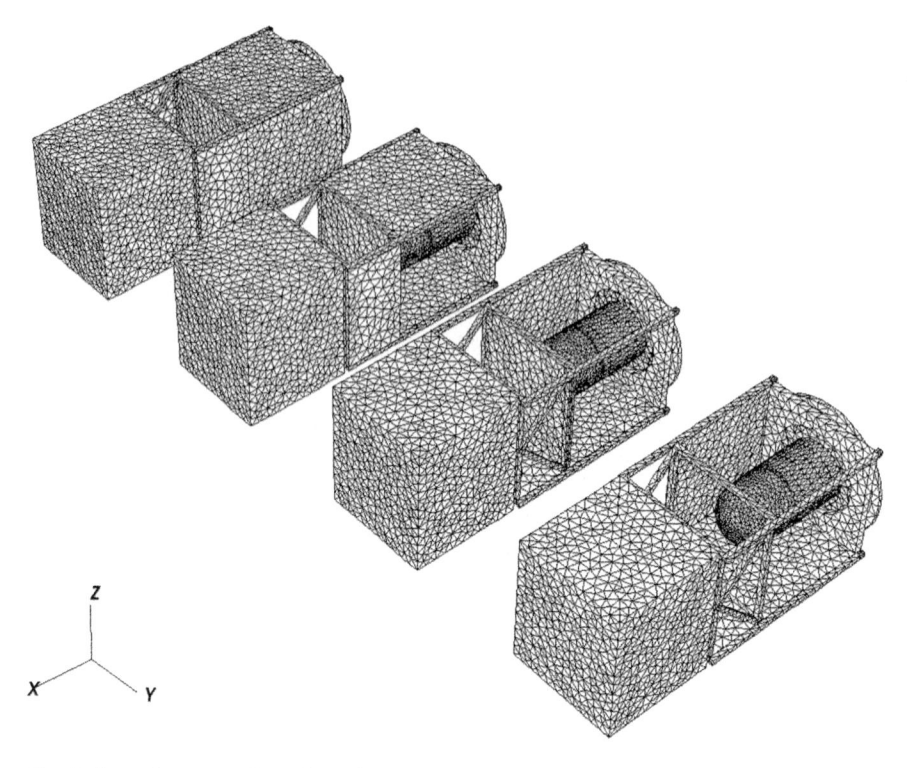

Fig. 2 Example spacecraft configurations

methodology is equivalent to that used in SCARAB, but the limited number of geometric configurations provides increasing benefit the more simulations are run due to the existence of repeat configurations. Pre-computation of aerodynamics and aerothermodynamic databases can also be performed in parallel for efficiency. On a serial machine, each new configuration requires about 30 min to compute, but if no new configurations are required, then the full destructive re-entry simulation takes approximately one minute. Therefore, this approach is also well suited for Monte Carlo type studies. The SAM code has a Monte Carlo capability for initial conditions, joint failure criteria and aerothermodynamic heating uncertainties, which has been used to provide a statistical representation of the fragmentation in more recent studies [4].

3 Fragmentation Analysis

Spacecraft oriented simulations are often highly sensitive to the initial orientation of the spacecraft. Therefore, to ensure that the demise techniques are robust to modelling choices, a set of 36 simulations at a range of initial orientations and spin

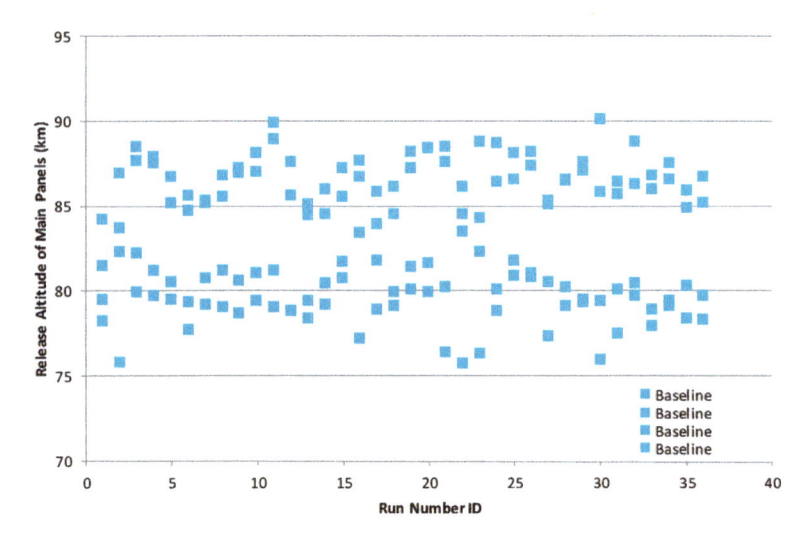

Fig. 3 Release altitude of main panels

rates were performed. A true sensitivity study would require more simulations, 1000 are used in practice with a wider range of uncertainties in an ongoing study, but this initial study of 36 is a similar number to that of an equivalent uncertainty study using a spacecraft oriented code [5]. The set of simulations show that the four main side panels of the spacecraft are removed, on the assumption of aluminium melt temperature of the structure resulting in fragmentation, between 75 and 90 km with little sensitivity to orientation or spin rate within this band (Fig. 3). It is interesting to note that due to the different rotational behavior of the spacecraft, the order in which the panels are released changes, which can lead to significant differences in the predicted casualty risk.

Given this consistency in the main panel release altitudes, design for demise techniques aimed at promoting the fragmentation of the vehicle can be assessed. The simplest concept is to consider the release of the external panels through joint failure, which is physically more reasonable than the standard melt-based models. This has been simulated by assuming that the epoxy inserts will fail once the potting material has reached a given temperature. This failure temperature is varied from 450 K to the initial assumption of the aluminium melt temperature. It is worth noting that recent tests [3] have shown this is likely to be an over-optimistic failure criterion for potted inserts, but could be a good criterion for edge inserts. As shown in Fig. 4, the altitude of the panel release is substantially increased with the reduction in the failure temperature of the joints. This has a substantial impact on the release altitude of the equipment and a clearly positive impact on the casualty risk.

In order to improve demise, the side panels can be opened to allow the hot shock layer gases access to the internal components. Activation of this at the spacecraft end-of-life necessitates that the panels remain connected to the spacecraft in order not to generate further orbital debris. From a modelling viewpoint, this requires

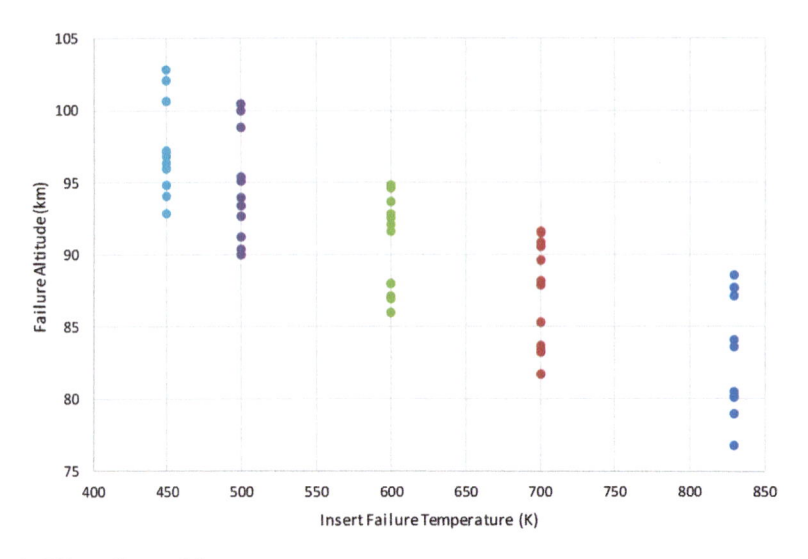

Fig. 4 Effect of insert failure temperature

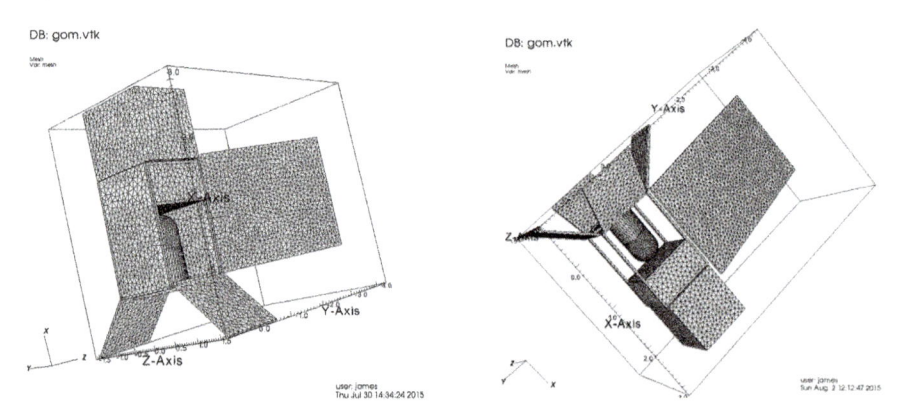

Fig. 5 Example panel opening geometries

a different set of initial geometries to be considered, which will have different aerodynamics and mass properties. Three cases have been run, with two, three and four of the side panels opened. Examples are shown in Fig. 5.

The improvement in the release altitude of the main panels can be seen in Fig. 6, with a gradual improvement as the number of open panels is increased. The difference in the last panel release at 75 km, increasing to 85 km, is substantial.

Further design-for-demise techniques have also been assessed. It has been found that the release altitude of the propellant tank and support structure is greatly enhanced by the jettisoning of the vehicle baseplate, but that the impact of aerodynamic stabilisation and the use of launch locks is limited.

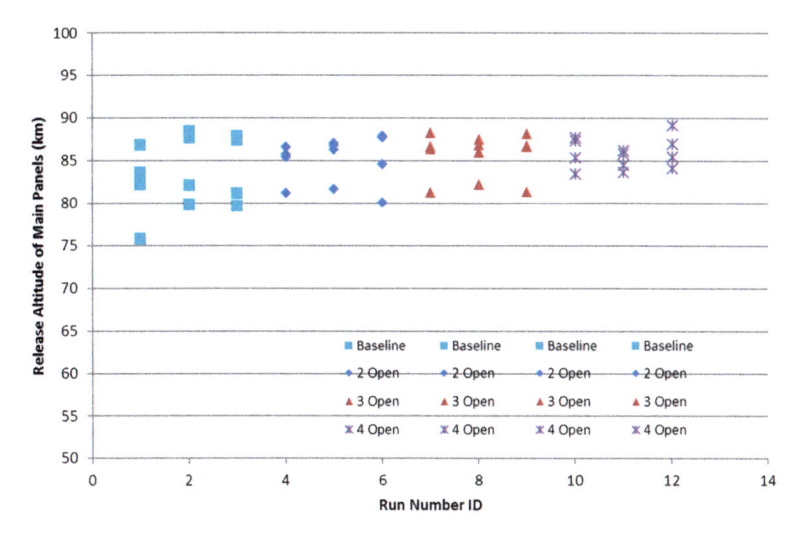

Fig. 6 Panel release altitude improvement from open panels

During the execution of this work, an important caveat on the quality of any spacecraft oriented predictions has been observed. Due to the small radius of curvature of the frame structure, the heating to this would be expected to increase substantially on failure of the panels. However, in the Lees heating correlation, which is the basis of the heating model is almost all current spacecraft oriented codes (Pampero being a notable exception), an overall vehicle equivalent radius is used, which is not appropriate for a frame-type structure. This is essentially the same effect which produces the underprediction of the heat to the magnetic torquers which is discussed later. Although this cannot be confirmed without some considerable work on improved heating models, this would be expected to fail the frame structure much earlier, resulting in more heating reaching the tank cone structure and an earlier failure (Fig. 7).

This observation suggests that the heating of individual, component level objects, is much more relaiably performed using appropriate correlations for the given shape. Therefore, in practice, the SAM code uses a 3dof analysis for components as this provides improvements in accuracy as well as computational cost.

4 Component Analysis

In most simple object-oriented codes it is often difficult to model a realistic demise pathway due to the limited range of modelling choices available to the user. SAM has been specifically designed to have a phenomenological approach intended to be able to model the demise of the object in a manner which is consistent with the physical processes encountered. A clear danger is that a model may be of a

Fig. 7 Persistent frame structure highlighting spacecraft oriented heating issues

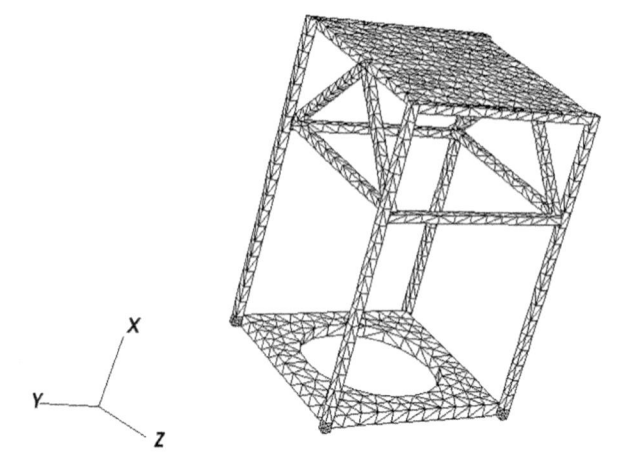

high granularity but not be representative of the correct phenomenology thereby having the potential to produce less accurate results than a far simpler model. The SAM simulations are set up such that it can be clearly understood both *whether* a particular understanding of demise is effective and *why* it is effective.

The components considered in this paper are batteries, reaction wheels and magnetic torquers.

4.1 Batteries

The battery layout considered is based on the Sentinel 5 Precursor (S5P) battery. The cells are in four separate bricks, which are separated by intermediate aluminium walls. The bricks are encased in GFRP and epoxy glues are used to bond the cells, GFRP plates and aluminium plates. As the aluminium case is not enclosed, the most likely demise mechanism is that the GFRP encased bricks will be released from the aluminium structure, and that battery will quickly disintegrate into small components, mainly cells dependent upon the temperature dependent failure of the GFRP and the epoxy bonds.

The battery is fixed to the spacecraft by metallic screws in epoxy inserts. Therefore, it is likely that the mechanism which causes the battery to be disconnected to the main structure is the failure of the inserts. Care is required here as there may be some delay in release as there is the possibility that the screws will become snagged within the insert holes, but it is strongly likely that the release will occur before the metal structure reaches melting point and that the battery component will be released cleanly as a separate item without remaining attached to a melting panel. Therefore, the assumption that the battery can initially be considered in isolation is good.

The cells are constructed of a steel can, with a copper and aluminium coil pack. It is worth noting that the can is open. This means that with the small dimension of

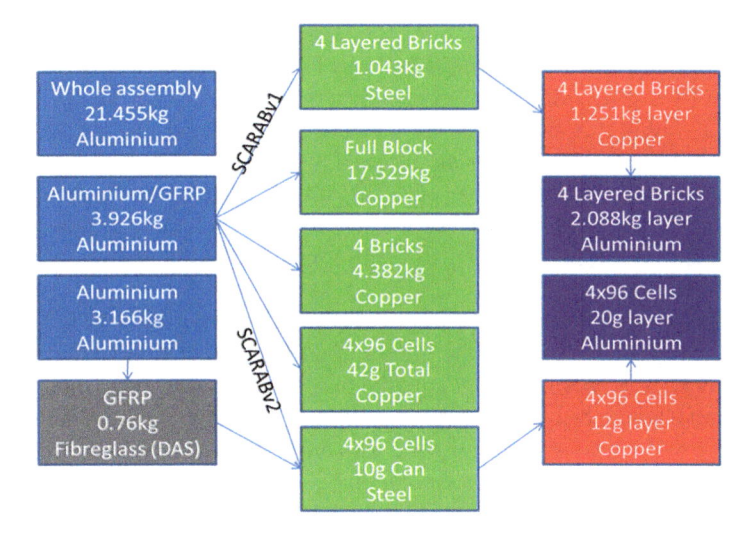

Fig. 8 Map of battery models

the cell, the temperature gradients are likely to be small and the coil pack will melt prior to the steel can. As there is an exit for the molten coil pack, it is likely that this molten material will escape prior to the melting of the steel can.

The battery has been modelled at a range of granularity levels, from a whole block, through four bricks, to an individual cell model. The layered bricks account for the cell material layers, but at brick level. Effects of the GFRP insulation have also been included. The map of battery models is shown in Fig. 8.

Simulations were run at a range of initial altitudes, reflective of the findings from the spacecraft breakup models. A range of heating correlations was also used. A standard object-oriented approach of 78 km was considered as a baseline, with a ± 20% variation. The simulations showed a very clear dependence on the modelling of the battery, with the single block and brick models suggesting that the battery does not demise in re-entry.

In contrast, all the simulations where the batteries were considered to fragment to the cell level demonstrated complete demise. This story was consistent with all the heating correlations and provided very good agreement with 6dof assessments at a range of initial attitudes and spin rates. As this is considered to be the most likely phenomenology of failure, this assessment suggests that batteries are not critical items.

4.2 Reaction Wheels

The reaction wheel is based on the S5P reaction wheel. An airtight aluminium housing, which is sealed with solder, but also welded together, protects the steel shaft and ball-bearing unit. The shaft supports the flywheel, which is the major mass, and critical part of the component. The ball-bearings are coated in Titanium Carbide (TiC) which might be expected to survive re-entry, but they are not considered within this analysis as they are small, and thus below the 15 J energy threshold. The separate elements are screwed together such that the joints between the components are expected to be reasonably strong. Therefore the expected demise phenomenology is that the aluminium housing will melt, releasing the internal parts. The shaft and flywheel would be expected to separate, and be the key parts which could survive re-entry.

As with the battery, a hierarchy of models is used to understand the behaviour of the reaction wheel demise, and the impact of different modelling and construction assumptions. It is worth noting initially that there is some uncertainty in the aerothermodynamic heating to very short cylinders as this is not well covered in the literature. The major source of good data on cylindrical objects, Klett [6], does not cover this range of shapes, although this is the source used for the heating in most object oriented codes, and is the basis of the model in SAM.

For the initial assessment, the complete reaction wheel is modelled as a single component, and the sensitivity to the material properties tested, by using steel, aluminium, and an equivalent material accounting for the energy required to demise a representative material mixture. Beyond this, the aluminium housing and steel components are modelled separately, with the flywheel being released at the point that the housing is demised. Flywheels of 3.5 and 5 kg are tested, with thicknesses of 42 and 53 mm. A map of the models is given as Fig. 9.

Use of a spoked flywheel instead of a solid flywheel is considered as a possibility for increasing demise. By assessing an unshadowed model of primitive shapes, the drag on the thinner spoked flywheel is seen to be somewhat lower than on the solid flywheel, but the thicker flywheels have similar drag. The composite shape demonstrates a reasonable level of increased heating due to the smaller radii of curvature. Using a more standard spacecraft oriented approach, with the model

Fig. 9 Map of reaction wheel models

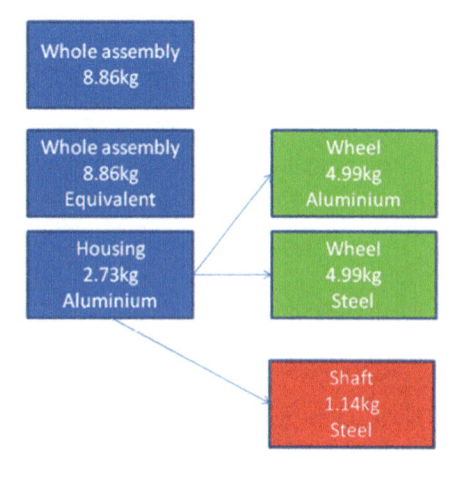

Fig. 10 Spacecraft-oriented style spoked reaction wheel model

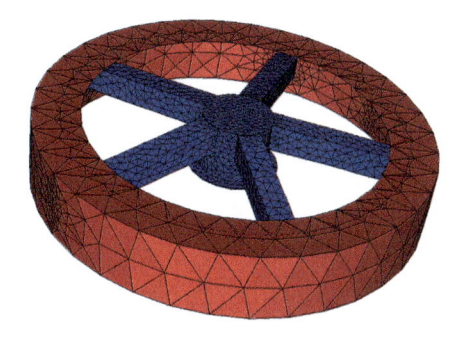

shown in Fig. 10, a reduced drag is predicted due to the shadowing. However, use of the primitives model demonstrates that the heating on these spoked structures is significantly underpredicted using spacecraft oriented models, in this case by approximately 35%. Therefore, a higher heating is used in this analysis.

As with the batteries, simulations at each level of granularity and the same range of initial altitudes are performed. In this case, the reaction wheels are seen to partially demise, but the flywheels reach the ground in almost all cases. The spoked wheels improve the demisability, but only sufficiently to allow complete demise in extreme cases. Therefore, reaction wheels are confirmed as a critical item in destructive re-entry assessments.

4.3 Magnetic Torquers

The magnetic torquer is essentially coils of insulated copper wire wrapped around a ferromagnetic cylinder, all within a cylindrical aluminium housing. It is worth noting that there is a layer of elastomer between the housing and the coil, which is a polyesterimide/polyamidimide high temperature insulated wire.

The demise phenomenology of the magnetic torquer is expected to be reasonably straightforward. The housing would be expected to fail initially, and at the aluminium temperature, the elastomer should already have no strength and is thus likely to have little bearing on the demise. The copper coil could then either remain in place, in which case it will then heat to its (higher) melt point and then melt to expose the iron core. There is a possibility that this is slightly conservative as the copper coil may well be removed, at least partially, from the core as it is reasonable to expect some level of unwinding of the coils. If the copper coils can become detached easily, then demise is more likely.

The melting of the solid iron core is expected to be the critical part of the demise process as this is the main monolithic part, and it is well shielded. As there is a layer of elastomer insulation, and the copper wire is also insulated, it is not clear that there will be significant heat conduction of the core prior to its exposure.

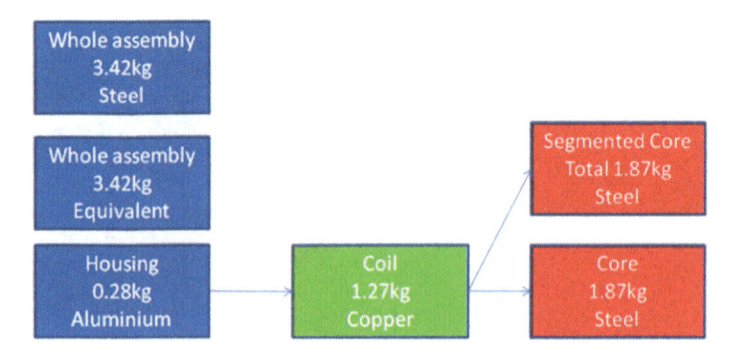

Fig. 11 Map of magnetic torquer models

The simplest way to capture the construction of the magnetic torquer is to consider a set of coaxial cylinders around the iron core. Again, a hierarchy of models is used. As an initial assessment, the component is modelled as a steel cylinder. This assessment is improved by using an equivalent material, which has the same heat to melt per unit mass, and latent heat as the compound multi-material component. Finally a fully layered approach is used, consisting of the aluminium housing, copper coils and iron core. The insulating material is not considered. The baseline model does not consider conduction between the layers, but an isothermal model considering perfect conduction, and a 1D conduction simulation have both been performed to assess this sensitivity. As a demise technique, a segmented core is also simulated. A map of the magnetic torque models is given as Fig. 11.

One of the key points of interest for magnetic torquers is that they have been identified as critical items in spacecraft oriented simulations, but are assessed as demisable in object oriented tools. This has often been argued as being due to the late release of the equipment, a hypothesis which is tested here through the use of a range of heating correlations and a range of initial altitudes. 6dof simulations are again performed at a range of initial orientations and spin rates.

For each heating correlation, the results are very clear. Using the Klett [6] formulation of most object oriented codes results in demise of the magnetic torquer for all assumptions from all initial altitudes. This is also true for the modified heating algorithm used in SAM, and is confirmed in 6dof simulations with this formulation. Use of the Lees correlation, the standard model in spacecraft oriented codes, results in a substantially reduced heat flux to the long cylindrical shape, and thus the magnetic torquers can survive when modelled in both 3dof and 6dof. The reduction in heating from the Lees correlation is approximately 40% for cylinders of this 16:1 aspect ratio, which results from the use of an equivalent length scale based on the windward projected area at the current orientation. This strongly suggests that the identification of magnetic torquers as critical items is based on a heating correlation being used outside its range of applicability rather than the release altitude. It is also evident from this work that the results of the object oriented codes are more valid than the results from spacecraft oriented codes (where the Lees model is used) for high aspect ratio shapes.

5 Conclusions

The SAM destructive re-entry code has been used to model spacecraft breakup in 6dof and critical component demise in both 3dof and 6dof, in order to assess a range of design for demise techniques at component and spacecraft level. During the performance of the work, a number of key issues with the fragmentation modelling and the applicability of the spacecraft oriented heating model to compound shapes and long cylinders have been identified. These are the source of some potentially erroneous conclusions in previous work.

The SAM tool has been successfully employed in the assessment of design for demise techniques for spacecraft fragmentation, and for the assessment of both the modelling, correlations in use and demise improvement techniques for three components which had been identified as critical. This work suggests that the concern over both batteries and magnetic torquers is likely to be unfounded.

This work has also demonstrated the utility of using a range of both modelling techniques and exploring uncertainties within a simulation campaign. The confidence which can be gained from this assessment is significantly greater than in some previous campaigns as many thousands of simulations have been performed with a large number of initial conditions, model granularities and correlations to produce a significant database. As a result of this, it is highly recommended to apply uncertainties and to perform statistical analyses in destructive re-entry simulations, whether object-oriented or spacecraft-oriented as this provides a substantially improved understanding of the casualty risk posed from spacecraft.

Acknowledgements This work has been funded by the ESA Clean Space Initiative.

References

1. Lips, T.: Equivalent break-up altitude and fragment list. 6th European conference on space debris, Darmstadt (2013)
2. Lips, T., et al.: Comparison of ORSAT and SCARAB re-entry survival results. 4th European conference on space debris, Darmstadt (2005)
3. Beck, J., Merrifield, J., Spel, M..: Joint fragmenation: phenomena testing and assessment. ESA clean space industry days, Noordwijk (2016)
4. Beck, J., Holbrough, I., Merrifield, J., Joiner, N.: Progress in hybrid spacecraft/object oriented destructive re-entry modelling using the SAM Code. 7th European conference on space debris, Darmstadt (2017)
5. Lips, T., Koppenwallner, G., Bianchi, L., Klinkrad, H.: Risk assessment for destructive re-entry. 5th European conference on space debris, Darmstadt (2009)
6. Klett, R.: Drag coefficients and heating ratios for right circular cylinders in free molecular and continuum flow from Mach 10 to 30, SANDIA report SC-RR-64-2141 (1964)

Low-Fidelity Modelling for Aerodynamic Characteristics of Re-Entry Objects

Gianluca Benedetti, Nicole Viola, Edmondo Minisci, Alessandro Falchi, and Massimiliano Vasile

Abstract This work presents the principal improvements and results of the Free Open Source Tool for Re-Entry of Asteroids and Debris aerodynamic module. The aerodynamic routines are based on the hypersonic local panel formulations, and several innovations to improve performances, in terms of computational time and accuracy across the hypersonic flow regime for re-entry of space vehicles and objects, have been introduced in the new version. A graphic-based preprocessing phase to reduce the computational time has been introduced and tested. New bridging functions, based on logistic regression model, aiming at providing a better estimate of aerodynamic outputs in the transitional flow regime have been introduced. The routines have been validated on different test cases, such as: spheres, STS orbiter, Orion capsule and ESA's IXV. In addition, the tool has been applied to perform the aerodynamic analyses of the cFASTT-1 spaceplane conceptual model and to compute the aerodynamics of GOCE during its re-entry phase. GOCE aerodynamic results have also been compared to DSMC high-fidelity simulations.

G. Benedetti (✉) · N. Viola
Department of Mechanical and Aerospace Engineering (DIMEAS), Politecnico di Torino, Torino, Italy
e-mail: gbenedetti92@gmail.com; nicole.viola@polito.it

E. Minisci · M. Vasile
Aerospace Centre of Excellence, Department of Mechanical and Aerospace Engineering, University of Strathclyde, Glasgow, UK
e-mail: edmondo.minisci@strath.ac.uk; massimiliano.vasile@strath.ac.uk

A. Falchi
Aerospace Centre of Excellence, University of Strathclyde, Glasgow, UK
e-mail: alessandro.falchi@strath.ac.uk

1 Introduction

Hypersonic aerodynamic and aerothermodynamic analyses play a key role in the design of space and aerospace vehicles, in particular for re-entry trajectory prediction and optimization. The correct estimation of these parameters can be obtained by using high fidelity methods such as Direct Simulation Monte Carlo (DSMC) or Computational Fluid Dynamics (CFD). These simulation methods are capable of reliably describing the challenging scenario of the hypersonic flow field, such as the interaction between the inviscid flow behind the shock waves, the viscous boundary layer near the wall of the vehicle and the aerothermal model uncertainties through the transitional flow regime.

However, these methods are usually characterized by extremely long computational times, even for performing aero-thermodynamic simulations for a single point along the vehicle's trajectory. For this reason, such tools cannot be extensively used in the preliminary design phase, when less computationally expensive approaches should be used even if less accurate results are to be expected. Several tools for re-entry analysis (such as ORSAT [1], DRAMA [2] and SCARAB [3]) have been already developed. However, they are either not open source nor easily available. Furthermore, most of them are capable to predict only a specific flow regime.

The Free Open Source Tool for Re-Entry of Asteroids and Debris (FOSTRAD), developed at University of Strathclyde [4], already includes tools for the aerothermal analysis across all hypersonic flow-density regimes. In this work all the latest improvements to FOSTRAD are reported, that is:

- Atmosphere model updated to the NRLMSISE-00.
- Integration of the back-face culling [5], through the pixelator [6] approach.
- New bridging function for transitional regime.

The improved version of FOSTRAD has been used to evaluate the aerodynamics characteristic of different space vehicles and objects: STS orbiter, Orion CEV, Intermediate Experimental Vehicle, cFASTT-1 spaceplane, Gravity field and steady-state Ocean Circulation Explorer (GOCE), and a simple sphere. With the analyses presented in this work, it has been possible to test the computational improvements, increase the confidence level on the tool aerodynamic module, and highlight possible future developments.

2 Introduced Innovations

The newly introduced changes led to the new aerothermal module in FOSTRAD (Fig. 1).

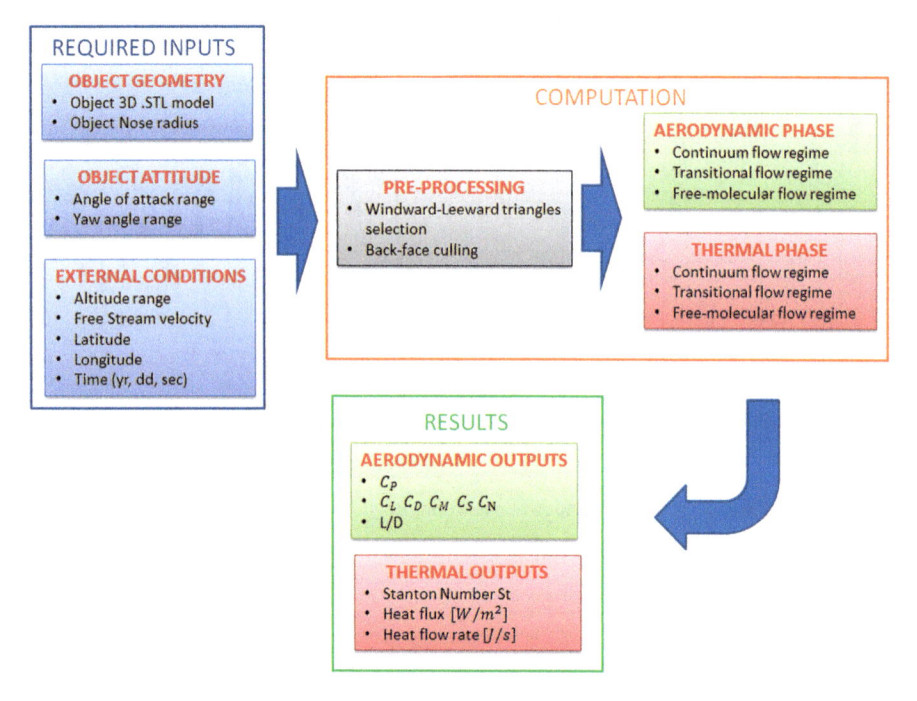

Fig. 1 Flow chart of the Aero-thermal model in FOSTRAD

2.1 Atmosphere model

The code has been updated to use the NRLMSISE-00 atmosphere model [7] for what concerns the estimation of total mass density and atmospheric temperature. The atmosphere model has been introduced in FOSTRAD by using the dedicated Matlab© 2014a Aerospace Toolbox atmosnrlmsise00 function [8]. The model extends from the ground to the exosphere, by increasing FOSTRAD capability to predict the atmospheric parameters up to 1000 km.

Compared to U.S. 1976 model, the NRLMSISE-00 model performs the total mass density thanks to the extensive use of drag and accelerometer data. Moreover, the O^+ isotope and hot atomic oxygen effects are taken into account for mass density computation.

Thanks to data obtained by Solar Maximum Mission [9], the model also considers the solar activity dependence on atmosphere number density and temperature. As example, Atmosphere temperature and gas density predictions above Turin (Italy) for the maximum and minimum 2010 solar activity are shown and compared to U.S. 1976 model, to better appreciate the dependence of atmospheric parameters from daily variation of solar activity (Figs. 2 and 3).

Significant differences between the NRLMSISE-00 and U.S 1976 model occurs in high atmosphere layers, such as the thermosphere and exosphere. Thus, it has

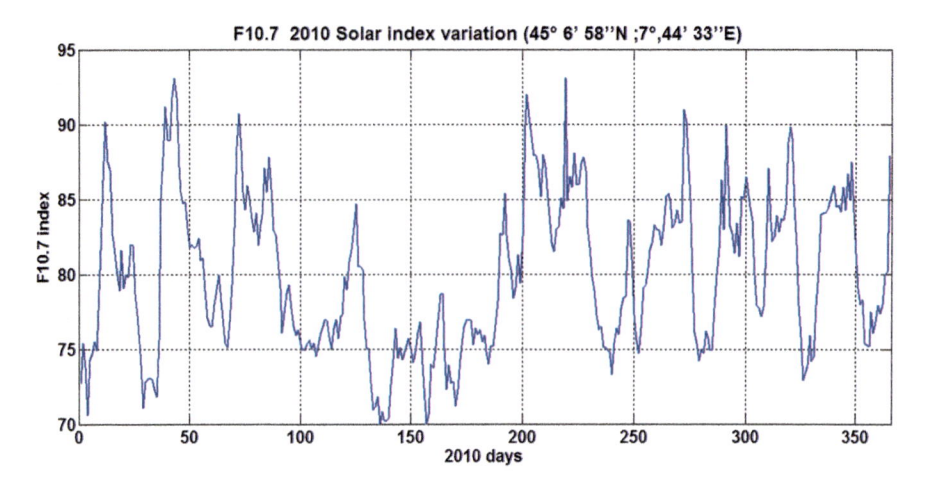

Fig. 2 F10.7 2010 Solar index variation above Turin, Italy

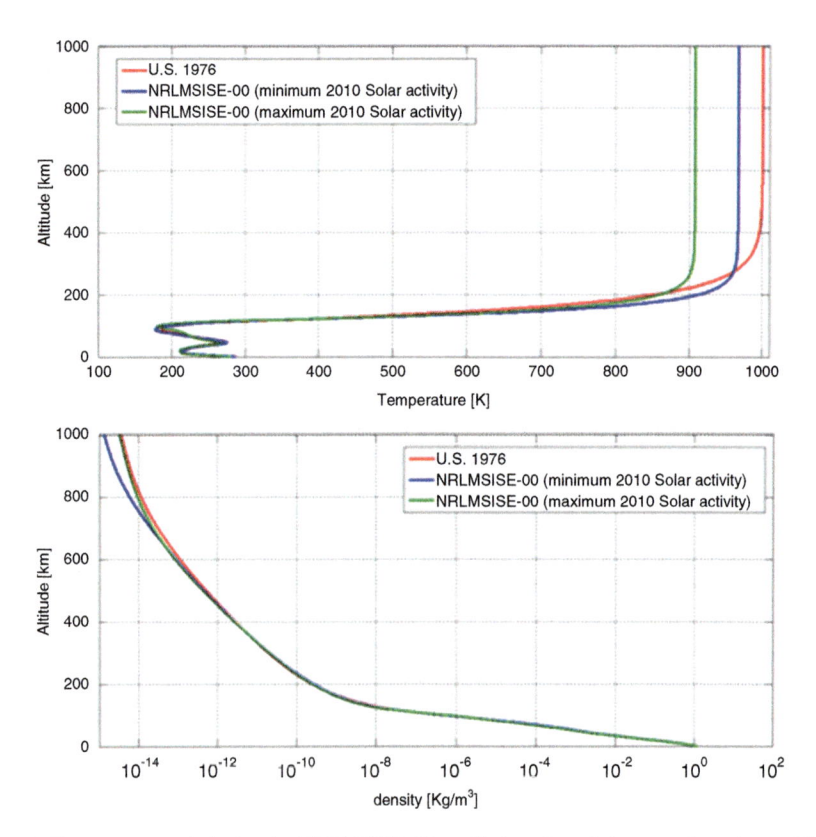

Fig. 3 Temperature and air density NRLMSISE-00 predictions for minimum and maximum 2010 solar activity compared with U.S 1976 data

been necessary to consider the more realistic NRLMISE-00 atmosphere-scenario for studying aerodynamics of re-entry vehicles, especially as concerns satellite re-entry studies.

2.2 Pre-processing graphic phase

The introduction of a new pre-processing graphic phase, based on two different rendering techniques: the back-face culling and occlusion culling. The second one has been implemented according to the approaches proposed by Wuilbercq [5] and Mehta (*Pixelator* [6]). The additional back-face culling step resulted in a quicker computation and more accurate aerodynamic estimation when compared to the original FOSTRAD. The pre-processing phase identifies the triangles that are directly facing the flow, and the ones that are shadowed by the body shape (example shown in Fig. 4). The developed algorithm follows the steps below:

1. The spacecraft geometry is modeled by using a stereolithographic CAD file format. The *.stl* file describes the surface geometry of the spacecraft by using a raw unstructured surface made by triangles, which are characterized by their unit normal vector, area, vertices and incenter.
2. The triangular mesh is correctly oriented in the direction of the oncoming flow, characterized by the free stream velocity vector V_∞:

$$V_\infty = [V_\infty \cdot \cos\alpha \cdot \cos\beta, -V_\infty \cdot \sin\beta, V_\infty \cdot \sin\alpha \cdot \cos\beta]$$

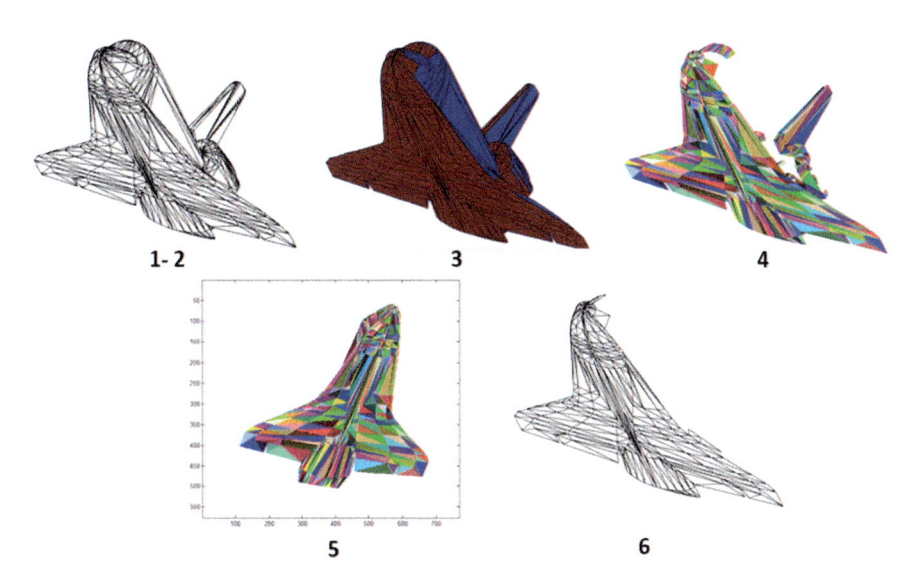

Fig. 4 FOSTRAD pre-processing phase applied on STS-Orbiter

Table 1 FOSTRAD results for pre-processing phase applied on STS Orbiter

Case	Mesh triangles	Air-wetted triangles	% time reduction
1	1936	844	51.246
2	7744	3376	55.516
3	17,424	7596	51.851
4	30,976	13,504	51.512

where α and β are respectively the attack and sideslip angles of the considered attitude.

3. The back-face culling process is computed in order to select the flow-facing triangular surfaces with the following inequality, evaluated on the i-th panel:

$$\frac{V_\infty}{||V_\infty||} \cdot \widehat{n}_i \begin{cases} < 0 \; flow-facing \\ \geq 0 \; back-facing \end{cases}$$

where $\widehat{n}_i$ is the normal vector of the local mesh panel. This step is capable of successfully identifying only the triangles on convex surfaces, while for concave geometries; this method does not properly distinguish the flow-wetted panels. It should be noted that the introduction of this step is not strictly necessary for the correct execution of the whole algorithm. However, by introducing this step, the required computational time for back-face culling algorithm has been reduced by 20–40%, depending on object attitude and mesh refinement.

4. Unique RGB colors are randomly generated and assigned to each remaining triangular facet. The use of this color model limits the pre-processing algorithm application to objects that have fewer than 16,777,216 triangular facets, which is the maximum number of unique RGB colors (256^3). A list of used colors is generated and saved in CPU memory.
5. A raster image is then generated by taking a screenshot of the colored object displayed according to the flow direction. The image is saved and stored in order to create a color list of the visible triangles.
6. The list of the complete RGB codes (point 4) and the visible facets one (point 5) are compared to find the visible facets IDs. With this simple procedure it is possible to obtain the list of triangular surfaces which are directly facing the flow and are not shadowed by other facets. The identified visible facets will be used for the object's aerodynamics computation.

The reduced number of triangles led to a 52.53% computational time reduction on the STS-orbiter simulation (averaged on four different refined meshes). Table 1 summarizes back-face culling time data on STS for the four analyzed cases.

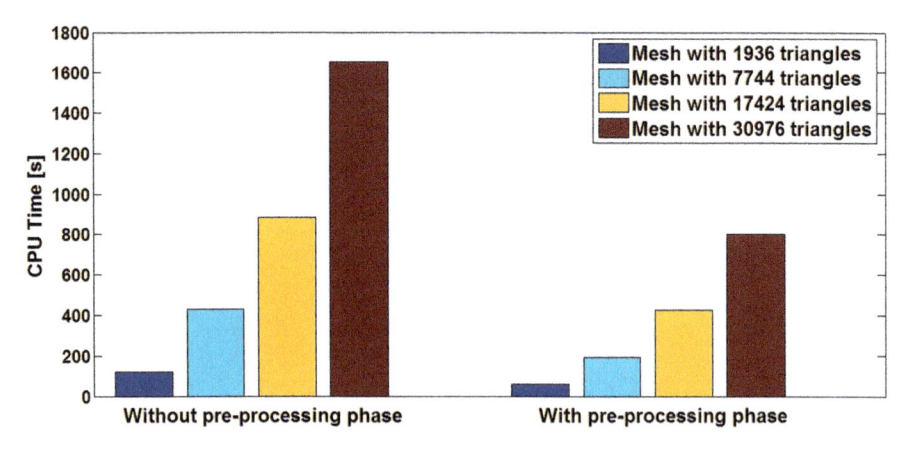

Fig. 5 FOSTRAD CPU time for the computation of STS Orbiter Aerodynamic characteristics

Figure 5 shows the computational time of *FOSTRAD* Aerodynamics module applied on STS Orbiter, at following fixed conditions:

- Altitude range: 50–220 km
- Attitude conditions: $\alpha = 20°, \beta = 20°$
- Number of iterations: 170 (1 iteration/km)

It is particularly interesting to observe the extremely low computational time required by FOSTRAD for a single step aerodynamic evaluation, which is just ∼4 s for a very fine mesh (∼30 k triangles). This is an extremely short computational time if compared to a high-fidelity simulation performed for example with a DSMC (free molecular and transitional regime) or CFD (continuum regime), which could require up to hours and days.

The other advantage obtained by introducing a graphical pre-processing phase is characterized by obtaining more precise aerodynamic results. As example, the drag coefficient estimation for Space Shuttle Orbiter, obtained with FOSTRAD and compared with flight data [10] is reported in Fig. 6. As it can be seen, a better estimation of drag coefficient is obtained by using FOSTRAD aerodynamic module with the graphical pre-processing phase.

2.3 Bridging Functions

Aerodynamic parameters (C_X) in the transitional flow region ($10^{-4} \leq \text{Kn} \leq 10$) are now computed using new bridging functions, based on 5PL weighted least squares regression method proposed by Baud [11], which is generally expressed as:

$$y = D + \frac{A - D}{\left[\left(1 + \left(\frac{x}{C}\right)^B\right)\right]^E}$$

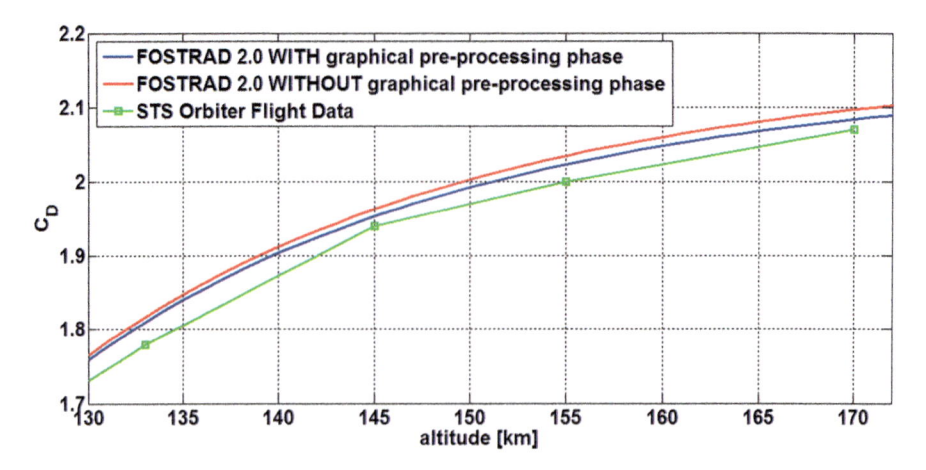

Fig. 6 STS Orbiter drag coefficient estimation versus altitude

where:

- A is the minimum asymptote.
- B is Hill's slope, which refers to the steepness of the curve. It could be either positive or negative.
- C is the inflection point, defined as the point on the curve where the curvature changes direction or signs.
- D is the maximum asymptote.
- E is the asymmetry factor

In this specific case, A and D represent a general aerodynamic coefficient in continuum and free-molecular regime respectively. For this reason, the newly implemented bridging functions are based on three fitting parameters (B, C, E), that allow a good match of the performance of different spacecraft and objects, as it will be shown in the validation section. All considered aerodynamic parameters are thus computed in this flow regime as:

$$C_{x_{trans}} = Cx_{fm} + \frac{C_{xcont} - C_{x_{fm}}}{\left[1 + \left(\frac{Kn}{C}\right)^B\right]^E}$$

where $C_{x_{trans}}$ refers to a generic aerodynamic parameter in the transitional flow region.

3 Validation of the Aerodynamic Models

The new models have been validated on different objects in order to prove FOSTRAD reliability for studying hypersonic re-entry aerodynamics.

3.1 Sphere

The evaluation of the drag coefficient for a sphere of 1.6 meter diameter has been performed via FOSTRAD and compared to DSMC data [12]. As it can be seen from Fig. 7, FOSTRAD matches satisfactory the high-fidelity literature data, with a maximum percentage error of 9.28%. Simulation conditions are reported in Table 2.

3.2 Orion Crew Exploration Vehicle

The second validation case has been performed by using the geometry shape of the Orion Crew Module, following the geometric indications of Moss et al. [13]. The lift/drag ratio variation with Knudsen number and angle of attack has been computed

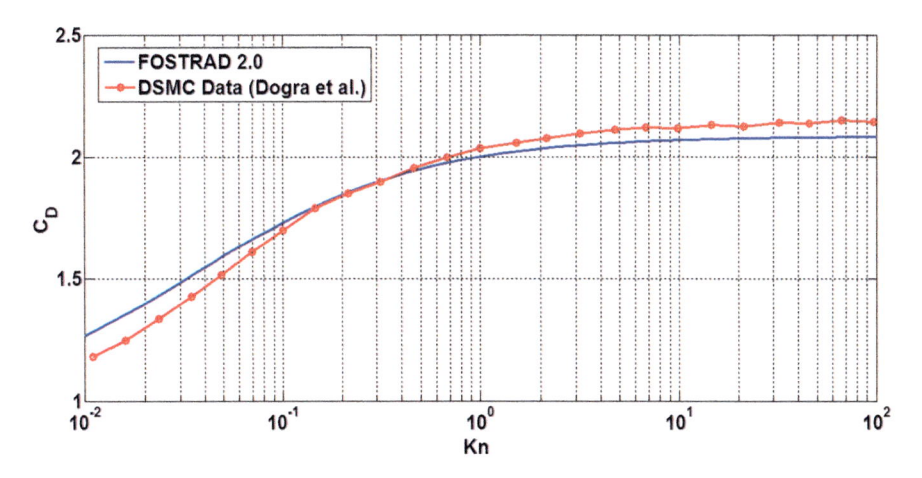

Fig. 7 1.6 meter-diameter sphere drag coefficient FOSTRAD estimation

Table 2 Simulation conditions for 1.6 m diameter sphere

Altitude range [km]	90–200
Free stream velocity [km/s]	7.5
Reference length [m]	1.6
Wall temperature [K]	350

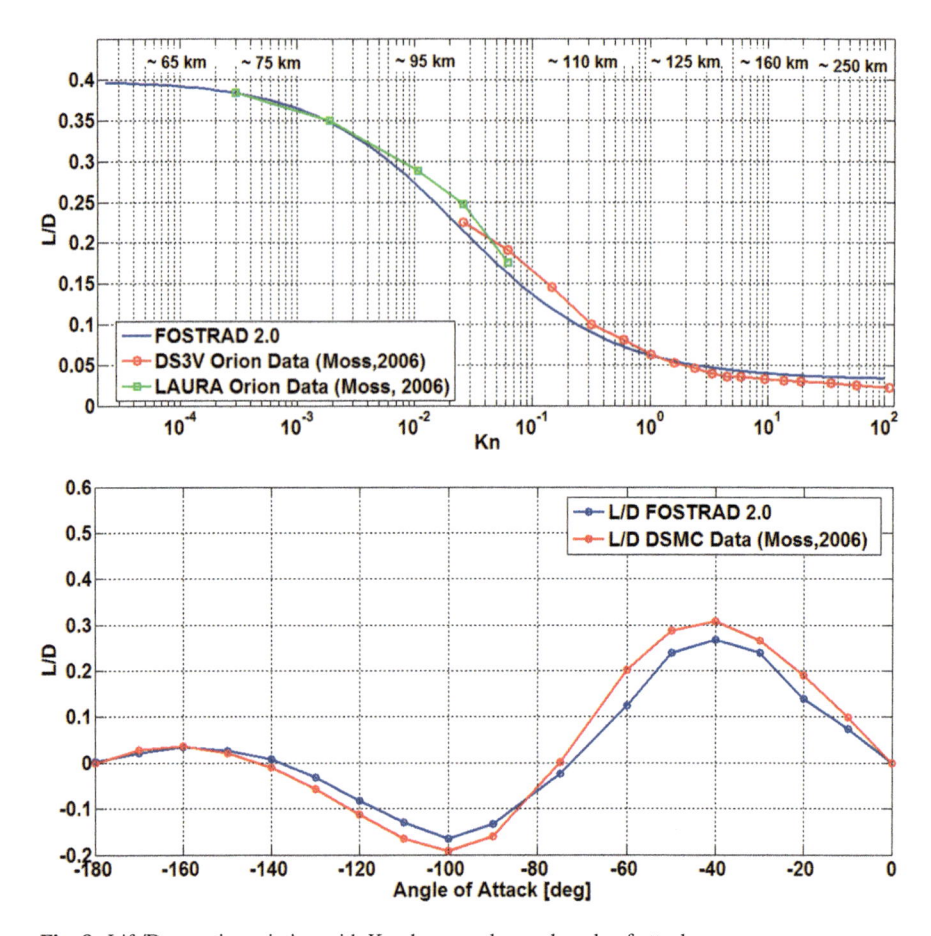

Fig. 8 Lift/Drag ratio variation with Knudsen number and angle of attack

by using FOSTRAD and compared with higher fidelity results [13], obtained with DSMC and CFD methods. The results are shown in Fig. 8.

3.3 Space Shuttle Orbiter

The validation of the aerodynamic analysis for the STS Orbiter case against the information available in literature, which have been obtained during the Space Shuttle Missions [10], or have been derived using different high-fidelity methods [14] is shown in Fig. 9.

The maximum percentage errors are, respectively 2.74%, 5.30% and 0.58% for $C_D, C_L, L/D$. The flow conditions are reported in Table 3.

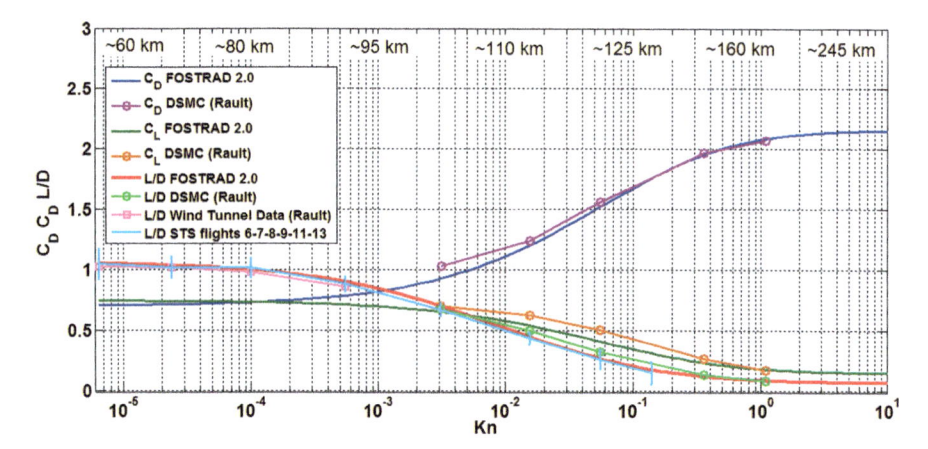

Fig. 9 Space Shuttle aerodynamic parameters comparison among FOSTRAD, DSMC simulations, Wind tunnel and Flight data

Table 3 Simulation conditions for STS orbiter

Altitude range [km]	60–170
Free stream velocity [km/s]	7.5
Reference length [m]	12.06
Nose radius [m]	0.719
Wall temperature [K]	300

3.4 Intermediate Experimental Vehicle (IXV)

The last validation case of FOSTRAD aerodynamic model has been done on the Intermediate Experimental Vehicle, the ESA suborbital re-entry prototype, developed by Thales Alenia Space. The validation of FOSTRAD in comparison with aerodynamic data provided by Pezzella et al. [15], has been performed by using the IXV stereolithographic CAD file (Fig. 10, Table 4).

The implemented changes allow a computation of aerodynamic parameters with a relative error ≤11% for the considered cases. This percentage error, typical of a low-fidelity tool, is due to the simplified models that do not take into account phenomena described by Anderson [16], such as viscous and high-temperature effects, gas-surface interactions, chemical reactions and solar pressure effects.

4 Application Cases

A selection of application cases has been reported, highlighting FOSTRAD capabilities on different shaped objects and scenarios.

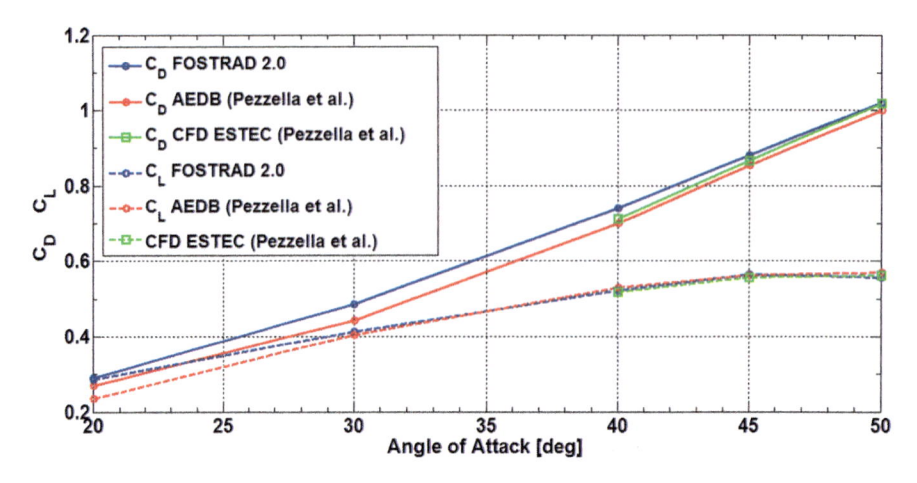

Fig. 10 IXV drag and lift coefficients evaluation with FOSTRAD, AEDB and CFD data

Table 4 Simulation conditions for IXV

Altitude [km]	50
Free stream velocity [km/s]	3.85
Reference length [m]	4.4
Nose radius [m]	1.329
Wall temperature [K]	300

Table 5 CFASTT-1 aerodynamic coefficients computation with FOSTRAD

AoA (deg)	h (km)	V (km/s)	C_D (–)	C_L (–)	C_M (–)	L/D (–)
74	120	7.86	3644	0.376	−0.433	0.103
74	105	7.86	2141	0.449	−0.347	0.209
74	91.4	7.81	1661	0.472	−0.320	0.284
74	82.6	7.34	1588	0.476	−0.315	0.299
74	80	6.46	1574	0.476	−0.315	0.302
74	74.3	5.39	1556	0.477	−0.314	0.306
74	65.1	3.21	1543	0.476	−0.313	0.308

4.1 CFASTT-1

FOSTRAD has been used to compute the CFASTT-1 aerodynamic characteristics of its hypersonic re-entry. The CFASTT-1 is a spaceplane project under development at the University of Strathclyde. The performed simulations are based on the vehicle expected atmospheric re-entry data [5]; the obtained results are presented in Table 5 and showed in Fig. 11. A stereolithographic CAD file, composed by 18,408 facets, has been used for the aerodynamic analysis.

The cases reported in Table 5 are all characterized by hypersonic flight conditions (shown in Fig. 11 with solid blue lines). Below 60 km, the flight conditions are not hypersonic anymore; as it can be clearly seen from Fig. 11 (dotted red line), the

Low-Fidelity Modelling for Aerodynamic Characteristics of Re-Entry Objects 259

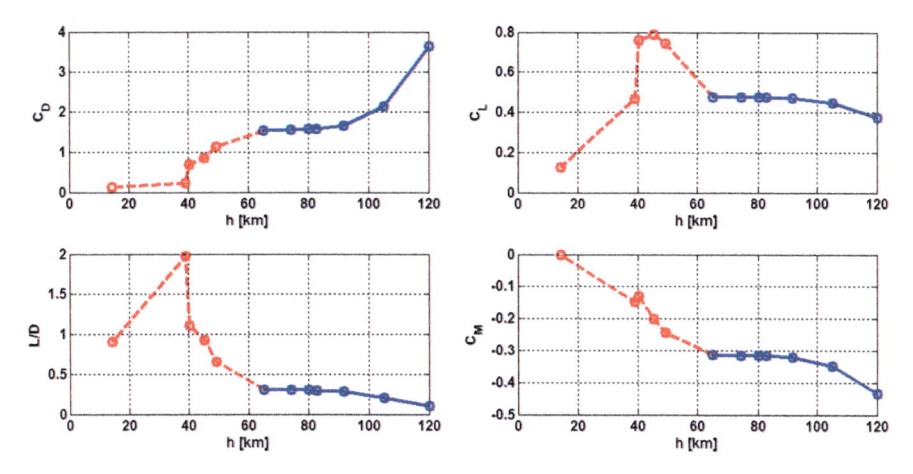

Fig. 11 CFASTT-1 aerodynamic parameters evaluation with FOSTRAD for its re-entry phase

software is capable of providing realistic trends for aerodynamic coefficients only for hypersonic flight conditions. For trans-supersonic cases the software shall not be used for aerodynamics estimation, because the modified Newtonian theory, which has been used for the C_P evaluation in continuum regime, can be applied only for hypersonic flow regime ($M \geq 5$).

4.2 Gravity Field and Steady-State Ocean Circulation Explorer

The Gravity Field and Steady-State Ocean Circulation (GOCE) was an ESA's satellite, manufactured by *Thales Alenia Space* and *EADS Astrium* [17]. On the 21st of October 2013, GOCE began its re-entry phase until its controlled re-entry on the 11th of November 2013. During GOCE de-orbiting and re-entry, all instruments maintained their functionality until ∼130 km, making this dataset extremely interesting for its aerodynamic and aerothermodynamic re-entry analysis. The performed aerodynamic analysis has been based on the data related to the attitude, altitude and their variations during the satellite end-of-life phase [18]. The attitude conditions are represented by variations in terms of pitch and sideslip angle, as shown in Fig. 12. Three different values have been considered for every re-entry day:

- Minimum daily value
- Maximum daily value
- Average daily value

Particular attitude instability due to the magnetic torquer saturation had been registered in the days 294, 303, 305, 312 and 314. During the last days of the re-entry phase, the attitude was not stable as it was during the drag-free flight;

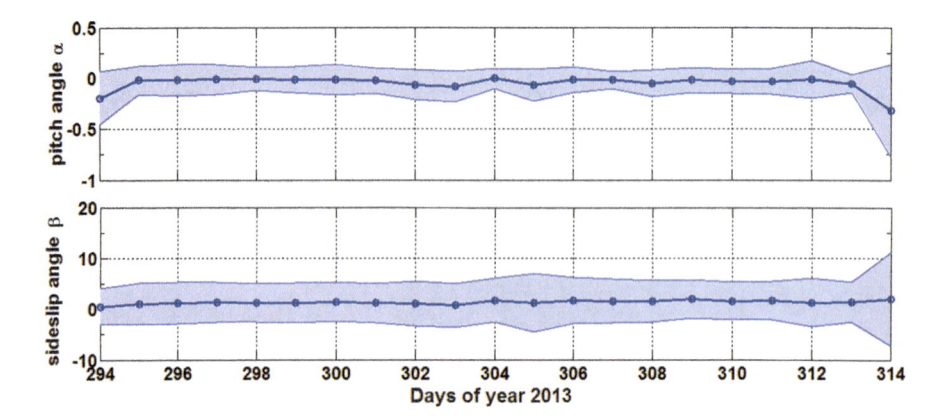

Fig. 12 GOCE re-entry attitude conditions

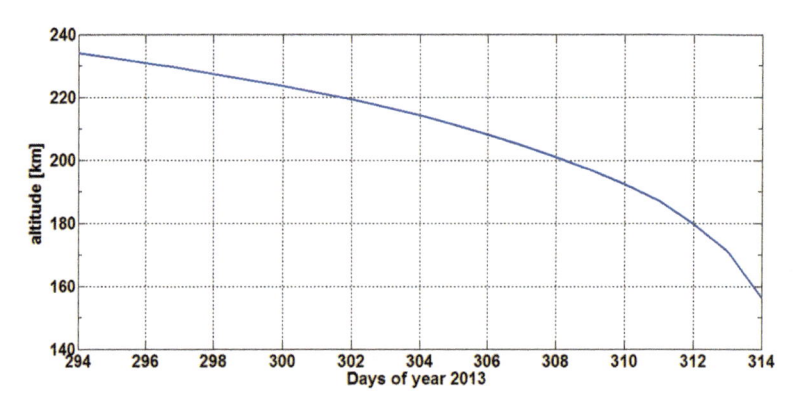

Fig. 13 GOCE re-entry altitude variations

indeed, the thrusters were turned off on the 21st of October 2013. Concerning the altitude variation, at start of re-entry time, *GOCE* had the perigee at 215 km and the apogee at 233 km [19] progressively decreasing until its re-entry, which ended with a splashdown in the south-eastern side of South America (Fig. 13).

The analyses have been performed with a simplified geometric model composed of 1772 facets (Fig. 14). The aerodynamic coefficients have been evaluated by using *FOSTRAD* using the presented attitude and altitude variations (Fig. 12). Three different curves have been computed for each different coefficient (drag and lift), according to the minimum, maximum and average daily values of α and β angles; the results are shown in Figs. 15 and 16.

The computed drag coefficients have been compared to literature data obtained with ESA's SCARAB and NAPEOS LSI tools (Fig. 15). It can be observed that the C_D dramatically increases during the last day before the destruction; this is more likely caused by the steep increase of the sideslip angle.

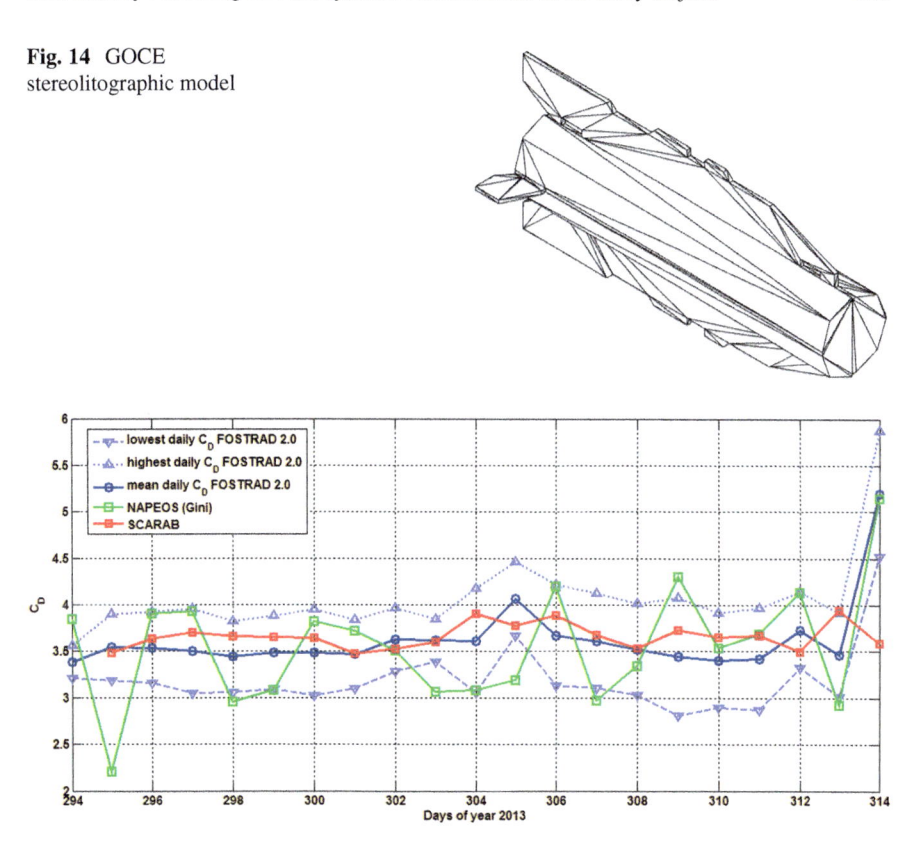

Low-Fidelity Modelling for Aerodynamic Characteristics of Re-Entry Objects 261

Fig. 14 GOCE stereolitographic model

Fig. 15 GOCE Drag Coefficient estimation during the re-entry phase

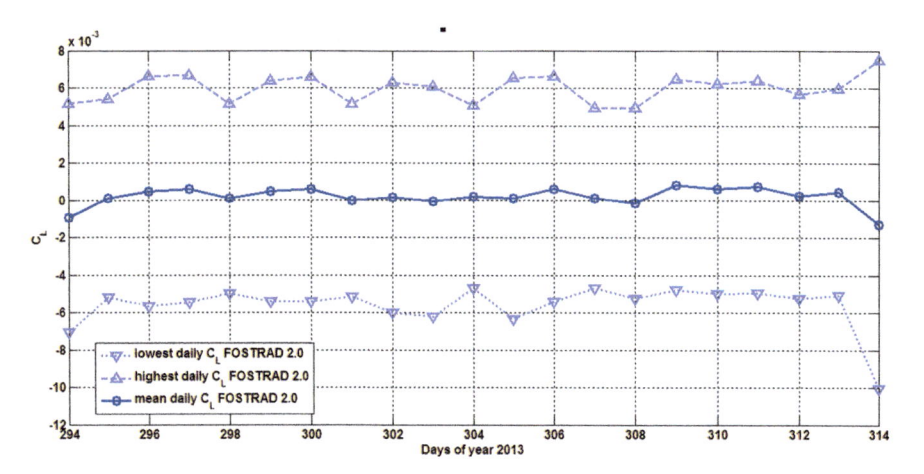

Fig. 16 GOCE Lift Coefficient estimation during the re-entry phase

Table 6 GOCE analysis conditions for CD- sideslip angle variation

AoA range (deg)	H (km)	V_∞ (km/s)	T_{wall} (K)	Ref. area (m^2)
[−10; 10]	260	7.76	350	1.13

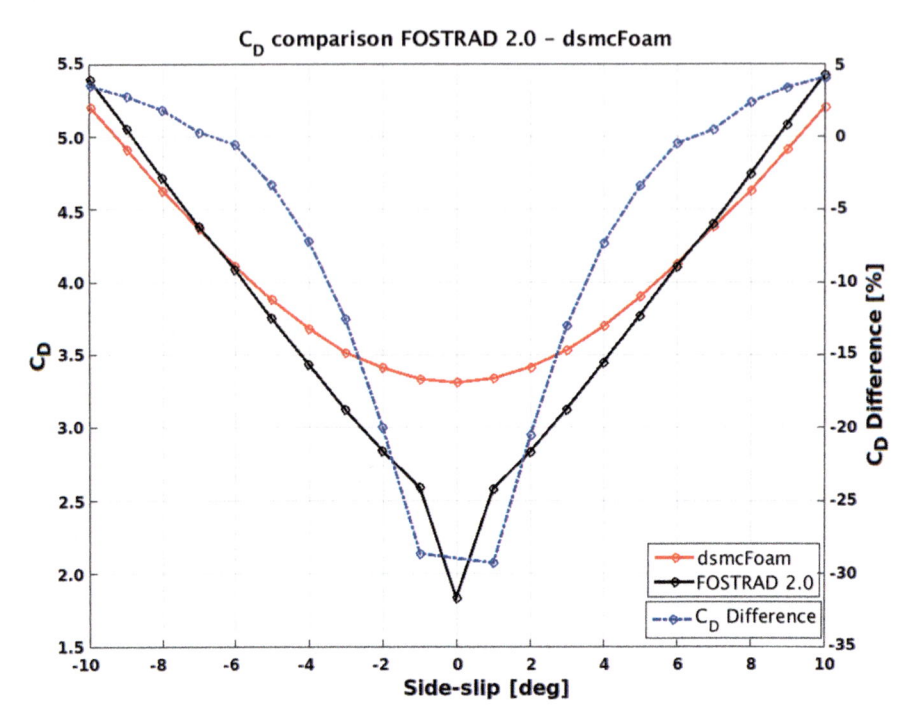

Fig. 17 GOCE Drag coefficient variation versus sideslip angle and altitude

The lift coefficient (Fig. 16) is characterized by very low values; this fact can be easily explained by considering the non-lifting shape of GOCE.

As shown in Fig. 12, the main variations for GOCE's attitude have been registered on the sideslip angle. To better understand the sideslip influence on the drag coefficient estimation, the $C_D - \beta$ curve has been analyzed at the conditions of Table 6. The results have been compared to a DSMC analysis performed at the same conditions with the same 3D model; the high-fidelity simulations have been obtained with the *dsmcFoam* [20] code.

As expected, the C_D rapidly increases when the sideslip angle raise. According to the simulations results shown in Fig. 17, the LSI (Local Surface Inclination) approach seems to underestimate the drag coefficient. The low-fidelity method error decreases as the exposed surface increases, i.e.: the attitude changes from the head-on flow condition. The relative difference goes from approximately 5% up to −30%. It must be highlighted that the difference with respect to the DSMC analysis rapidly falls in the range of ±10% and ±5% as the attitude moves away from ±3° and ±5° respectively.

Thanks to additional tests, it has been observed that simplifying the geometry by removing the big later fins from GOCE geometry greatly reduces the differences. This is due to the fact that the DSMC takes into account the viscosity forces acting on the lateral surfaces. On the other hand, FOSTRAD does not take into account the viscosity forces on the facets parallel to the flow. Thus, objects (or parts) having a high lateral to front surface ratio could cause a major deviation for attitudes nearby the head-on flow aerodynamic condition.

5 Conclusions

A low fidelity code for aerodynamics and aerothermodynamics analysis has been further developed; the work has focused on its aerodynamic module, which has improved with the aim of providing reliable prediction of hypersonic aerodynamics. The recently introduced innovations have been presented in this paper.

Among all the improvements, the introduction of a new pre-processing graphic phase, also known as back-face culling algorithm, resulted in a quicker computation and more accurate aerodynamic estimation when compared to the original code [6]. The aerodynamic simulation is performed only on the panels directly facing the flow and not shadowed by other facets. The reduced number of stereolithographic mesh triangles leads to a significant computational time reduction on every investigated object simulations. Moreover, the introduced pre-processing phase has led to a more accurate prediction on aerodynamics.

Aerodynamic parameters in the transitional flow region have been computed using new bridging functions, based on 5PL weighted least squares regression using the Baud method. The new tool has been validated on different object shapes, and has shown a good agreement with high fidelity codes or experimental data.

Furthermore, the improved code has been used to predict the re-entry aerodynamic characteristics of two re-entering objects: GOCE and the CFASTT-1. The tests have highlighted the software applicability on hypersonic re-entry study for two completely different shapes: GOCE, which is characterized by a sharp-edged cylindrical body with a high lateral surface and the CFASTT-1 geometry which is mostly hemispherical.

In light of the recent studies and results, the considered code (implemented models, and general architecture) is undergoing a continuous development aimed at increasing its accuracy and flexibility on both the aerodynamics and aerothermodynamics modules.

Concerning the graphic pre-processing phase, a further possible development on the back-face culling process may regard the development of an algorithm capable of considering only the effective wetted part of a partial shadowed facet instead of considering its entire surface. Such a development would further increase the tool aerodynamic accuracy.

Acknowledgments The authors would like to thank Roberto Destefanis, Lilith Grassi and Simone Bianchi of *Thales Alenia Space Italy*, who kindly provided the IXV's stereolitographic model, which was used during the FOSTRAD validation tests.

References

1. Dobarco-Otero J., Smith N., Bledsoe K. J., Delaune R. M., Rochelle W. C. and Johnson N. L.: The Object Reentry Survival Analysis Tool (ORSAT)–Version 6.0 and its Application to Spacecraft Entry, American Institute of Aeronautics and Astronautics (2005)
2. Lips, T., Fritsche, B.: A comparison of commonly used re-entry analysis tools. Acta Astronaut. **57**, 312–323 (2005)
3. Koppenwallner, G., et al. SCARAB–a multi-disciplinary code for destruction analysis of spacecraft during re-entry. Proceedings of the 5th European symposium on aerothermodynamics for space vehicles, Cologne, Germany (2005)
4. Mehta, M., et al.: An open-source hypersonic aerodynamic and aerothermodynamic modeling tool. 8th European symposium of aerothermodynamics of space vehicles (2015)
5. Wuilbercq, R.: Multi-disciplinary modeling of future space-access vehicles. s.l. PhD dissertation, University of Strathclyde, UK (2015)
6. Mehta, P., et al.: Computer graphics for space debris. 6th international conference on astrodynamics tools and techniques (ICATT). Darmstadt: s.n. (2016)
7. Picone, J.M., et al.: NRLMSISE-00 empirical model of the atmosphere: statistical comparison and scientific issues. J. Geophys. Res. **107**, 1468 (2002)
8. MathWorks. Aerospace toolbox (2016). https://it.mathworks.com/help/aerotbx/ug/atmosnrlmsise00.html#responsive_offcanvas
9. Aikin, A.C., et al.: Thermospheric molecular oxygen measurements using the ultraviolet spectrometer on the solar maximum mission spacecraft. J. Geophys. Res. **98**(A10), 17607–17613 (1993)
10. Rault, D.: Aerodynamics of the shuttle orbiter at high altitudes. J. Spacecr. Rocket. **31**, 944–952 (1994)
11. Baud, M.: Data analysis, mathematical modeling. Methods Immunol. Anal. **1**, 656–671 (1993)
12. Dogra, V.K., Wilmoth, R.G., Moss, J.N.: Aerothermodynamics of a 1.6-meter-diameter sphere in hypersonic rarefied flow. AIAA J. **30**(7), 1789–1794 (1992)
13. Moss, J., Katie, B., Greene, F.: Orion aerodynamics for hypersonic free molecular to continuum conditions. 14th AIAA/AHI international space planes and hypesonic systems and technologies conference, Canberra, Australia (2006)
14. Moss, J., Bird, G.: Direct simulation of transitional flow for hypersonic reentry conditions. J. Spacecr. Rocket. **40**, 338–360 (2003)
15. Pezzella, G., Marino, G., Rufolo, G.C.: Aerodynamic database development of the ESA intermediate experimental vehicle. Acta Astronaut. **94**, 57–72 (2014)
16. Anderson, J.D.: Hypersonic and high temperature gas dynamics, 2nd edn. AIAA, Reston, VA (2006)
17. Agency, European Space. GOCE. [Online] http://www.esa.int/Our_Activities/Observing_the_Earth/GOCE
18. Gini, F., et al.: Precise orbit determination of the GOCE re-entry phase. 5th International GOCE user workshop, Paris, France (2014)
19. Virgili, B.B., et al.: GOCE re-entry campaign. 5th international GOCE user workshop, Paris, France (2014)
20. Scanlon, T.J., Roohi, E., White, C., Darbandi, M., Reese, J.M.: An open source, parallel DSMC code for rarefied gas flows in arbitrary geometries. Comput. Fluids. **39**, 2078–2089 (2010)

Re-entry Predictions of Potentially Dangerous Uncontrolled Satellites: Challenges and Civil Protection Applications

Carmen Pardini and Luciano Anselmo

Abstract Currently, nearly 70% of the re-entries of intact orbital objects are uncontrolled, corresponding to about 50% of the returning mass, i.e. approximately 100 metric tons per year. In 2015, 79% of the mass was concentrated in 40 upper stages and the remaining 21% mostly in about ten large spacecraft. The average mass of the sizable objects was around 2 metric tons. Predicting the re-entry time and location of an uncontrolled object remains a very tricky task, being affected by various sources of inevitable uncertainty. In spite of decades of efforts, mean relative errors of 20–30% often occur. This means that even predictions issued 3 h before re-entry may be affected by an along-track uncertainty of 40,000 km (corresponding to one full orbital path), possibly halved during the last hour if further tracking data is available. This kind of information is not much useful and manageable for civil protection applications, often resulting in confusion and misunderstandings regarding its precise meaning and relevance. Therefore, specific approaches and procedures were developed to provide understandable and unambiguous information useful for civil protection planning and applications, as shown in practice for recent re-entry prediction campaigns of significant satellites (UARS, ROSAT, Phobos-Grunt, GOCE, and Progress-M 27 M).

1 Introduction

As of mid-January 2017, and since the decay of the Sputnik 1 launch vehicle core stage on 1 December 1957, more than 24,000 cataloged orbiting objects have re-entered into the Earth's atmosphere, with a total mass of about 30,000 metric tons. Of these, approximately 71% were orbital debris, while the remaining 29% was represented by intact objects, where most of the mass (>99%) was concentrated. Currently, approximately 70% of the re-entries of intact orbital objects

---

C. Pardini (✉) · L. Anselmo
Space Flight Dynamics Laboratory, ISTI-CNR, Pisa, Italy
e-mail: carmen.pardini@isti.cnr.it; luciano.anselmo@isti.cnr.it

are uncontrolled, corresponding to about 50% of the returning mass, i.e. ~100 metric tons per year. On average, there is one spacecraft or rocket body uncontrolled re-entry every week, with an average mass around 2000 kg [1].

Detailed computer simulations and the analysis of retrieved spacecraft and rocket body components have suggested that, also in the case of objects not specifically designed to survive the severe mechanical and thermal loads, 5–40% of the mass of sufficiently massive bodies is able to reach the Earth surface [2, 3, 4]. In terms of mass, number and component survivability, the uncontrolled re-entries of spent upper stages generally present a higher risk on the ground compared to spacecraft and, apart from uncommon accidental cases, as the tragic loss of the Columbia Space Shuttle orbiter (2003), or the demise of Skylab (1979), the bulk of the re-entry fragments recovered so far on the ground comes from rocket bodies.

No case of personal injury caused by re-entering orbital debris has yet been confirmed. Nonetheless, uncontrolled re-entries of sizable space objects are becoming of growing concern due to the increase of space activities around the Earth and population on the ground. The ground casualty risk, even if still small compared to other commonly accepted risks linked to the lifestyle or the workplace and household safety, will presumably show a tendency to grow in the coming years.

Therefore, specific guidelines to minimize the risk to human life and property on the ground have been defined. For instance, single re-entries compliant with the NASA standard 8719.14 must have a world-wide human casualty risk not exceeding 0.0001. In other words, the chance for anybody anywhere in the world of being injured by a piece of falling debris from a single uncontrolled re-entering object must be lower than 1:10,000 [5]. Such alert threshold is now adopted by several organizations and countries around the world, even though only for a relatively small number of spacecraft and upper stages detailed breakup studies are being carried out, or disclosed to the public, in order to estimate their casualty expectancy [2, 6].

Hence, every week or two, on average, an uncontrolled re-entry violating the alert human casualty risk threshold of 1:10,000 probably occurs, unknown to most of the governments and safety authorities around the world.

2 Re-entry Statistics

Since 1957, re-entered on average in the atmosphere 54 payloads, 63 upper stages and 272 debris per year, i.e. 2–3 intact objects per week. During the last decade (2007–2016), re-entered on average 45 payloads, 40 rocket bodies and 354 debris per year, i.e. 1–2 intact objects per week (Fig. 1).

Considering, for instance, the uncontrolled re-entries of intact objects occurred in 2015, 62% (64) were payloads and 38% (40) were upper stages. However, while the latest accounted for nearly 79% of the mass (i.e. 82 metric tons), the payloads, consisting of small satellites with a mass below 50 kg in 83% of the cases, contributed with the remaining 21% (i.e. 22 metric tons) to the uncontrolled re-entered mass [7].

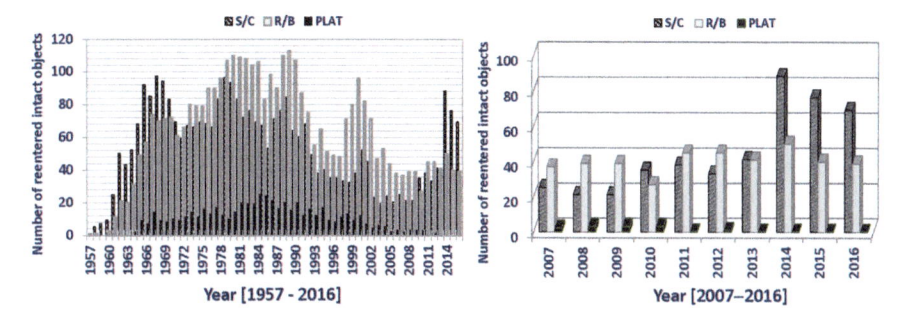

Fig. 1 Intact space objects (S/C: spacecraft; R/B: rocket bodies; PLAT: platforms) re-entered since 1957 (left) and from 2007 to 2016 (right)

The decay rate of intact objects was mainly driven by the launch activity, with a lower, but not trivial, contribution linked to the solar activity cycle and the corresponding change in the magnitude of the drag perturbation. A much more strong correlation with the launch activity was highlighted in [1], by excluding from the tally all spacecraft associated with human spaceflight, i.e. manned spaceships and capsules, space stations, man-tended modules and cargo vehicles. The top decay rates, observed between the mid-1960s and the end of the 1980s, were followed by a declining trend, at the beginning of the 1990s, in consequence of the breakup of the Soviet Union. The last decade was instead characterized by an increase in the number of commercial launches, also from emerging countries, typically consisting of lightweight payloads.

A systematic decreasing trend in the decay rate of intact objects was observed during the last four decades. Fig. 2, based on the re-entry statistics carried out in [1], between December 1957 and April 2013, shows the yearly (in the top) and the weekly (in the bottom) decay rate of intact objects over the last 5, 10, 20, 30, 40, 50 and 55 years with respect to the end of 2012, by excluding all spacecraft associated with human spaceflight (typically performing controlled re-entries) from the tally. The decay rate passed from a maximum of about 130 re-entries per year, corresponding to nearly 2–3 re-entries per week, over 50 years [1963–2012], to a minimum of approximately 57 re-entries per year, i.e. just a re-entry per week, during the 2008–2012 five years period.

Such a progressive decline of the average decay rate of intact objects might most likely reflect changes in the number of launches, as well as in the mission profiles, for which a considerable number of payloads and upper stages would have been placed in high Earth orbits. Moreover, a notably long and deep minimum characterizing the solar activity cycle 24 (the smallest sunspot cycle in over a century) undoubtedly could have played a significant role.

While the decay rate of all cataloged intact objects can be directly inferred from the information included in the US Strategic Command (USSTRATCOM) catalog, made available to registered users through the Space-Track Organization (www.space-track.org), the actual number (and mass) of spacecraft and upper

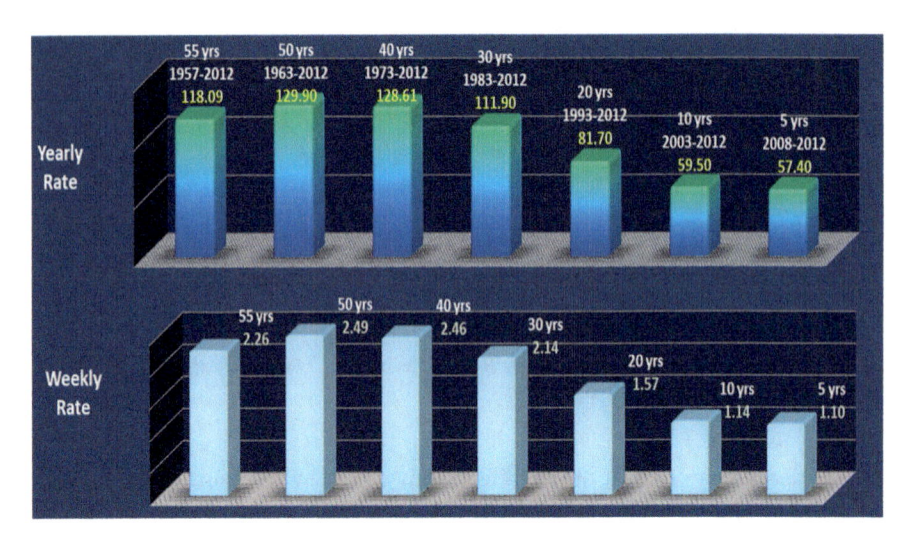

Fig. 2 Average decay rate (yearly in the top, weekly in the bottom) of intact objects, excluding the spacecraft associated with human spaceflight programs (statistics based on re-entries in between 1957 and 2013 [1])

stages re-entered so far into the Earth's atmosphere without control cannot be easily attained. As a matter of fact, in addition to vehicles associated with manned space programs and controlled re-entries carried out for safety reasons, there have been several classified military reconnaissance satellites, which have systematically performed a controlled re-entry to recover and/or to prevent the retrieval of secret spacecraft components. Moreover, also for a small number of upper stages, a controlled re-entry has been accomplished during the last few years.

Notwithstanding, thanks to a detailed and massive research based on the information available in the open literature, it was possible to identify, with an adequate level of confidence, all spacecraft performing controlled re-entries [1]. It was found that during the Space Age, on average, one out of five re-entries of intact objects was controlled in some way. The total mass associated with the uncontrolled re-entered objects was assessed to be around 11,000 metric tons, mainly concentrated (>98%) in almost 5700 intact objects, as of April 2013. Currently, more than 11,300 metric tons of man-made materials are expected to have re-entered into the Earth's atmosphere without control. On average, there is one sizable spacecraft or rocket body uncontrolled re-entry every week, with an average mass around 2000 kg.

An uncontrolled satellite can re-enter anywhere on a large portion of the Earth surface, putting all locations within the latitude band defined by the orbit inclination into the area potentially affected by the surviving debris fall. The knowledge of the approximate re-entry points for a representative sample of decayed objects might be of value to investigate the potential impact of some factors related to the launch pattern, the mission profile and the lifetime in orbit on the distribution of re-entries over the Earth.

A distribution of the re-entry locations was obtained [7] by analyzing the uncontrolled re-entries of intact objects occurred from 2004 to 2015 and for which a USSTRATCOM post-re-entry time assessment with a claimed error of ±1 min was available (for 287 rocket bodies and 59 payloads/platforms). For this sample of objects, and in the interval of time considered, it was found a slight prevalence (53.5% vs. 46.5%) of re-entries in the Southern Hemisphere. Moreover, the Southern Hemisphere bias was smaller for upper stages (51.6% vs. 48.4%) and higher for payloads (62.7% vs. 37.3%).

In a previous study carried out by Nicholas Johnson in 1997 [8], by analyzing the uncontrolled re-entries of 331 objects occurred from September 1992 to December 1996, it was instead found a prevalence of re-entries in the Northern Hemisphere (56.5% vs. 43.5%). Moreover, while objects staying in orbit less than 1 month showed a more marked bias (62% North vs. 38% South), those with a longer orbital lifetime displayed a nearly symmetrical distribution (51% North vs. 49% South).

From the results of both analyses, it can be reasonably concluded that the varying north-south asymmetry observed in the two cases was mainly driven by the different launch pattern, mission profile and residual lifetime of the objects put in orbit in different historical periods.

Considering all the uncontrolled re-entries of intact objects (i.e. not only those characterized by a post-event assessment claimed error of ±1 min) occurred from 2004 to 2015 (a total of 722 uncontrolled re-entries, including 447 rocket bodies and 275 payloads/platforms), it was found that the mean flux of decaying intact objects over the Earth was approximately 1.05×10^{-7} km^{-2} per year [7]. This would imply, on average, an uncontrolled re-entry over Italy every 28 years and over Europe every 10 months.

3 Re-entry Risk Evaluation

If an average surviving fraction of 15–20% is applied to the total amount of mass suspected to have re-entered the Earth's atmosphere without control so far (i.e. about 11,300 kg), around 1695–2260 metric tons of manmade debris would have likely survived re-entry and hit the ground, with no case of personal injury confirmed heretofore.

Nonetheless, due to an expanding use of space and to a consequent rise in the amount of space hardware, the number of uncontrolled re-entries is doomed to remain significant in the foreseeable future. Moreover, if the concurrent increase of the population is taken into account, the ground casualty risk, even if still relatively small, will probably raise in the coming years.

This is the reason why specific guidelines to minimize the risk to human life and property on the ground have been defined and are now adopted by several organizations and countries around the world. The case of NASA has been already mentioned in the introduction [5], but also for the European Space Agency the human casualty risk should not exceed 1 in 10,000 for any re-entry event, either controlled or uncontrolled (ECSS-U-AS-10C/ISO 24113) [9, 10].

The main factors affecting the estimation of the risk of human casualty from uncontrolled re-entries include the number of debris expected to reach the surface of the Earth, the kinetic energy of each surviving fragment and the amount of the world population potentially at risk. A kinetic energy threshold of 15 J is typically accepted as the minimum level for potential injury to an unprotected person [5], while a probability of fatality of 50% corresponds to a kinetic energy of 103 J [11].

A crucial metric used by NASA [5] to represent and to evaluate the potential risk from re-entering debris is the so-called total debris casualty area (A_C), which for a re-entry event is the sum of the casualty areas of all debris pieces able to survive re-entry. It is computed as follows:

$$A_C = \sum_{i=1}^{n} \left(\sqrt{A_h} + \sqrt{A_i} \right)^2 \qquad (1)$$

where $A_h = 0.36$ m^2 is the projected cross-sectional area of a standing human and A_i is the cross-section of each individual fragment reaching the ground. A_C is de facto a simple and effective method to combine in a single figure all information on the breakup process of a re-entering space object.

The casualty area (Eq. 1) and the impact location of the surviving fragments for a re-entry event are usually computed by means of specific re-entry analysis tools, which can be grouped in two main families, named the object-oriented tools, e.g. the NASA's Debris Assessment Software (DAS) and the Object Re-entry Survival Analysis Tool ORSAT (orbitaldebris.jsc.nasa.gov/re-entry/orsat.html), and the spacecraft-oriented tools, like the ESA's SpaceCraft Atmospheric Re-entry and Aerothermal Break-up software tool SCARAB (www.htg-hst.de/1/htg-gmbh/software/scarab).

The total human casualty expectation, better known as the casualty expectancy (E_C), is obtained as the product of the total debris casualty area and the total average population density (P_D) in the area overflown by the re-entering object, i.e.: $E_C = A_C \times P_D$. For instance, for mid-inclination orbits, it can be shown that a world-wide casualty expectancy of 1:10,000 can be currently exceeded in a single uncontrolled re-entry event if the total casualty area of the surviving debris is greater than approximately 8 m^2.

The re-entry casualty risk can be determined through the probability to cause serious injury or death. For a re-entry event with surviving fragments, and inside a given latitude belt, the probability of debris fall is one, but the expected consequences, at least for people in the open, are not particularly adverse with respect to the common risks accepted in the everyday life. As an example, the risk of being hit by falling orbital debris amounts to about one part per trillion per human per lifetime, i.e. it is of the order of 10^{-12} [12]. Instead, the risks in our daily life are comparatively huge: the risk of being killed in a car accident amounts to about 1/100 in industrialized countries, of death by fire is about 1/1000, of being hit by a lightning is approximately 1/1,500,000 [12].

4 Re-entry Prediction Uncertainty

After nearly six decades of space activity, predicting the re-entry time and location of an uncontrolled satellite remains a very tricky task. There is considerable uncertainty in the estimation of the re-entry epoch due to sometimes sparse and inaccurate tracking data, complicate shape and unknown attitude evolution of the re-entering object, biases and stochastic inaccuracies affecting the computation of the atmospheric density at the altitudes of interest, magnitude, variability and prediction errors of solar and geomagnetic activity, and mismodeling of gas-surface interactions and drag coefficient.

All these uncertainty sources combine in a complex way, depending on the specific properties of the re-entering object considered and on the particular space environment conditions experienced during the final phase of the orbital decay. Therefore, even applying the same (best) models, methods and procedures, the overall relative re-entry prediction errors may be quite different for various objects and in diverse epochs.

The experience accumulated worldwide shows that a relative prediction error of ±20% should be adopted to compute the uncertainty windows associated with nominal re-entry epoch predictions, in order to reasonably cover all possible error sources. However, in specific cases, more conservative prediction errors, up to ±30%, should be considered, in particular during the last 2–3 days of residual lifetime. In support of this, it was found that for the recent USSTRATCOM last re-entry predictions before decay, from December 2014 to January 2016, a re-entry time uncertainty of ±30% was able to include nearly 90% of the events [7]. Moreover, these predictions were based on orbit data with a mean epoch at nearly 5 h before decay (only in 2% of the cases the last available orbit was at less than 1 h before re-entry, in 20% of the cases at less than 2 h, in 33% of the cases at less than 3 h).

Therefore, if an uncertainty of ±30% is applied, for instance, to a residual lifetime of 5 h, the re-entry time prediction error is ±1.5 h, corresponding to a couple of orbit tracks around the Earth for a re-entering satellite in near-circular orbit. Anyway, also when the flux of orbit determinations is steady and optimal, there is an unavoidable processing and communication delay of at least 2–3 h between the orbit determination epoch and the release of the corresponding re-entry prediction, so the final forecasts issued during the last hour or minutes preceding the actual re-entry are based on a state vector with a 2–3 h old epoch. Therefore, even predictions issued around 3 h before re-entry have a typical along-track uncertainty of approximately one orbit (i.e. ∼40,000 km), while those issued immediately before re-entry present a typical along-track uncertainty of half an orbit (i.e. ∼20,000 km).

As a consequence, for uncontrolled re-entries driven by thermospheric drag, it is not possible to predict a re-entry location, which remains quite undetermined until the end, along the satellite trajectory, but it is only possible to identify the areas of the planet where the re-entry may no longer occur, with a given confidence level.

5 Re-entry Prediction Process

Following the accidental re-entry of the nuclear-powered satellite Cosmos 954, in 1978, independent re-entry prediction capabilities were established and maintained at the facilities of the Italian National Research Council (CNR) in Pisa (formerly CNUCE-CNR, now ISTI-CNR), to provide support to the Italian civil protection authorities in case of new emergencies. The criterion for the activation of a re-entry prediction campaign of national concern is in theory met whether an uncontrolled re-entering satellite, apparently exceeding the casualty expectancy alert threshold of 1:10,000, overflies the Italian territory. If this is the case, the goals of the appointed ISTI-CNR team are those to monitor the orbital decay, to provide re-entry predictions with uncertainty time windows and to predict possible passes over Italy, together with the related sub-satellite tracks, during the last phases of the flight.

The purpose of a re-entry prediction process is to determine the time interval (or re-entry window) in which the natural re-entry of a satellite can be foreseen, taking into account all the uncertainties affecting the re-entry predictions. The definition of appropriate re-entry uncertainty windows is obviously a critical aspect of the prediction process and is typically based on past experience. Re-entry windows amplitudes in between $\pm 15\%$ and $\pm 25\%$ of the residual lifetime may be adequate in 90% of the cases [2], depending on satellite characteristics, decay phase, solar activity level and atmospheric model. However, residual lifetime errors well in excess of 30% cannot be completely excluded, due to unpredicted geomagnetic storms in the last few days of flight, or to ballistic parameter and atmospheric density mismodeling in the hours preceding the re-entry.

Among the main ISTI-CNR activities carried out during the uncontrolled re-entry of a potentially dangerous space object there are the following: (1) Acquisition of the orbital elements of the re-entering object; (2) Updating of data files including the environmental conditions, i.e. the observed and forecasted solar and geomagnetic activity indices; (3) Determination of the object's ballistic parameter; (4) Propagation of the last available orbital state up to the final plunge down to the altitude of 80 km (nominal re-entry epoch); (5) Evaluation of the global uncertainty time window around the nominal re-entry epoch; (6) Representation of the sub-satellite ground track corresponding to the current global uncertainty time window, during the last 2–3 days preceding the final decay (Fig. 3).

5.1 Civil Protection Applications

However, typical re-entry prediction standard products, such as those of points 4 to 6, are of no, or very limited, use for civil protection applications. As a matter of fact, the nominal decay forecast is absolutely useless for civil protection planning, due to its intrinsic large uncertainty. The global uncertainty window provides relevant information, identifying the time interval in which the re-entry should be expected

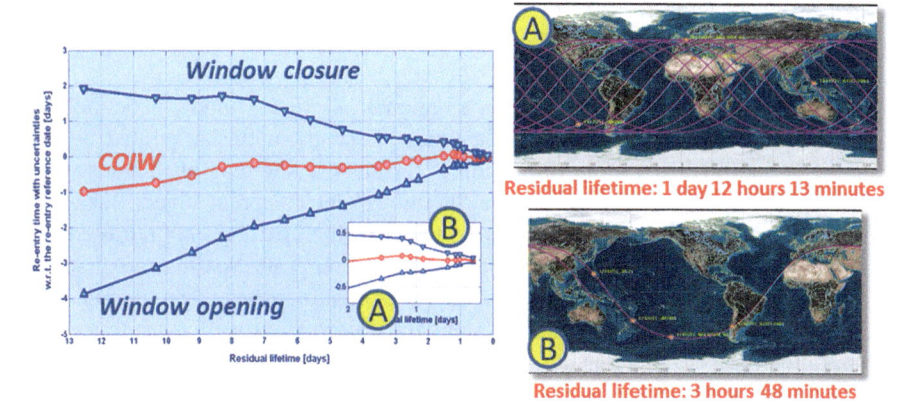

Fig. 3 Example of the representation of the sub-satellite ground track corresponding to two global uncertainty time windows computed 1 day 12 h 13 m (case A) and 3 h 48 m (case B) before re-entry (the uncertainty window evolution is shown on the left)

somewhere in the world. However, this interval remains too large until re-entry, so it is not possible to devise and apply practical precautionary civil protection measures based on it. Finally, the re-entry location inside the global uncertainty window remains quite undetermined, along a varying number of orbital sub-satellite tracks, themselves possibly affected by a considerable cross-track error.

Therefore, the locations possibly at risk in a given area, for instance in Italy, cannot be identified reasonably ahead of re-entry using the information from points 4 to 6. For these reasons, a new approach was devised and applied in Italy to real re-entry prediction campaigns since the orbital decay of the BeppoSAX spacecraft in 2003 [13]. It was firstly based on the attempt to answer the question: "Given a certain global uncertainty time window, where and when a re-entering satellite fragment might cross the airspace and hit the ground on a specific area of the world overflown by the falling uncontrolled object?", and then on the following reasoning: "For each location inside the global uncertainty time window, the re-entry and debris ground impact is possible but not certain. However, the eventual re-entry or impact may occur in each place only during a specific and quite accurate risk time window, which can be used to plan risk mitigation measures on the ground and in the overhead airspace".

Hence, the solution of the problem should consist, in general, in identifying the risk time window for each overflown location of the planet inside the global uncertainty window, and in computing, in particular, the "regional risk time window" corresponding to each pass over an area of interest, e.g. Italy [14]. The procedure adopted at ISTI-CNR to assess the regional risk time windows for a finite area embracing Italy starts 3–4 days before the final decay (this is to focus the attention on a relatively low number of sub-satellite tracks overflying the target area), by simulating a re-entry opportunity for each pass over the area of interest (that is obtained by slightly modifying the nominal predicted trajectory through small changes of the re-entering object's ballistic parameter, in order to obtain simulated re-entries over the target area in the time interval corresponding to the current global uncertainty window).

Then, for each re-entry opportunity, a regional risk time window is defined by accounting for: (1) The different flight times of the fragments generated by the satellite breakup (their timing dispersion is typically a few tens of minutes wide and includes the time of flight of the "fictitious" intact parent object, taken as reference to set the absolute scale of time, and also small particles not representing a hazard on the ground, but possibly dangerous for aircraft crossing the airspace during the re-entry); (2) The variation of the initial conditions leading to nominal re-entries in different parts of the country along the trajectory, as well as the trajectory propagation errors (a few minutes); (3) The finite size of the area of interest around the simulated re-entry opportunity (± 2 min for areas of about 2000×2000 km^2, i.e. ten times those of a country like Italy). Considering the re-entries of typical spacecraft or upper stages, the amplitude of the risk time window for such areas should be around 30–40 min, including the airspace up to an altitude of 10–20 km.

It is worth mentioning that the ground tracks crossing the area of interest are much more stable and less affected by uncertainties, being computed with the "right" times (in fact, as a direct consequence of the almost exact synchronization between the satellite dynamical evolution and the Earth's rotation, the potential re-entry time over specific locations of the planet can be estimated with a reasonable accuracy already a few days before the final decay) and approximately including the re-entry dynamics down to ground impact.

Finally, a cross-track safety margin, with respect to each re-entry ground track for the area of interest, should be defined to obtain the volume of airspace and the surface on the ground associated with the regional risk time window. Its definition depends on: (1) The expected dispersion of the fragments perpendicularly to the satellite trajectory (this is a function of the breakup nature and of the endo-atmospheric dynamics of the fragments, amounting to as much as several tens of km); (2) The cross-track trajectory uncertainty due to the mismodeled evolution of the orbital decay (it might amount to a few tens of km 3–4 days before decay, progressively decreasing as the re-entry is approaching); (3) The effects of the prevailing or predicted winds in the stratosphere and troposphere (the cross-track drift of macroscopic fragments exposed to winds during the final phase of nearly vertical fall is less than a few tens of km).

Limiting the attention to the relevant fragments and depending on the specific nature of the re-entering parent object, the cross-track safety margin may be ± 90–200 km around 3–4 days before re-entry, and ± 80–120 km during the last 24–48 h. In conclusion, the volume of airspace which could be potentially affected by the debris fall is the region of space extended up to the relevant geodetic altitude (e.g. 15 km), centered on the re-entry ground track and with a cross-track swath of ± 100–200 km, which might progressively drop to ± 100 km, or less, as the re-entry is approaching.

6 Recent Significant Re-entries

During 5 months, from September 2011 to January 2012, three massive satellites re-entered without control in the atmosphere: UARS, ROSAT and Phobos-Grunt [6].

UARS had a dry mass of 5668 kg, a diameter of 4.6 m, a length of 9.7 m and a quite complex shape, with booms, appendages, protruding structures and a big solar array.

ROSAT had a dry mass of 2426 kg, dimensions of 2.2 × 4.7 × 8.9 m, a quite compact shape and solar array configuration, and just one boom aligned with the longitudinal axis.

The Phobos-Grunt vehicle, trapped by a failure in orbit around the Earth, was a complex spacecraft whose main mission was a soil sample return from the major moon of Mars, i.e. Phobos. The failed probe had a total mass at launch of 13,525 kg and dimensions of 3.76 × 3.76 m (7.97 m with the solar arrays deployed) × 6.32 m. Historically, it was the 12th most massive space object re-entering the atmosphere uncontrolled, but more than 82% of the total mass, i.e. about 11,150 kg, consisted of very toxic liquid hypergolic propellants. The dry mass was therefore around 2350 kg, a value not uncommon among spacecraft and upper stages usually re-entering without control.

Another engrossing and uncommon re-entry was that of the ESA's GOCE satellite, having a dry mass of 1002 kg and a roughly cylindrical shape of 1 m diameter and 5.3 m length, with wing-shaped fins spanning 2 m. Following the automatic shutdown of its depleted propulsion system, on 21 October 2013, the satellite re-entered on 11 November 2013 [15].

Finally, after encountering severe problems immediately after launch, the cargo ship Progress-M 27 M was declared officially lost on 29 April 2015 and re-entered uncontrolled nine days later, on 8 May 2015. It had a launch mass of 7289 kg, a dry mass in excess of 5 metric tons and carried on board 1373 kg of highly toxic propellants [7].

As in previous similar occurrences, the Space Flight Dynamics Laboratory of ISTI-CNR was in charge of the re-entry predictions for the Italian civil protection authorities and space agency. The five mentioned satellites are shown in Fig. 4.

6.1 UARS: Upper Atmosphere Research Satellite

The NASA's satellite UARS was deployed into a 580 km circular orbit by the space shuttle Discovery, on 15 September 1991. After 14 years of mission, the residual propellant was used to lower the satellite orbit with a series of eight maneuvers, for the purpose of reducing its residual lifetime, according to the space debris mitigation guidelines. UARS was decommissioned by NASA in December 2005, leaving the tanks completely empty in order to complete the satellite passivation. Since then, the orbit of UARS continued to decay up to the final re-entry into the Earth's atmosphere, on 24 September 2011.

UARS   **ROSAT**   **Phobos-Grunt**

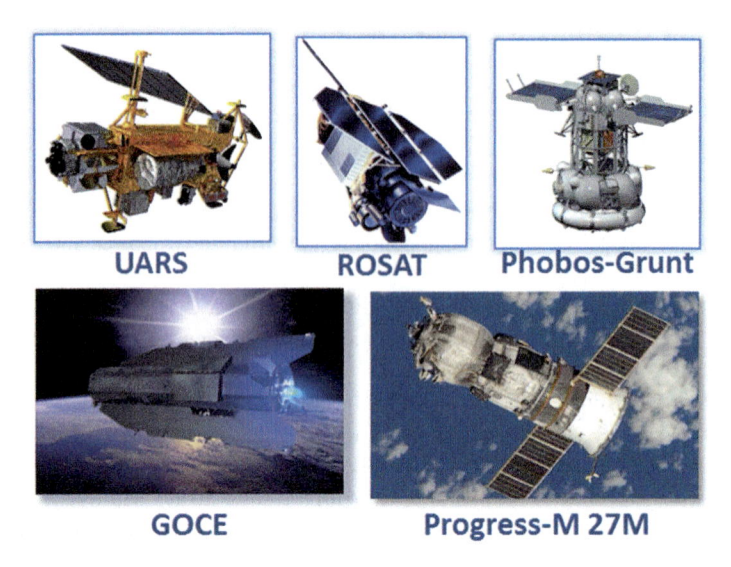

GOCE   **Progress-M 27M**

Fig. 4 Artistic representation of five significant objects recently re-entered without control in the Earth's atmosphere

A re-entry survivability analysis of UARS had been performed by NASA in 2002, using the software tool ORSAT and assuming a breakup altitude of 78 km. 26 fragments with a total mass of 532 kg were expected to survive re-entry and distribute along a debris footprint 788 km long. According to NASA, the total debris casualty area was assessed to be 22.4 m^2 and the risk of human casualty was about 1:3200 in the latitude belt between $\pm 57°$.

The prediction activity was carried out at ISTI-CNR during the last 12 days of residual lifetime, marked by the solar flux on the rise and a couple of geomagnetic storms in the first half. The re-entry uncertainties windows for UARS were obtained by varying its ballistic parameter by $\pm 20\%$. The mean relative residual lifetime error was close to 15% over the re-entry campaign and about 20% during the last 2 days. The maximum absolute errors were instead close to 28% around one day before re-entry.

Information concerning the possible cross-track dispersion of the fragments was not available. Therefore, on the basis of the experience of past re-entries and on the expected trajectory inaccuracies, a ground swath of ± 100 km around the nominal track was assumed. The risk zones and time windows for Italy were issued about 64 h ahead of re-entry. At that time, the satellite re-entry tracks possibly affecting the Italian territory were two in a global uncertainty window 30 h wide. The risk time window associated with each possible re-entry track was 38 min wide, including the airspace up to a geodetic altitude of 10 km. The last remaining risk zone over Italy fell out of the global uncertainty window 5 h before re-entry [6]. After the event, the US Joint Space Operation Center (JSpOC) assessed that the re-entry at the altitude of 80 km had occurred at 04:00 UTC ± 1 min, on 24 September 2011.

6.2 ROSAT: ROentgen SATellite

The satellite ROSAT of the German aerospace center (DLR) was launched from Cape Canaveral with a Delta II rocket on 1 June 1990, and placed into a 575 km circular orbit to study astronomical sources in the extreme ultraviolet and X-ray bands of the spectrum. After 8 years of data collection, the orbit of the abandoned satellite was left to progressively decay due to the action of air drag.

A re-entry survivability analysis of the satellite had been performed by DLR using SCARAB. Eighteen fragments with a total mass of 1700 kg were expected to survive re-entry and distribute along a debris footprint 1200 km long. The largest fragment would have had a mass of 1500 kg and the total debris casualty area was estimated to be around 20 m^2. The risk of human casualty from surviving debris was about 1:3000 (DLR) in the latitude belt between $\pm 53°$.

Re-entry predictions were carried out at ISTI-CNR during the last 11 days of satellite lifetime, marked by solar activity on the rise and relatively quiet geomagnetic conditions. The re-entry uncertainties windows for ROSAT were obtained by varying its ballistic parameter by $\pm 25\%$. The maximum relative absolute error was about 8% and occurred 19 h before decay. The mean prediction errors were about 3% over the re-entry campaign and 5% during the last 2 days.

Based on the information issued by DLR, implying a maximum cross-track dispersion of the fragments of ± 40 km, and on the estimated re-entry trajectory error, an initial and very conservative ground swath of ± 90 km around the nominal track was assumed, successively reduced to ± 85 km to account for the decreasing trajectory propagation uncertainties. The risk zones and time windows for Italy were issued about 88 h ahead of re-entry. At that time, the uncertainty window was still wide (51 h), but the satellite re-entry tracks possibly affecting the Italian territory were five. The risk time window associated with each track was 30 min wide, including the airspace up to a geodetic altitude of 10 km. The last "surviving" risk zone fell finally out of the global uncertainty window about 18.5 h before re-entry [6]. According to JSpOC, the re-entry of ROSAT at 80 km occurred at 01:50 UTC ± 7 min, on 23 October 2011.

6.3 Phobos-Grunt

The Roscomos planetary probe Phobos-Grunt was launched on 8 November 2011 from the Baikonour cosmodrome with a Zenit-2 rocket. Initially placed into a 208 × 344 km orbit, the spacecraft, directed towards the main moon of Mars, remained unfortunately trapped in orbit around the Earth, probably due to a malfunction of the on-board computer. Any attempt to regain control from the ground was unsuccessful and, since 22 November 2011, its orbital decay was essentially compatible with natural perturbations alone.

According to various estimates carried out in Russia and in Germany, from 8 to 30 fragments, with a total mass of 200–1000 kg, were expected to survive re-entry and distribute along a debris footprint 800–1300 km long. The risk of human casualty on the ground due to debris impact was assessed to be about 1:5000–1:3000 in the latitude belt between ±51.5°.

The re-entry campaign was carried out at ISTI-CNR during the last 13 days of residual lifetime, characterized by a declining solar activity and relatively quiet geomagnetic conditions. The re-entry uncertainties windows for Phobos-Grunt were obtained by varying its nominal residual lifetime by ±25%. Overall, the maximum absolute error was about 8%, and 6% during the last two days.

Conservatively assuming a quite improbable propellant tank explosion at high altitude during re-entry, and taking into account the estimated trajectory error, a ground swath of ±120 km around the nominal track was considered. The risk zones and time windows for Italy were issued about 57 h ahead of re-entry. At that time, the satellite re-entry tracks possibly affecting the Italian territory were three in a global uncertainty window 28 h wide. The risk time window associated with each track was 30 min wide, including the airspace up to a geodetic altitude of 10 km. The last "surviving" risk zone remained in play until the end. Later on it was assessed that the probe had re-entered before crossing the Atlantic Ocean and then Europe. According to JSpOC, the re-entry at 80 km had occurred at 17:46 UTC ±1 min, on 15 January 2012.

6.4 GOCE: Gravity Field and Steady-State Ocean Circulation Explorer

The ESA's satellite GOCE was launched on 17 March 2009 from the Plesetsk cosmodrome on a Rokot launcher. After mapping the geopotential for four years from an extremely low circular orbit, on 21 October 2013 the low thrust ion propulsion motor, used to contrast the atmospheric drag, was automatically shut down. Then the satellite entered in a Fine Pointing Mode (FPM), with an attitude control minimizing the aerodynamic drag effect. According to pre-launch specifications, the FPM state would have been maintained up to the reaching of an average drag force along the orbit of 20 mN. But contrarily to any expectation, the attitude control system remained operational until re-entry, with drag forces perhaps exceeding 2000 mN. Therefore, even if the casualty expectancy for this re-entry was estimated to be slightly above the alert threshold, i.e. 1:5000, it presented a number of challenges and opportunities, from the prediction and risk evaluation point of view, by reason of its peculiar nature.

A pre-launch destructive analysis of GOCE had been carried out for ESA by HTG (Hyperschall Technologie Göttingen) using SCARAB. Overall, 43 macroscopic fragments, totaling approximately 270 kg, were expected to survive re-entry and hit the ground along a 900 km footprint, with a time dispersion of 17 min.

Re-entry Predictions of Potentially Dangerous Uncontrolled Satellites... 279

No.	Day	Opening	Closure
1	9 Nov. (descending)	06:37 UTC	07:17 UTC
2	9 Nov. (ascending)	17:55 UTC	18:35 UTC
3	10 Nov. (descending)	07:26 UTC	08:06 UTC
4	10 Nov. (ascending)	18:44 UTC	19:24 UTC
5	11 Nov. (descending)	06:48 UTC	07:28 UTC
6	11 Nov. (ascending)	18:10 UTC	18:50 UTC

Exclusion of the last surviving risk zone at ~14^h before reentry, thanks to the contraction of the ground safety swath

Fig. 5 Risk time windows (table on the right) and satellite re-entry ground tracks possibly affecting the Italian territory for cases 2–4 associated with the GOCE re-entry. The last surviving risk zone (No. 4), no longer interesting Italy due to a contraction of the ground safety swath, is shown on the bottom right

The orbital evolution of GOCE was monitored at ISTI-CNR since 23 October 2013. Following an initial period of test and analysis, in which only the opening of the re-entry uncertainty window was of importance to exclude the re-entry before a given epoch, reasonably and conservative criteria, mainly based on the uncertainty affecting the duration of the FPM phase, were elaborated and applied, with good and consistent results through the end of the campaign [15].

The possible cross-track dispersion of the fragments was assumed to be ±200 km around the nominal track. The risk zones and time windows for Italy were issued about 61 h ahead of re-entry. At that time, the satellite re-entry tracks possibly affecting the Italian territory were six in a global uncertainty window 67 h wide. The risk time window associated with each track was 40 min wide, including the airspace up to a geodetic altitude of 12 km. The last "surviving" risk zone fell out of the global uncertainty window about 14 h before re-entry, thanks to a significant contraction (to ±120 km) of the ground safety swath (Fig. 5). The re-entry at 80 km occurred at 00:16 UTC ±1 min, on 11 November 2013 (JSoPC). The GOCE fragments eventually plunged into the Southern Atlantic Ocean between 00:24 and 00:40 UTC.

6.5 Progress-M 27 M

The Russian cargo ship Progress-M 27 was launched on 28 April 2015 from the Baikonour cosmodrome aboard a Soyuz 2-1A rocket. After nearly 9 min of powered flight, the cargo was separated in orbit. Shortly afterwards, during the crucial phase of the Soyuz third stage shutdown and the separation of the Progress spacecraft, there was, unfortunately, a loss of telemetry from the upper stage and the spacecraft.

After vain attempts to regain control of the vehicle, the mission was declared lost by the Russian space agency on 29 April 2015. The third stage, with a mass of 2300 kg, re-entered the same day. With no ability to command Progress for a safe return, the out of control supply vessel was irremediably doomed to an uncontrolled re-entry.

No detailed fragmentation and demise analysis was available. However, considering the Progress launch mass of 7289 kg, and the presence on board of 1373 kg of highly toxic propellants, this uncontrolled re-entry was expected to violate the alert casualty expectancy threshold of 1:10,000. A number of parts were presumed to survive re-entry, like the docking mechanism and the spherical pressurant tanks. An additional unknown was represented by the tanks holding the very toxic propellants. However, due to the very short permanence in orbit of the vehicle, it was unlikely that propellants had time to completely freeze. Their dispersion at high altitude during re-entry was instead considered the most probable scenario.

The first ISTI-CNR re-entry prediction was issued on 30 April. In the morning of 7 May, the only potentially risky re-entry trajectory over central Italy was identified and in the afternoon of the same day, about 12 h before the actual re-entry, any residual risk for Europe and Italy was finally excluded. According to JSoPC, the re-entry of Progress at 80 km occurred on 8 May 2015, at 02:20 UTC ± 1 min, over the Southern Pacific Ocean.

7 Conclusions

Uncontrolled re-entries of sizable spacecraft and upper stages still represent a negligible risk compared with other hazards widely accepted in the everyday life of developed countries. However, this risk will probably increase in the future, due to a rapidly growing use of circumterrestrial space for a plethora of commercial, civilian, military and scientific applications, and the high media visibility of space activities and accidents makes the prevention of human casualties and property damages on the ground an important goal.

Re-entry predictions of uncontrolled space objects are still affected by significant uncertainties, and this situation cannot be easily improved, in general, because some of the main error sources involved are outside the control of analysts and modelers. For example, better atmospheric density models for the lower layers of the atmosphere might be developed, but without reliable predictions of solar and geomagnetic activities such hypothetical improved models would not offer substantial advantages for predictive orbit propagations. The detailed modeling of the attitude evolution of complex re-entering bodies might also contribute, in principle, to a reduction of the uncertainties, but would request a thorough knowledge of the objects design, which in most cases is not, and probably will not be, available. Moreover, the use of complex tools for dynamical modeling very close to re-entry might result quite tricky, for the lack of enough time to reliably check the input parameters and the results. Another obvious area open to improvement would be the availability of more sensors sharing their tracking data, possibly reducing the observation gaps and improving the orbit determinations and ballistic parameter estimations. This would probably represent the development with the highest progress potential for re-entry predictions, but its implementation would be expensive and dependant from a high level of international cooperation.

However, in spite of the existing limitations and independently from the envisaged progresses, re-entry predictions for civil protection authorities can be managed in effective way with specific approaches and procedures, as those described in the previous sections. They are able to provide understandable and unambiguous information useful for civil protection planning on selected areas, with the definition of a few narrow risk windows, in terms of space and time. The process may be started two or three days before re-entry, then discarding the risk windows that are left out of the progressive shrinking of the global uncertainty window. In such a way, the civil protection authorities can focus their attention, and any appropriate measure for mitigating the potential risks, on specific "ground corridors" and the overlying airspace, over time intervals of just 20–40 min, at least until one of these re-entry opportunities is still possible.

References

1. Pardini, C., Anselmo, L.: Re-entry predictions for uncontrolled satellites: results and challenges. In: Ouwehand, L. (ed.) Proceedings of 6th IAASS conference "safety is not an option", ESA SP-715 (CD-ROM). European Space Agency, Noordwijk (2013)
2. Anselmo, L., Pardini, C.: Computational methods for re-entry trajectories and risk assessment. Adv. Space Res. **35**(7), 1343–1352 (2005)
3. Ailor, W., Hallman, W., Steckel, G., et al.: Analysis of reentered debris and implications for survivability modeling. In: Danesy, D. (ed.) Proceedings of the 4th European conference on space debris, ESA SP-587, pp. 539–544. ESA Publication Division, European Space Agency, Noordwijk (2005)
4. Klinkrad, H., Fritsche, B., Lips, T., et al.: Re-entry prediction and on-ground risk estimation. In: Klinkrad, H. (ed.) Space debris–models and risk analysis, pp. 241–288. Springer Praxis Publishing, Chichester (2006)
5. NASA: Process for limiting orbital debris. NASA-STD-8719.14, revision A with change 1, NASA technical standard, p. 44. National Aeronautics and Space Administration, Washington (DC) (2012)
6. Anselmo, L., Pardini, C.: Satellite re-entry predictions for the Italian civil protection authorities. Acta Astronaut. **87**(1), 163–181 (2013)
7. Pardini, C., Anselmo, L.: The uncontrolled re-entry of progress-M 27M. J. Space Safety Eng. **3**(3), 117–126 (2016)
8. Johnson, N.: The re-entry of large orbital debris. In: Heath, G.W. (ed.) Space safety and rescue Science and technology series, vol. 96, pp. 285–293. Univelt Inc, San Diego (1997)
9. European Space Debris Safety and Mitigation Standard Working Group: European code of conduct for space debris mitigation. Issue 1.0, pp. 7 (2004)
10. ESA Space Debris Mitigation Working Group: ESA space debris mitigation compliance verification guidelines. Issue 1, European Space Agency, pp. 28 (2015)
11. Cole, J.K., Young, L.W., Jordan-Culler, T.: Hazards of falling debris to people, aircraft, and watercraft. Sandia Report, SAND97-0805-UC-706, Sandia National Laboratories, Albuquerque (1997)

12. Nicollier, C., Gass, V. (eds.): Our space environment, opportunities, stakes and dangers. EPFL Press, Lausanne (2015)
13. Anselmo, L.: Risk analysis and management of the BeppoSAX re-entry. ISTI Technical Report 2003-TR-23, Pisa (2003)
14. Pardini, C., Anselmo, L.: Satellite re-entry prediction products for civil protection applications. In: Sgobba, T., Rongier, I. (eds.) Proceedings of the 7th IAASS conference "space safety is no accident", pp. 453–462. Springer, Cham (2015)
15. Pardini, C., Anselmo, L.: GOCE re-entry predictions for the Italian civil protection authorities. In: Ouwehand, L. (ed.) Proceedings of the 5th international GOCE user workshop, ESA SP-728 (DVD). European Space Agency, Noordwijk (2015)

Uncertainty Quantification for Destructive Re-Entry Risk Analysis: JAXA Perspective

Keiichiro Fujimoto, Hiroumi Tani, Hideyo Negishi, Yasuhiro Saito, Nobuyuki Iizuka, Koichi Okita, and Akira Kato

Abstract In order to improve the accuracy of the expected casualty risk prediction for the destructive re-entry, the related uncertainty factors are identified and the uncertainty quantification approach are proposed. Based on our studies and experience, the identified dominant uncertainty factors are the model accuracy, the attitude stability mode, the shape complexity and shape change, and the initial conditions at the entry start and the break-up. High-fidelity numerical simulations play the important role to model complex multi-disciplinary physics and to cover the wide range of environmental conditions with small numbers of model validation data. The real shape and the deformation effects are initialy modeled by the high-fidelity numerical simulations and finally modeled by the reduced empirical physics-based models. Flight data acquisition also plays important the role to quantify the uncertainties of the attitude stability mode, the break-up altitude, and the temperature distributions and to validate the integrated risk analysis model. Some of the key findings obtained by the high-fidelity simulation are discussed. It is also shown that various types parameter dependencies such as Mach number, wall temperature, surface curvature, and the effect of the turbulent flows on the aerodynamic characteristics and the heat flux distributions should be considered in the empirical models.

1 Introduction

Space debris problem is a growing concern to be tackled internationally to keep our space activity sustainable. Numerous pieces of debris are tracked on the orbit around the Earth, those number will be rapidly increased due to the growing

---

K. Fujimoto (✉) · H. Tani · H. Negishi · Y. Saito · N. Iizuka · K. Okita
Research and Development Directorate, Japan Aerospace Exploration Agency (JAXA), Tokyo, Japan

A. Kato
Safety and Mission Assurance Department, Japan Aerospace Exploration Agency (JAXA), Tokyo, Japan

satellite demands such as mega-constellations. The increasing number of space debris directly results in the increasing potential risk to all spacecraft. Therefore, a comprehensive research and development (R&D) efforts should be made to understand the space debris environment, assess its risk, mitigate its number growth, and control the risk. The importance to ensure a sustainable space environment has been firstly stated in space basic act of Japan in 2008. One of the major plans on the space policy is to address the R&D activities for the space debris problems. Reorganizations of the regulation issues and R&D activities for space debris of Japan Aerospace Exploration Agency (JAXA) were initiated in 2016. Four major activities were identified and re-organized, those are (1) formulation of space debris international standards and guidelines, (2) debris situational awareness and defense, (3) low cost active debris removal, and (4) debris mitigation [1, 2].

Space debris mitigation includes a set of strategies to reduce the number of orbital debris in the critical orbital zones for long term sustainability. It is commonly recommended both in the world space debris mitigation guidelines and standards, that the un-operational spacecraft and launch vehicle orbital stages in low earth orbits should be removed by controlled re-entry or reduction of orbital lifetime to be shorter than 25 years followed by natural re-entry. Risk should be minimized as much as possible by the comprehensive consideration on the design and the disposal operation based on the mission analysis and the risk assessment. The destructive re-entry of satellites and rockets and the related ground risk due to the fragments reaching the ground are getting increased interest in recent years. Survived fragments derived from the launch vehicle's ascent and the spacecraft's re-entry have been found and retrieved in many places all over the world. Fatal accident due to artificial space debris has not happened yet, however, survived fragments have been found at the place near populated areas. It is truly the strong reason to emphasize the importance to predict the risk quantitatively, and minimize the risk by the detailed consideration on the design and the disposal operations.

Therefore, main focus of this study lies in the significant uncertainty reduction of the destructive re-entry risk analysis to realize the detailed design and operation considerations. The uncertainty quantification factors and related research topics are identified, and effective uncertainty quantification strategy based on the high-fidelity numerical simulations and the flight data acquisition is proposed. Finally, an effectiveness of the high-fidelity numerical simulations is shown, including the discussion on the key findings obtained by the computed results.

2 Re-entry Risk Analysis Methods

In order to predict the ground risk due to the survived fragments, the multi-disciplinary physics should be considered to describe the complicated destruction process. After an initiation of re-entry into the atmosphere, the spacecraft experiences severe aerodynamic and aero-thermal loads. Those loads rapidly increse with the altitude decrease due to the larger atmospheric density. Finally, the spacecraft is broken into numerous fragments due to the mechanical and thermal destructions.

Most of the fragments are demised, but some fragments can survive and reach the ground. Consequently, the survived debris strike, the blast-wave over-pressure and thermal radiation due to the explosive debris, and the toxic emissions can result in human injury and property damage.

Therefore, it is essential to predict the likelihood of the fragment survivability and the consequences based on quantitative phyics-base models. There are dominant three uncertainty factors: (A) multi-disciplinary nature of the related physics, (B) significant non-linearity of the off-nominal physics, and (C) the effect of the complicated shape and deformation.

For the survivability analysis, the multi-disciplinary physics-based models such as the trajectory, attitude motion, aerodynamics, aerothermodynamics, and thermal transfer should be considered. In terms of the free-stream conditions, wide range of flow speed regime from hypersonic to subsonic flows, and wide range of atmospheric density and pressure conditions from the free molecular flow to the continuum flow should be considered. In the hypersonic and supersonic flow conditions, the non-linear effects such as the chemical reactions due to the high temperature behind the shockwave and the shockwave interactions on the aerodynamic characteristics and the heat flux should be considered. In the free molecular flow regime at the high altitude, the effects of the rarefied gas dynamics such as non-equilibrium of thermodynamics and chemical reactions should be considered.

While, in the destruction limit state predictions, both mechanical and thermal destruction should be considered. The spacecraft is not designed to be protected from the severe heating, so that the resulting heat path and the heat flux distributions become complicated. Therefore, the resulting temperature distributions is also complicated, and its variation range becomes very large. In addition, the thermal conductivity and the strength of material significantly depend on the temperature. Such temperature dependencies are main factors of the difficulty of non-linear relationship characterization.

Effect of the complicated shape and its deformation are also the dominant uncertainty factors. Shape of the spacecraft is generally complicated, so that the resulting aerodynamic chracteristics and the heat flux distributions are complicated as well. Furthermore, the spacecraft's shape is changed in time due to the break-up and the destructions, so that the trajectory becomes different due to changes of the mass, the moment of inertia, the aerodynamic characteristics. The aerodynamic heat flux distributions and the heat path are also changed. All of these uncertainty factors should be included and related uncertainties should be reduced in order to achieve the accurate fragment's survivability analysis.

Then, the technical detail of the re-entry analysis method is introduced, and two approaches practically used in the space agencies are compared to identify the research scope of this study. The break-up altitude is evaluated by a break-up altitude model, while the fragment dispersion area, and the number of the fragments, with related kinematic energies are evaluated by fragment dispersion model as shown in Fig. 1. There are mainly two approaches for the destructive re-entry risk analysis, the one is "object-oriented" approach and the other is "spacecraft-oriented" approach as compared in Table 1. Main difference is how to take into account the object's shape

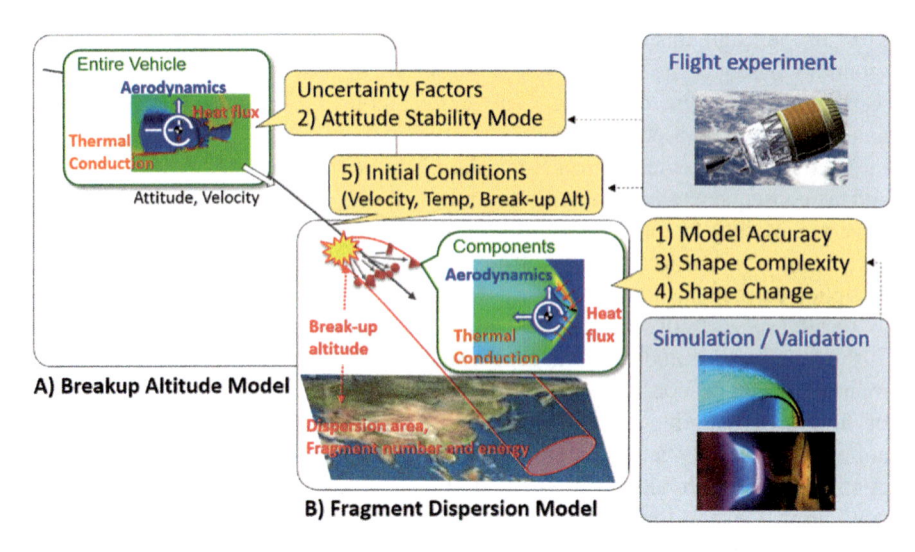

Fig. 1 Uncertainty factors for the destructive re-entry risk analysis

Table 1 Comparisons of re-entry risk analysis approaches

	Object-oriented – ORSAT (NASA) [3] – ORSAT-J (JAXA) – DEBRISK (CNES) [4]	Spacecraft-oriented – SCARAB (ESA) [5] – PAMPERO (CNES) [6] – LS-DARC (JAXA) [7]
Object shape	– Basic shapes – Shape change is not considered, except for that by break-up	– Real shape is considered – Shape change in time including the change by material fusion
Breakup altitude	Decided based on past flight data or modelled based on thermal and/or stress analysis	
Equation of motion	3 DOF – Random tumbling is assumed – No lift considered	6 DOF – Attitude variation is evaluated – All aerodynamic forces and moments are considered
Thermal analysis	– 1D or 2D model	– 3D model
Aerodynamic model	Predetermined function based on base shape, Knudsen number and Mach number.	Evaluated based on the local panel method
Heat flux model		

change in time. In the object-oriented approach, all of the spacecraft's components are considered as the simple base geometries such as sphere, box, and cylinder. While in the spacecraft-oriented approach, the complicated real shape is directly considered including the break-up and the demise of the structures. In terms of the attitude and the trajectory modeling, the random tumbling attitude is assumed in the object-oriented approach. While in spacecraft-oriented approach, 6 degree-of-freedom (DoF) motion of the fragment is directly considered by solving 6

DoF equations of motion. In order to evaluate the aerodynamic characteristics and the heat flux distributions, the pre-determined database or regression equation is used in the object-oriented approach. While in the spacecraft-oriented approach, the aerodynamic characteristics and the heat flux are evaluated by the surface integrations for each time step based on the empirical correlation model [5, 7].

Orbital re-entry survivability analysis tool of Japan (ORSAT-J) is re-entry risk analysis tool used in JAXA, which is derived from NASA-ORSAT ver. 4.0 and implements an object-oriented approach. The drag coefficient C_D is determined based on the shape, the flow regime, the attitude motion, and the free-stream Mach number. Convective heating rate in ORSAT are mainly evaluated based on the object shape, Stanton number, and Knudsen number. In hypersonic continuum flow, the conventional empirical correlations for a spherical stagnation point such as the Detra-Kemp-Riddell correlation [8] is used as the primary basis. Based on the experimental results of the stagnation point heating rate and the heat flux distributions obtained for various geometries, the shape and motion dependencies are modeled. To obtain spatial average of the heating rate, the stagnation point heating rate is multiplied by factors that account for the heat flux distributions.

Both approaches have unique advantages. The object-oritented approach is attractive in terms of the simplicity and the validated accuracy for the base geometries such as spheres and cylinders. On the other hand, the spacecraft-oriented approach is attractive because it can be applied to the design for demise considerations and real shape effect can be considered. The heating surface area generally increases by considering real complicated shape. Although it is highly depending on the shape type and the flow conditions, the fragment might be aerodynamic stable rather than being random tumbling, as it is basically assumed in the object-oriented approach.

In order to exploit the full advantages of both approaches, the spacecraft-oriented re-entry risk analysis tool is under the development at JAXA, which is named as LS-DARC (Destructive Atmospheric Re-entry Code) [7]. However, there is still discussions on the difference of the results of two approaches, mainly due to the difficulty of the model validations to cover all of the physics-based models, the analysis conditions, and the geomerty type. It is clear that the technological gap which should be filled in this study is an establishment of the efficient uncertainty quantification methodology for the destructive re-entry risk analysis. Detailed discussions on the efficient uncertainty quantification strategy will be given in the next section.

3 Strategy for Uncertainty Quantification

There exist various uncertainty factors in the re-entry risk analysis. Major uncertainty factors identified in this study are shown in Fig. 1, those are (1) model accuracy, (2) attitude stability mode, (3) shape complexity and shape change in time, and (4) initial conditions at the entry start and the break-up.

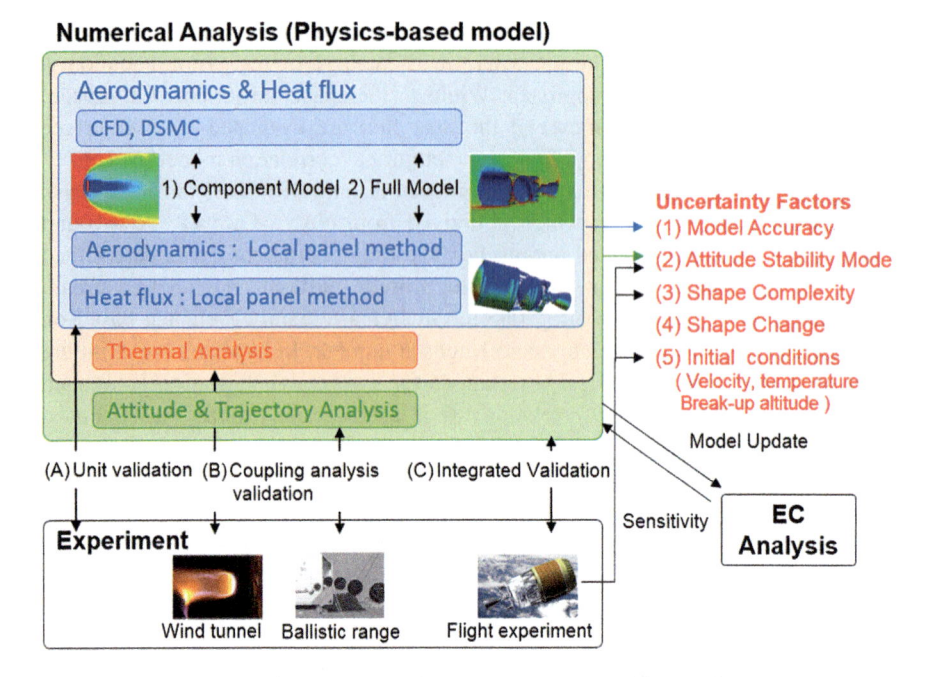

Fig. 2 Uncertainty quantification strategy for destructive re-entry risk analysis

Overview of the uncertainty quantification strategy is also shown in Fig. 2. Brief description of each uncertainty factors and its quantification strategy is given respectively.

3.1 Model Accuracy

In order to predict the fragment's survivability, the multi-disciplinary complicated physics such as the trajectory and attitude, aerodynamic characteristics, heat flux distributions, and heat transfer should be modelled simultaneously. In addition, wide range of analysis conditions such as flow speeds and the various geometries should be covered in the model validations. Furthermore, it is generally difficult to obtain the validation data for the multi-disciplinary coupling analysis, which is mainly due to the capability limitation of the experimental facilities. These are main cause of the difficulty of the uncertainty quantification for physics-based models.

3.1.1 Aerodynamics and Heat Flux Models

Since uncertainties on the aerodynamic and the heat flux models are dominant factors on the resulting expected casualty, the uncertainty quantification strategy is described for them.

In the proposed strategy, the model validation is started from unit validation as shown in Fig. 2, in which each physical model is validated without considering the interactive effect from the other discipline. Wind tunnel experiments to obtain the aerodynamic characteristics and the heat flux distributions are the typical examples of the unit validation.

Then, the coupling analysis validation is carried out to validate the coupling models such as the coupled heat flux and heat transfer analysis, and the coupled 6DoF motion and aerodynamics analysis. High enthalpy wind-tunnel and the ballistic range experiments are typical examples. Finally, the integrated validation is carried out to validate the integrated multi-disciplinary model. Flight data acquisition is a typical example, which will be discussed in the following sections. Based on this approach, both high-fidelity numerical simulations and the empirical correlation models are validated by comparing with the experimental data at the representative flow conditions. To cover a wide range of flow conditions, the empirical correlation model used in the re-entry risk analysis is validated by comparing with the results of high-fidelity numerical simulations, instead of the comparison with the experimental data. Selection of the geometry and the experimental conditions should be carried out carefully to obtain effective experimental data for the model validations, in order to assess the prediction accuracy for all of the important flow structures.

3.1.2 Mechanical and Thermal Destruction

Due to severe aerodynamic and aero-thermal loads, the spacecraft can be demised. The uncertainties to predict the temperature and stress distributions and those to predict the destruction limit state functions are also dominant factors on the resulting expected casualty. As already discussed above, accuracies to predict the complicated temperature and mechanical and thermal stress distributions are essential. Although an establishment of the practical destruction model is ultimate goal, the high-fidelity numerical simulations such as thermal transfer and structural analysis based on the finite element method are firstly employed.

Since related numerical simulation algorithms has been matured and the available computing capability is becoming significantly larger, the remained research needs are only on the grid resolution requirements and the material property data acquisition such as the Young's modulus, strength, thermal conductivity, heat capacity, and radiation emissivity with temperature dependency.

There are two categories of the uncertainties related to the limit state functions. One is that of its formulation and the other is that of the related material properties. The formulation of limit state function should be decided carefully to describe the interaction mechanism between the mechanical and thermal destructions. There have been comprehensive research efforts to describe the mechanical destructions depending on the failure modes, such as the ductile fracture, the plastic collapse and so on. There is also established formulation to decide the thermal destruction which is based on the melting temperature and the latent heat of vaporization. However, further research effort should be made to describe the destruction state under the interaction effect between the mechanical and thermal destructions. In addition, the re-entry risk analysis by the probabilistic analysis of high-fidelity numerical simulation is not practical, further research effort should be made for the model reduction with keeping sufficient accuracy.

3.2 Attitude Stability Mode

The assumptions on the spacecraft's attitude motion have significant influence on the resulting fragment's survivability. In the object-oriented approaches, the fragment's attitude motion is generally assumed to be the random tumbling, meaning that the object surface is heated equally. However, if the aerodynamic stability is available and the heavier components are in the wind-ward side, it is resulting in the higher heat flux on the larger heat capacity components. As it will be discussed in following sections, the rocket upper stage might be stable aerodynamically during its re-entry. The consideration of the real complicated shape is essential for such an aerodynamic stability analysis, so that the development of the spacecraft-oriented analysis tool is under the way at JAXA. In addition, the destructive re-entry flight data acquisition planned to be conducted in this research to understand the real behaviour of the attitude motions.

3.3 Shape Complexity and Shape Change

It is clear that the shape has a significant influence on the aerodynamic characteristics and the heating amount and distributions, and the survivability of resulting fragments. As already discussed above, the fragment's shape is simplified especially in the object-oriented approach, generally resulting in a reduction of the heating area.

Conversely, if the real complicated shape can be considered in the fragment's survivability analysis, the predicted heating amount will become larger. Therefore, an intensive research efforts have been made at JAXA to validate the empirical correlation models for the aerodynamic and heat flux predictions for the complicated geometries. Heat flux increases by the small curvature effect, the shock wave impingement are the typical example of this issue.

As it will be discussed in the following sections, high-fidelity numerical simulation tools to handle 6DoF motion of the fragments have been developed. By the effective complemental use of the spacecraft-oriented analysis and the high-fidelity numerical simulations, the real shape effect on the fragment's survivability and its mechanism have been investigated in this study. As also discussed above, shape change effect has significant influence on all the related physics such as aerodynamics, heat flux, trajectory, and attitude motions. The effects of the shape change on the fragment's survivability will be investigated by the spacecraft-oriented analysis tools.

3.4 Initial Conditions

Uncertainties on the initial conditions such as the break-up altitude, the temperature level of the system component structures at the break-up, and the fragment's radial velocity at the break-up have significant influence on the structure demisability and the dispersion area. The initial conditions of such parameters are generally decided based on the design analysis and the flight data, and the consideration of the worst-case scenario. As discussed above, the temperature level has significant effects on the fragment's survivability since the heating amount required for the thermal destruction is significantly depending on the initial temperature level, and the material strength decreases at the elevated temperatures. Low-temperature and high pressure gas storage tanks inside main tank are often found to survive. It is also true from the facts for the retrieved survived fragments on the ground, many of low temperature and high pressure thick wall tanks are found to be survived. It is the strong reason of the importance of the temperature level of the components.

Since it is still unpractical to predict all the destruction process from the re-entry initiation until the ground impact accurately without any assumptions, the flight data acquisition is essential. If sufficient amount of the flight measurements is available, it is straightforward to quantify the uncertainties of the initial conditions and to justify the validity of analysis assumptions. As it will be discussed in following section, the development of the destructive re-entry data acquisition system is under the way at JAXA. By using this system, the break-up altitude, the temperature and the pressure histories can be obtained, which would significantly contribute to the uncertainty reduction and the justification of the assumptions.

4 Re-entry Flight Data Acquisition

In order to reduce the uncertainties on the attitude stability mode, the break-up altitude, and the temperature distributions on the system components, the re-entry flight experiments have been conduced in JAXA, and new re-entry flight data acquisition system is under the development. The schematic overview of the system is shown in Fig. 3. It has the capability to measure the pressure and temperature

▷ Acceleration, Pressure and Temperature Sensors (Wired, Wireless is future option)
▷ GPS / Iridium radio interference
▷ Reasonable cost and size

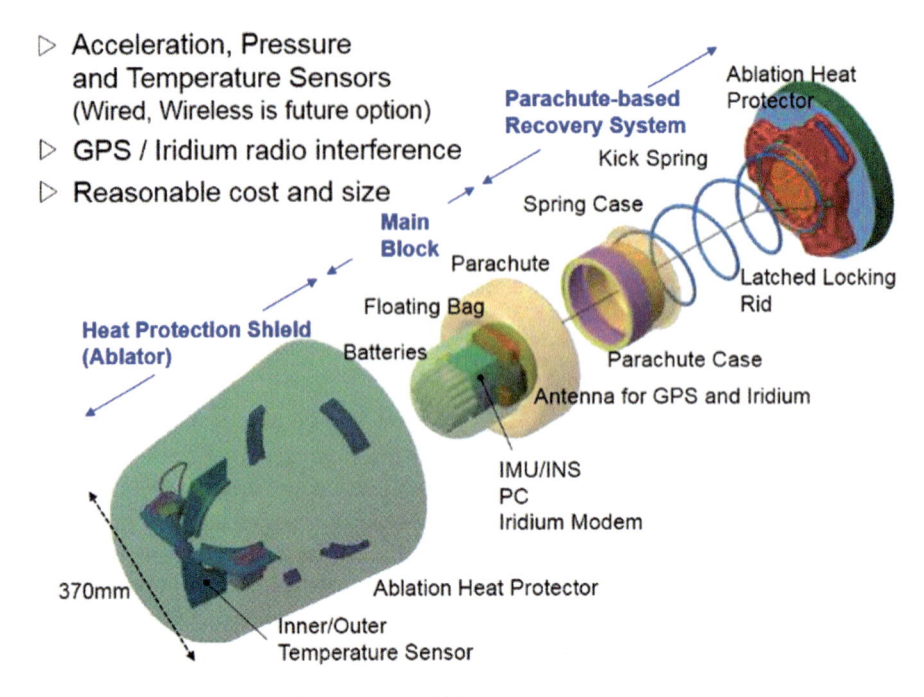

Fig. 3 Overview of re-entry flight data acquisition system

data, and has GPS / Iridium radio interface. The system consits of three components, the ablative heat protection sield, the main block, and the parachute-based recovery system. Currently, the measurement system consists of the wired sensors, and the wireless sensing is one of the future options. Schematic overview of the sequence of events for H-IIB launch vehicle is shown in Fig. 4. Acceralation, pressure, and temperature data will be acquired at various locations for each system components during entire re-entry phase before the break-up. After the break-up, the system will be separated passively and recovered by the parachute-based water landing.

By the obtained acceralation data, the vehicle's attitude motion can be identified for the entire period before the break-up. In addition, by the detailed investigation based on the obtained flight data and the corresponding simulation results, the destruction process and its mechanism can also be clarified. As will be discussed in the following sections, the sensor positions have been selected based on the predicted surface pressure and the heat flux distributions, which will be used not only for the destruction process investigations but for the model validations. Furthermore, the important initial conditions such as the temperaure level at the break-up can be updated to the less uncertain values.

In the past H-IIB No 4, HTV No 3 and 4 missions, the re-entry flight data had already been obtained by the re-entry flight data acquisition system named i-ball [9]. Research activities for the destruction process clarification and the model validation have already been conducted in this study.

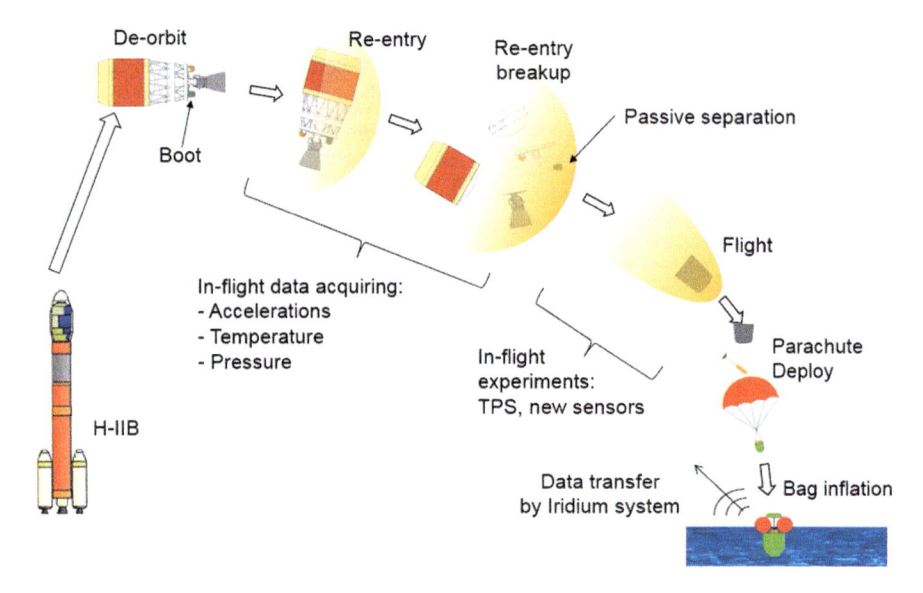

Fig. 4 Sequence events for the re-entry flight data acquisition

5 High-fidelity Numerical Simulations

High-fidelity numerical simulation is the essential technology to achieve high-order accuracy in the re-entry risk analysis. If the physical models used in the simulation are validated by comparing with the experimental data throughout the possible environmental condtions, the significant uncertainty reduction can be achieved. Based on the detailed investigation of the computed results, the mechanism of the demise process can be understood, and the empirical correlation models used in the re-entry risk analysis can be improved by comparing with the high-fidelity numerical simulation results.

Since the accuracy of the aerodynamics and the heat flux model has significant influence on the evaluation of the expected casualty, the accuracy of the used CFD method for the destructive re-entry analysis is essential. In the last two decades, the comprehensive and intensive research efforts have been made in order to improve CFD schemes and the computational speed especially for the aerodynamic design of the aircrafts, rockets, re-entry vehicles. Consequently, the aerodynamic and the heat flux analysis for the complicated geometry have now mature and available.

However, there are still technological gaps to establish an accurate CFD for the destructive re-entry, those are (1) model validation and accuracy improvements for the non-smooth bodies, (2) development of the capability to handle deforming complicated bodies, and (3) development of the heat flux model including important effects such as the surface chemical reactions and the shape deformations.

In this study, two state-of-the-art in-house CFD codes, named as UNITED and LS-FLOW have been employed. Those codes have advantages such as the capability to handle complicated geometry, and the capability of the solution adaptive mesh refinement (AMR). These two codes have been employed to predict various types of the aerodynamic characteristics and the heat flux predictions, and intensive research efforts for the accuracy validaton studies have been made. Technical details on the computational fluid dynamics are described here below.

5.1 Rarefied Flow Solver (UNITED)

At the high altitude, roughly over 90 km, the effecs of the rarefied gas dynamics such as non-equilibrium of thermodynamics and chemical reacions are significant. For the prediction of the surface pressure and heat flux distributions, the Boltzmann equation of kinetic theory is solved by direct simulation Monte Carlo (DSMC) method [10]. In this study, University of Tokyo Evolution DSMC (UNITED) code is employed, which has been originally developed at the university of Tokyo and has been successively developed by JAXA.

UNITED has been developed and employed for the wide range of applications such as the thrustor plume interaction issues for the cargo transfer vehicle HTV [11], analysis of the aerodynamic fluctuations for the rocket upper stages.

The variable soft sphere [12] (VSS) and variable hard sphere (VHS) models [13] are available for the collision model. It also has the capability of the sub-cells [14] and chemical reactions of air [15].

Geometries/shapes of satellites and rocket upper stages are complicated and change in time. The required grid resolution is strongly depending on the important flow structures, and thus flow conditions. For the computational grid, Cartesian cut cell with AMR method is employed. In addition, the capability to handle multiple bodies in 6DoF motion is available. These are key features to analyze moving multiple complicaed objects within the practical turn-around time.

5.2 Continuous Flow Solver (LS-FLOW)

At the moderate altitude, roughly bellow 90 km, the flow can be treated as continuous. For the prediction of the surface pressure and heat flux distributions, the compressible Navier-Stokes flow solver for the arbitrary polyhedral unstructured grids, named LS-FLOW [16] is employed in this study. It is the core CFD code of JAXA, which has been applied to various flow problems including the aerodynamic analysis [17], high temerature reactive flows and the cryogenic flows [18] for the liquid rocket engines. It was initially developed by the present author [16] and has been improved continuously, various numerical schemes have been implemented [17]. LS-FLOW has been validated by comparing with the experimental data for the

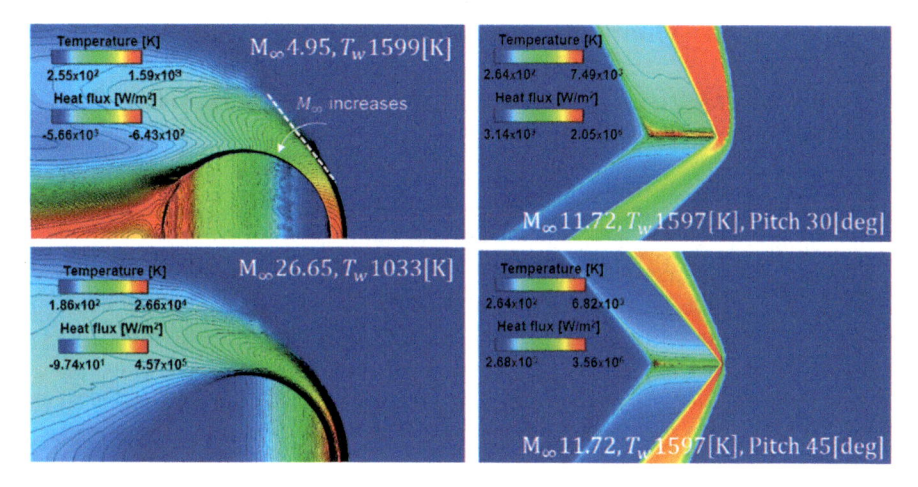

Fig. 5 Mach number and wall temperature dependency on the heat flux

various types of the geometris under the wide range of flow conditions [19, 20]. Then, it has been applied to various aerodynamic design problems such as the aerodynamic analysis for the expendable and re-usable rockets and the re-entry capsules, and the capsule outer shell separation in the chaotic freestream conditions.

The system of equations are discretized and solved by the cell-centerd finite volume method based on the arbitrary polyhedral unstructured grids. Various types of the numerical schemes have been empoyed such as for the gradient evaluation of flow variables, the gradient limter functions, Euler flux, and the viscous flux. Some of the schemes have been developed in JAXA in order to establish the robust and still spatially accurate aerodynamic analysis methodology for the turbulent high Reynolds number flows with including the strong shockwaves and the massive separated flows. Various types of the turbulence models and the high-order reconstruction scheme for the unstructured grids [21] are available. Further detailed descriptions on the key features can be followd in the previous publications [20].

For the capability to handle moving multiple complicated objects, LS-FLOW has the capability to handle the arbitrary polyhedral unstructured grid, and the overset grid method with including 6DoF motion. For the computational grid, body-fitted Cartesian grid generated by LS-GRID [22] is used in this study, in which Cartesian grid is used as the volumce cells and the body-fitted layer grid is used to resolve high Reynolds number boudary layers. Since volume cells are topologically Cartesian grid, and thus the AMR method is available, which is essential to achieve practical turn-around time to consider the various types of geometries under the wide range of flow conditions.

6 Uncertainty Quantification for Rocket Upper Stages

6.1 Parameter Dependency on the Aerodynamics and Heat flux

As already investigated in previous studies, the parameter dependencies of the flow conditions, such as Knudsen, Mach and Reynolds numbers, and the wall temperature, on the aerodynamic characteristics and the heat flux should be considered correctly in the re-entry risk analysis. Firstly, the aerodynamic characteristics and the heat flux are predicted by LS-FLOW for the base geometries used in ORSAT-J in order to investigate the accuracy of the present empirical correlation models and parameter dependencies such as free-stream Mach number and the wall temperature. Results of the two flow conditions are shown for the sphere whose diameter is 0.554 m (Fig. 5). One conditions are flow speed 7336.8 m/s, wall temperature 1033.2 K, and altitude 85.28 km, the other conditions are flow speed 1586.3 m/s, wall temperature 1599.7 K, and altitude 41.86 km. Results of two flow conditions are shown for the box, whose dimensions are 0.085 m, 0.085 m and 0.11 m. Flow conditions are flow speed 3821 m/s, wall temperature 1597 K, altitude 53.7 km, pitch angles are 30 and 45°. Those conditions are equivalent to the trajectory obtained in the re-entry risk analysis for the rocket upper stages by ORSAT-J. Computed surface heat flux and temperature distribution in the pitch symmetry plane is shown in Fig. 5. Since shock wave angle has significant effect on the flow variable distributions behind the shock wave, there is strong Mach number dependency on the resulting surface pressure and the heat flux distribution. As shown in the figure for Mach number 4.95, near-wall flow temperature is less than the wall temperature, therefore difference of the heat flux distribution is significant as comparing with that for the Mach number 26.65. It is clarified that the wall temperature effect can be dominant if the stagnation temperature becomes relatively close to the wall temperature. In addition, it is also clarified that the significant larger heat flux is observed near the sharp corners as shown in Fig. 5, which is mainly due to the smaller shockwave distance at the larger surface curvature area. Although it is not shown here, however, it has been confirmed that the accuracy of the conventional models is excellent especially for the smooth shapes such as the spheres, while there should be continuous research efforts to improve the accuracy for the non-smooth shapes. Technological difficulties lie in how to describe the governing flow mechanism based on small number of parameters such as the surface normal direction against free-stream and the surface local curvatures. Small curvature effect is considered in the empirical correlation model developed for LS-DARC, so that the significant increase of the heat flux at the sharp corner is quantitatively predicted [7].

6.1.1 Turbulent Flow Effect on the Heat flux

Turbulent flow effects on the heat flux level and its distribution is another key factor of the uncetainty. CFD analysis of the heat flux distribution for the flat-faced round disk whose diameter is 1.665 m is carried out by LS-FLOW in this study. Two

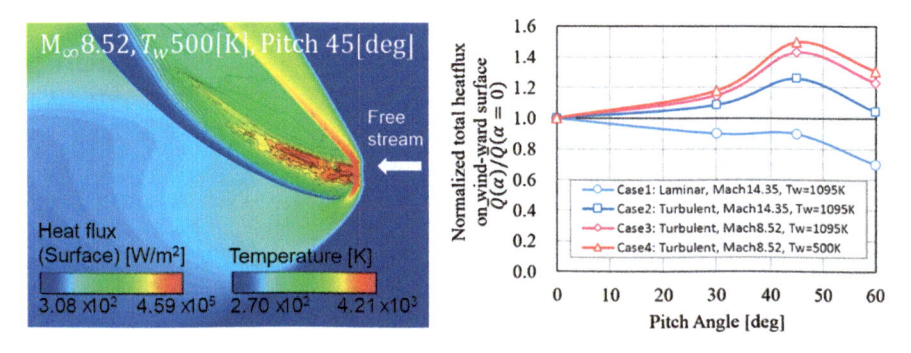

Fig. 6 Parameter depency on wind-ward mean heat flux for flat-faced body

flow conditions are considered, the one is free-stream static pressure 7.99 Pa, static temperature 227 K, and flow speed of 4337 m/s, the other is static pressure 93.87 Pa, static temperature 270.65 K, and flow speed of 2810 m/s. In addition, two wall temperatures are considered 1095 K and 500 K, respectively. Computed surface heat flux and temperature distribution in the pitch symmetry plane is shown in Fig. 6. Total heat flux variation on the wind-side flat surface normalized by the value at the zero pitch angle is also shown in Fig. 6. In the laminar flow computations Case 1 as shown in Fig. 6, total heat flux decreases with the pitch angle. While, in the turbulent flow computations Case 2, total heat flux increases with the pitch angle. It is not shown due to the restrictions of data disclosure, the similar trend has been observed in the experimental data. It is clarified that there is significant turbulent flow effect and the pitch angle dependency on the total heat flux. In most of the conventional empirical correlation models, the surface normal direction against the free-stream, the surface curvature, and the pressure are used to describe heat flux distributions. Thus, it is clear that this total heat flux increase with the pitch angle cannot be predicted by the conventional models. It is clear that there are also Mach number dependency and the wall temperature dependency on the total heat flux as shown in Fig. 6. In addition to consider the turbulent flow effect, these dependencies also should be considered. Therefore, the comprehensive research efforts have been made to improve the empirical correlation model in this study.

6.1.2 Real Shape Effects

As is already discussed in the previous sections, the shape of the satellites and the rocket upper stages are complicated and it also changes in time due to the thermal or mechanical failures. In order to obtain the validation data for the empirical correlation models, and to reconsider the assumptions of the re-entry risk analysis, high-fidelity numerical simulations for the rocket upper stages are carried out by using UNITED. Considered flow speed is 7900 m/s and the altitude is 95 km. Computed number density, the heat flux and the surface pressure distributions are

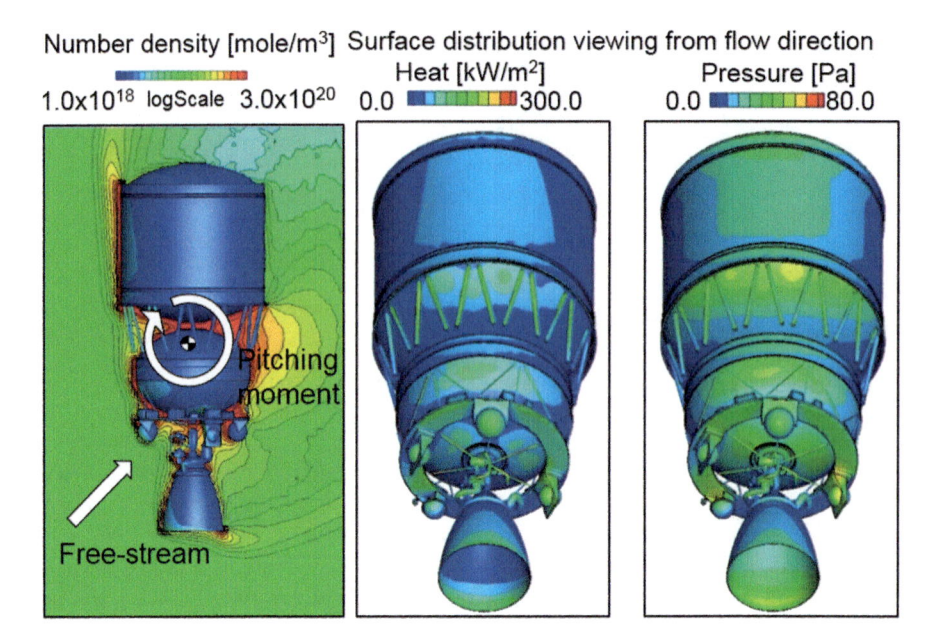

Number density [mole/m³] Surface distribution viewing from flow direction
                          Heat [kW/m²]   Pressure [Pa]
1.0x10¹⁸ logScale 3.0x10²⁰ 0.0 ▬▬▬▬300.0   0.0 ▬▬▬▬80.0

Fig. 7 Surface heat flux and pressure distributions for the rocket upper stage

shown in Fig. 7. In terms of the attitude stability mode, the axial location of the center of gravity tends to be wind-ward due to the heavier mass of the engines, and the projection area of the aft part is larger than that of the fore part. As a result, it is numerically shown that the vehicle might be aerodynamically stable since the pitching trim angle is available. By using high fidelity numerical simulations, the detailed distributions of heat flux and surface pressure can be obtained. This greatly contributes to the uncertainty reduction of the empirical correlation model by the comparison of results. In addition, in order to obtain the effective flight data to reduce the uncertainty, the sensor location has been selected based on the detailed analysis of the computed results. Since the heat flux distributions is complicated due to the flow interactions between the system components such as shock wave impingement, and it significantly changes depending on the vehicle's velocity and the attitudes, the sensor location should be carefully selected based on the sensitivity analysis for the uncertainty reduction. The optimum selection of the seonsor locations significantly contribute to the uncertainty reduction.

Aerodynamic interactions between the multiple objects such as shock-wave interaction and the low dynamic pressure wake effect on the down-stream objects and the aerodynamic characteristics of the moving objects are also the uncertainty factors. Although the investigation on this issue is under way in this study, it has been demonstrated that the 6DoF motion analysis for the multiple complicated ojects is applicable to practical problems by using UNITED. Based on this method, time variation of the pressure and the heat flux can also be predicted with including flow interaction between multiple system components. In this study, the aerodynamic characteristics and the heat flux distributions predicted by the high-fidelity numerical simulation have been compared with those obtained by the conventional empirical correlations. In addition, the trajectory and the attitude history predicted by the high-fidelity numerical simulations have also been compared with those obtained by the conventional models as coupling analysis validation.

Although the uncertainties on the high-fidelity numerical simulation should be still quantified, the high fidelity numerial simulation should play an important role to validate the empirical models and to understand the complicated physics of the destructive re-entry.

7 Conclusion

An improvement in the accuracy of re-entry risk analysis is getting increased interest in recent years. Important uncertainty factors have been identified, and those quantification methods have been proposed. Core approaches to achieve significant uncertainty reduction such as the high-fidelity numerical simulations and the flight data acquisition system were proposed. Some of the findings related to the identified uncertainty factors were discussed. It was shown that various types parameter dependencies such as Mach number, wall temperature, surface curvature, and the effect of the turbulent flows on the aerodynamic characteristics and the heat flux distributions should be considered for the uncertainty reduction. Finally, it was proposed and the demonstrated that the aerodynamic analysis and the heat flux evaluation for the realistic complicated geometry is available. In addition, 6 degree-of-freedom motions of the multiple moving fragments were also demonstrated.

By using high fidelity numerical simulations, it is possible to validate the current empirical correlation models and to re-consider the risk analysis assumptions such as the random tumbling and the surface area which is strongly related to the amount of the aerodynamic heat. It is clear that high-fidelity numerical simulation should play the important role to maximize the accuracy of the destructive re-entry risk analysis.

References

1. Fujimoto, K., Tani, H., Negishi, H., Saito, Y., Iizuka, N., Okita, K.: Update of aerodynamics and heat flux model for ORSAT-J. 8th IAASS conference (2016)
2. Fujimoto K., Saito Y., Shimizu R., Kawamoto S., Ohnishi M.: Overview and progress of research on space debris in JAXA. Stardust final conference on asteroids and space debris (2016)
3. Dobarco-Otero, J., et al.: Upgrades to object reentry survival analysis tool (ORSAT) for spacecraft and launch vehicle upper stage applications. In: AAS science and technology series Space Debris and Space Traffic Management Symposium, vol. 109, pp. 273–289 (2003)

4. Omaly, P., Magnin, V.C., Galera, S.: DEBRISK, CNES tool for re-entry survivability prediction: validation and sensitivity analysis. Proceedings of the 6th IAASS Conference, Montréal, Ca, 21–23 May 2013
5. Lips, T., Fritsche, B., Homeister, M., Koppenwallner, G., Klinkrad, H., Toussaint, M.: Re-entry risk assessment for launchers–development of the new SCARAB 3.1L. Proceedings of the 2nd IAASS conference (2007)
6. Annaloro, J.: Elaboration of new spacecraft-orientated tool: PAMPERO. 8th European symposium on aerothermodynamics for space vehicles, Lisbon, 2–6 March 2015
7. Fujimoto, K., Negishi, H., Saito, Y., Spel, M., Prigent, G.: Benchmark of JAXA and CNES reentry safety analysis tools for accurate heat-flux prediction. In: Proceedings of the 9th IAASS Conference, Toulouse, France, 18–20 October 2017
8. Detra, R.W., Kemp, N.H., Riddell, F.R.: Addendum to heat transfer to satellite vehicles reentering the atmosphere. Jet Propul. **27**(12), 1256–1257 (1957)
9. Sugimura, F., Moriya, A., Kishino, Y., Morisaki, H., Makino, T., Wada, K., Yamamoto, H., Sasaki, H.: Development of i-Ball. In: Proceedings of 29th International Symposium on Space Technology and Science, 2013-g-15, Nagoya, Japan, June 2–9, 2013
10. Bird, G.: Molecular gas dynamics and the direct simulation of gas flows. In: Oxford Engineering Science Series, vol. 42. Clarendon Press, Oxford (1994)
11. Tani, H., Ohmaru, T., Takata, S., Uematsu, H., Matsuura, M., Daimon, Y., Negishi, H.: Hybrid N-S/DSMC simulations of HTV main engine plume impingement to the ISS. J. Jpn. Soc. Simul. Technol. **34**(3), 186–192 (2015)
12. Koura, K., Matsumoto, H.: Variable soft sphere molecular model for inverse-power-law or Lennard-Jones potential. Phys. Fluids. **3**, 2459–2465 (1991)
13. Bird, G.A.: Monte-Carlo simulation in an engineering context. Prog. Astronaut. Aeronaut. **74**, 239 (1981)
14. Boyles, K.A., et al.: The use of virtual sub-cells in DSMC analysis of orbiter aerodynamics at high altitudes upon reentry. AIAA-2003-1030, 41st AIAA aerospace sciences meeting and exhibit, Reno, NV, January 2003
15. Bird, G.A.: Chemical reactions in DSMC. 27th international symposium on rarefied gas dynamics, Pacific Grove, CA, 10–15 July 2010
16. Fujimoto, K.: Study on the automated CFD analysis tools for conceptual design of space transportation vehicles. Ph.D. Dissertation, University of Tokyo, Tokyo, Japan (2006)
17. Kitamura, K., Nonaka, S., Kuzuu, K., Aono, J., Fujimoto, K., Shima, E.: Numerical and experimental investigations of epsilon launch vehicle aerodynamics at Mach 1.5. J. Spacecr. Rocket. **50**(4), 896–916 (2013)
18. Negishi, H., Aono, J., Shimizu, T., Sunakawa, H., Sezaki, C., Nagao, N., Nanri, H.: Mixing characteristics of transcritical hydrogen flows around a mixer in a liquid rocket engine (in Japanese). JSASS-2016-2117-A (2016)
19. Kitamura, K., Fujimoto, K., Shima, E., Kuzuu, K., Wang, Z.J.: Validation of arbitrary unstructured CFD code for aerodynamic analyses. Transac. Jpn. Soc. Aeronaut. Space Sci. **53**(182), 311–319 (2011)
20. Fujimoto, K., Nambu, T., Negishi, H., Watanabe, Y.: Validation of LS-FLOW for reentry capsule unsteady aerodynamic analysis. AIAA paper 2017-1411, 55th AIAA Aerospace Sciences Meeting, AIAA SciTech Forum (2017)
21. Haga, T., Kawai, S.: Toward accurate simulation of shockwave-turbulence interaction on unstructured meshes: a coupling of high-order fr and lad schemes. AIAA Paper 2013-3065 (2013)
22. Fujimoto, K., Fujii, K., Wang, Z.J.: Improvements in the reliability and efficiency of body-fitted cartesian grid method. AIAA Paper 2009-1173 (2009)

HDMR-Based Sensitivity Analysis and Uncertainty Quantification of GOCE Aerodynamics Using DSMC

Alessandro Falchi, Edmondo Minisci, Martin Kubicek, Massimiliano Vasile, and Stijn Lemmens

Abstract A sensitivity analysis of aerodynamic coefficients has been performed by coupling a Direct Simulation Monte Carlo method and a High Dimensional Model Representation based uncertainty quantification approach. The study has been performed on the Gravity Field and Steady-State Ocean Circulation Explorer satellite. The uncertainty on aerodynamics has been quantified with respect to atmospheric parameters, which have been obtained using the NRLMSISE-00 atmospheric model, within a free molecular flow regime in the Low Earth Atmosphere. The aerodynamic simulations have been performed with the dsmcFoam code, based on the open-source OpenFOAM platform.

1 Introduction

Aerodynamic simulations of satellites, space debris and asteroids are of utmost importance when performing re-entry analysis, impact prediction and satellite design optimization. The identification and characterization of the impact footprint of re-entering objects/fragments is largely dependent on the prediction of aerodynamic characteristics. Although most known tools performing aerodynamic simulations are deterministic, and then do not take into account the uncertainty associated with the provided inputs, in fact, atmosphere parameters, attitude, and in some specific case also the geometry may be affected by uncertainty. In addition,

---

A. Falchi (✉) · M. Kubicek
Aerospace Centre of Excellence, University of Strathclyde, Glasgow, UK
e-mail: alessandro.falchi@strath.ac.uk; martin.kubicek@strath.ac.uk

E. Minisci · M. Vasile
Aerospace Centre of Excellence, Department of Mechanical and Aerospace Engineering, University of Strathclyde, Glasgow, UK
e-mail: edmondo.minisci@strath.ac.uk; massimiliano.vasile@strath.ac.uk

S. Lemmens
ESA/ESOC, Robert-Bosch-Str. 5, 64293 Darmstadt, Germany
e-mail: Stijn.Lemmens@esa.int

also some coefficients employed by the methods, which are usually treated as conventional constants, are uncertain. The uncertainty in the inputs, coefficients, and model parameters, has a direct impact on the analyses outcome. Performing the uncertainty quantification (UQ) of the aerodynamic performance will provide the statistical distribution of aerodynamic coefficients.

In this paper, a preliminary sensitivity analysis (SA) and UQ of the Gravity Field and Steady-State Ocean Circulation Explorer (GOCE) satellite aerodynamics is presented. The SA is extremely useful to understand which parameters mostly influence the considered phenomena. The uncertainty quantification process characterizes the uncertainty of the output on the basis of the distributions which describes the input uncertainties. The free molecular aerodynamic UQ analysis is important in order to estimate the confidence on the coefficients commonly used; e.g.: the C_D assumed to be 2.2. As explained by Cook [4], the aerodynamic coefficients are highly influenced by the molecular speed ratio, which depends on atmospheric parameters as temperature, molecular composition, satellite velocity. In addition, as shown by Moe et al. [8, 9] the drag coefficient depends also on the diffusive-reflective energy accommodation coefficient, which is commonly assumed to be 0.93 and it is affected by uncertainty. Moreover, the aerodynamic coefficients are also affected by the wall to flow temperature ratio, as it has been reported by Bailey [1]. Instead, simplified aerodynamic tools [11, 13] usually employ constant aerodynamic coefficients that can affect the aerodynamic behavior of the re-entering object. Therefore, it is paramount to quantify the uncertainty on the previously cited parameters on the aerodynamics in order to perform robust re-entry analyses.

The SA and UQ in this work have been performed by coupling the Adaptive Derivative High Dimensional Model Representation (AD-HDMR), developed by Kubicek et al. [7], and the Direct Simulation Monte Carlo (DSMC) code dsmcFoam developed by Scanlon et al. [14] at the University of Strathclyde, and based on the platform OpenFOAM. The AD-HDMR algorithm can use a single-fidelity (SF) or a multi-fidelity (MF) approach, and both of them have been used to perform the SA and evaluate the efficiency of the technique.

Two different studies for GOCE's aerodynamics are presented focusing on the drag coefficient (C_D); the first one is a global sensitivity analysis study performed with five variables: molecular speed ratio (SR), side-slip angle (α), single atomic species N_2, wall temperature to free flow temperature ratio, and diffusive accommodation coefficient. For this simplified test case, the ranges of the variables were decided arbitrarily, on the basis of their influence over the aerodynamics known from the literature. In the second case, the UQ has been studied in accordance with GOCE's real orbital altitude and its orbital parameters, nine different variables have been examined in this case study: the 5 ones previously cited; three additional atomic species (O, O_2, Ar), and the vibrational relaxation coefficient. To determine the input variables distribution for the real case, the analysis of the Naval Research Laboratory Mass Spectrometer and Incoherent Scatter Radar (NRLMSISE-00) has been performed with respect to altitude, latitude, and longitude. The influence of atmospheric parameters on the input distribution will be shown.

2 Atmosphere Model Range Analysis

The atmosphere model analysis was required in order to understand the variability range of each considered parameter. The atmospheric variables considered in this work are the following:

- Four species atomic density [N_2, O, O_2, Ar]
- Atmospheric Temperature
- Altitude
- Latitude and longitude
- Time and date of the year

For the purpose of the performed analyses, it was decided to use a mean magnetic index. In this work, the atmospheric model developed by the U.S. naval research laboratories, the "Naval Research Laboratory Mass Spectrometer and Incoherent Scatter Radar Exosphere [12]" (NRLMSISE-00), was used. This atmosphere model is based on the older MSISE-90 and provides experimental-based data on the mass density and neutral temperature from ground to ~1000 km altitude (lower exosphere).

The DSMC code uses as direct inputs the atmospheric neutral temperature and atomic densities. All these parameters vary with respect to altitude, longitude, latitude, time and date of the year:

$$T_\infty = f(H, lat, long, time, day, year) \quad (1)$$

$$N_{O_2, O, N_2, Ar} = f(H, lat, long, time, day, year) \quad (2)$$

These inputs change sensibly with all the considered variables, in order to perform a C_D analysis of the orbiting satellite, it was decided to average the daily values over an entire year of GOCE's re-entry for each of the five parameters.

In addition to the atmospheric inputs, it has been required to use the following information on GOCE geometry and orbital parameters present in the literature [5, 16]:

- GOCE detailed geometric model (Fig. 1)
- Reference cross-section ($A_{ref} = 1.131$ m^2)
- Orbital velocity (V = 7760 m/s)
- Orbital altitude (~260 km)
- Outer Wall Temperature ($T_{Wall} = 350$ K)

As it is known from the literature [1, 17], in the rarefied transitional and free molecular flow, the aerodynamics are highly influenced by the SR ([4], Eq. (3)), wall to free flow temperature ratio (T_{wall}/T_∞), and the degree of rarefaction, which is defined according to the Knudsen number (Kn). Another important parameter for the Gas-Surface Interaction model, that cannot be derived directly from the atmospheric model, is the specular-diffusive accommodation coefficient (σ_{diff}, also known as energy accommodation coefficient [9]).

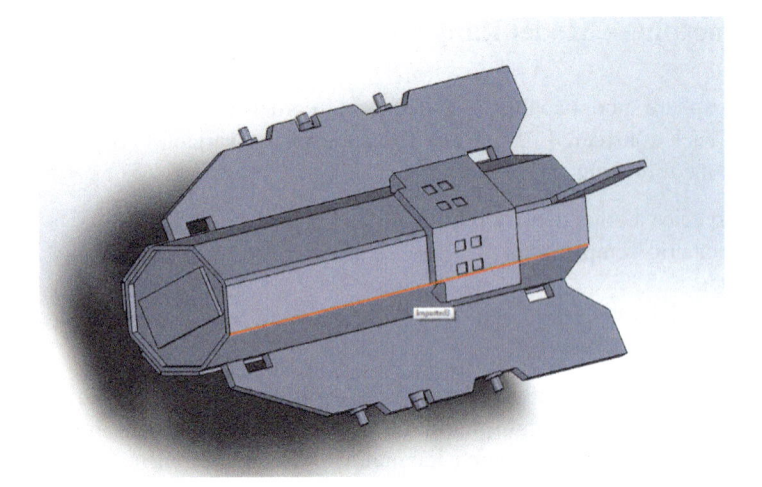

Fig. 1 GOCE satellite 3D model used for the aerodynamic simulations

$$SR = \frac{Satellite Velocity}{MostProbableMolecularSpeed} = \frac{V}{MPMS} \quad (3)$$

where the Most Probable Molecular Speed is defined as:

$$MPMS = \sqrt{2RT_\infty} \quad (4)$$

where R is the specific gas constant.

The diffusive accommodation coefficient is a parameter that influences the particle-surface energy exchange and it is a DSMC simulation parameter. Although, this parameter is subject to uncertainty but within the Low Earth Orbit (LEO) it is generally assumed $\sigma = 0.93$. In this work, the diffusive accommodation coefficient uncertainty has been characterized with a uniform distribution.

The investigation has been performed within the Free Molecular Flow (FMF) Regime. Even though, the Kn (Eq. (5)) changes with the altitude, solar activity, latitude and longitude. Therefore, it has been required to compute the Kn for the analyzed atmospheric range.

$$Kn = \frac{\lambda}{l_{ref}} \quad (5)$$

Where λ is the *Mean Free Path* (MFP: Eq. (6)), and the l_{ref} is the object reference length. The mean free path for a rarefied gas, according to the Chapman-Enskog viscosity coefficient and the Variable Hard Sphere (VHS) molecular model [2, 6] is defined as in Eq. (6).

$$\lambda_{VHS} = \left(\frac{2\mu}{15\rho}\right)(7-2\omega)(5-2\omega)\frac{1}{\sqrt{2\pi RT_\infty}} \quad (6)$$

Where μ is the viscosity coefficient and it is computed according to the Chapman-Enskog Hard Sphere model:

$$\mu = \frac{5m}{16\pi\delta^2}\sqrt{\pi RT_\infty} \quad (7)$$

The atmospheric model analysis has been performed in the following ranges:

- **Altitude:** the range is from 220 to 280 km
- **Latitude:** the range is from -90 to 90 deg
- **Longitude:** fixed to 0 deg

This investigation lead to the results shown in Figs. 2, 3, 4.

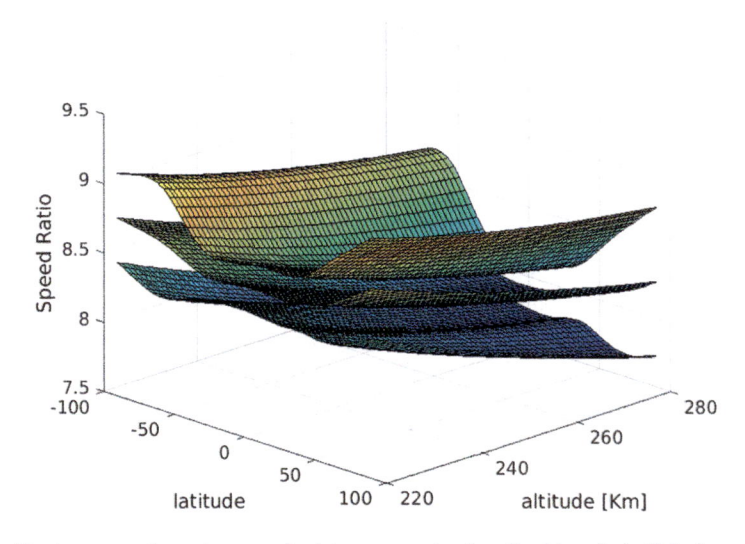

Fig. 2 Yearly averaged, maximum and minimum speed ratio—fixed longitude (0 deg)

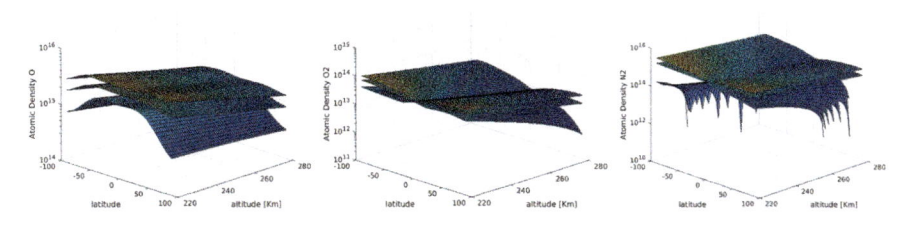

Fig. 3 Yearly averaged, max and min atomic density O (left), O_2 (center), N_2 (right)—fixed longitude (0 deg)

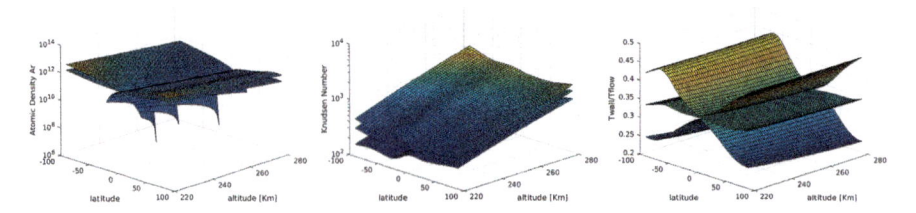

Fig. 4 Yearly averaged, max and min Ar atomic density (left), Kn (center) T_{ratio} (right)—fixed longitude (0 deg)

3 Direct Simulation Monte Carlo

In this work, the DSMC code developed at the University of Strathclyde, dsmcFoamStrath, based on the free and open-source CFD (C++) platform openFOAM [10], has been used. The DSMC solves the Boltzmann's equation by using a statistical representation of the particles present in a rarefied flow. Firstly developed by Bird [3], the method can provide a HF representation for molecular gas dynamics. A DSMC simulation can be described by three main phases:

1. Mesh generation and particles initialization
2. Particles tracking and statistical collisions simulation
3. Averaged field properties computation

Initially, the user defines the boundary conditions and geometry. The mesh must be defined in accordance with the Mean Free Path (MFP) in the free stream region. The maximum cell size has to be a fraction of the MFP. The number of particles per cell has to be defined in accordance to the level of precision sought. In the common DSMC practice [3] it is recommended to use a MFP to Cell size ratio between 2 and 3. Usually, the mesh is initialized with 10–40 particles per cell, the higher number of particles usually provides a better accuracy (Fig. 5).

In the DSMC simulation, different boundary conditions can be imposed to best simulate the environment and the flow. For the nature of this problem, a mixed specular-diffusive reflective GSI model, along with a variable hard sphere (VHS) model for the inter-particles collisions and viscosity models, has been employed. The code allows the use of a collision-less model, which is specifically implemented for the FMF, even if the VHS model was used, due to the number of inter-particles collisions being changing within the considered input distributions. In the dsmcFoam code a 5-species Quantum-Kinetic (QK) chemical reacting model [15] is available. Although, the influence of chemical reactions on the aerodynamics at the considered altitude is negligible. Therefore, in order to reduce the computational cost, the reacting model has not been activated.

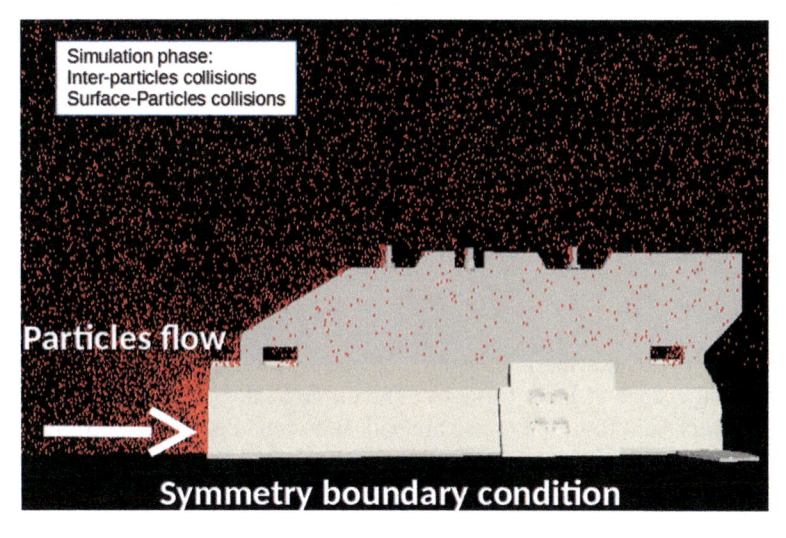

Fig. 5 GOCE free molecular flow, half-symmetry particles representation at steady-state

4 AD-HDMR Uncertainty Quantification

In this paragraph, the AD-HDMR methodology used to perform the UQ and the SA is briefly described.

4.1 High Dimensional Model Representation

The method is based on a cut-High Dimensional Model Representation (HDMR) decomposition, where each increment function (defined in Eq. (10)) is handled separately. This allows the use of various approximation methods for the model of interest. If $F(\mathbf{x})$ is a derivable and integrable function defined on a n-dimensional unit hypercube—$[0, 1]^n$ and $\mathbf{x} \in [0, 1]^n$, the ANOVA representation of $F(\mathbf{x})$ can be given as:

$$F(\mathbf{x}) = F_0 + \sum_{i=1}^{n} F_i(x_i) + \sum_{1 \leq i < j \leq n} F_{i,j}(x_i, x_j) + \ldots + F_{1,\ldots,n}(x_1, \ldots, x_n) \quad (8)$$

where F_0 is the constant term and represents the mean value of $F(\mathbf{x})$, the function $F_i(x_i)$ represents the contribution of variable x_i to function $F(\mathbf{x})$, the function $F_{i,j}(x_i, x_j)$ represents the pair correlated contribution to $F(\mathbf{x})$ by the input variables

x_i and x_j, which are defined as $1 \leq i < j \leq n$, etc. The last term $F_{1,\ldots,n}(x_1,\ldots,x_n)$ contains the correlated contribution of all input variables and the total number of summands for Eq. (8) is 2^n.

Each independently differentiable and integrable term in Eq. (8) is differentiated according to its generic variable x_i to obtain the infinitesimal increment to the function of interest. This leads to the following equation:

$$dF(\mathbf{x}) = \sum_{i=1}^{n} \frac{\partial F(\mathbf{x})}{\partial x_i} dx_i + \sum_{1 \leq i<j \leq n} \frac{\partial F(\mathbf{x})}{\partial x_i, x_j} dx_i dx_j + \ldots + \frac{\partial F(\mathbf{x})}{\partial x_1,\ldots,x_n} dx_1 \ldots dx_n \quad (9)$$

Equation (9) relates the infinitesimal change of the function of interest on the infinitesimal change of input variables. The differential equation as represented in Eq. (9) is very hard to use in practical applications and obtaining derivatives from a function of interest is in many cases a hard task and in some cases practically impossible. Therefore, an integral form of Eq. (9) is introduced as

$$F(\mathbf{x}) - F(^c\mathbf{x}) = \sum_{i=1}^{n} \int_{^cx_i}^{x_i} \frac{\partial F(\xi)}{\partial \xi_i} d\xi_i + \sum_{1 \leq i<j \leq n} \int_{^cx_i}^{x_i} \int_{^cx_j}^{x_j} \frac{\partial F(\xi)}{\partial \xi_i, \xi_j} d\xi_i d\xi_j + \ldots$$
$$+ \int_{^cx_1}^{x_1} \ldots \int_{^cx_n}^{x_n} \frac{\partial F(\xi)}{\partial \xi_1,\ldots,\xi_n} d\xi_1 \ldots d\xi_n \quad (10)$$

where $^c\mathbf{x}$ represents a central position in the stochastic space, called the central point considered to be the statistical mean of a given stochastic random variable. The terms in Eq. (10) that represent the integral of the derivative of each independent function is defined as the increment function.

The non-important stochastic spaces are neglected in order to decrease the number of function calls. The stochastic space reduction is done in two ways; the first predicts the importance of the increment function, and the second neglects the zero-th increment functions. The prediction is based on fundamental aspects of the integral form and an inverse logic. The importance of each interaction is predicted and iteratively added to the final model. The process is stopped when desired accuracy is reached. The neglect approach, as the name suggests, neglects all zeroth order increment functions that are passed through the prediction phase. After selecting the important increment functions, the method switches to an automatic sampling approach and interpolates each increment function separately. This leads to an optimal number of function calls for the problem of interest.

The combination of various interpolation techniques is used which includes the Lagrange interpolation, Kriging model, spline and Independent Polynomial Interpolation, which is in-house developed interpolation technique. The adaptive sampling process takes in account the position, behaviour, and input probability of the function of interest. The position and behaviour are captured using the Error Comparison (EC) function and is later modified to take into account the input probability distribution of the given stochastic random variable. The proposed EC function converges in the L2 sense; however, this convergence is not suitable for practical reasons. Therefore, the convergence criterion is based on the observation of the standard deviation and the expected value. The statistical properties are obtained using the Monte Carlo sampling on the created surrogate model (SM). This allows the visualization of each increment function separately (i.e. the probability distribution function (PDF) for each stochastic random variable.) as well as the final PDF.

4.2 Sensitivity Analysis

The sensitivity of each increment function represents the influence of the corresponding random variable on the model of interest. The orthogonal property of the increment functions allows for each function to be handled separately i.e. each increment function statistics is calculated independently. This allows the definition of the statistical properties of each increment function as follows:

$$\mu_i = \int_{-\infty}^{\infty} \int_{c_{x_i}}^{x_i} \frac{\partial F(\xi)}{\partial \xi_i} d\xi_i p_i(x_i) dx_i \quad (11)$$

$$\sigma_i^2 = \int_{-\infty}^{\infty} \left(\int_{c_{x_i}}^{x_i} \frac{\partial F(\xi)}{\partial \xi_i} d\xi_i - \mu_i \right)^2 p_i(x_i) dx_i \quad (12)$$

where μ_i represents the partial mean and σ_i^2 represents the partial variance. The mean and variance sensitivity indices for each increment function are defined as:

$$S_i^\sigma = \frac{\sigma_i^2}{\sigma^2} \quad (13)$$

$$S_i^\mu = \frac{\mu_i}{\mu} \quad (14)$$

where σ^2 and μ are computed as follows:

$$\sigma^2 = \int_{-\infty}^{\infty} (F(\mathbf{x}) - \mu)^2 p(\mathbf{x}) d\mathbf{x} = \int_{-\infty}^{\infty} \cdots \int_{-\infty}^{\infty} ((F(^c\mathbf{x}) + \sum_{i=1}^{n} \int_{c_{x_i}}^{x_i} \frac{\partial F(\xi)}{\partial \xi_i} d\xi_i$$

$$+ \sum_{1 \le i < j \le n} \int_{c_{x_i}}^{x_i} \int_{c_{x_j}}^{x_j} \frac{\partial F(\xi)}{\partial \xi_i, \xi_j} d\xi_i d\xi_j + \cdots$$

$$+ \int_{c_{x_1}}^{x_1} \cdots \int_{c_{x_n}}^{x_n} \frac{\partial F(\xi)}{\partial \xi_1, \ldots, \xi_n} d\xi_1 \ldots d\xi_n) - \mu)^2 p(\mathbf{x}) dx_1 \ldots dx_n \quad (15)$$

$$\mu = \int_{-\infty}^{\infty} F(\mathbf{x})p(\mathbf{x})d\mathbf{x} = F(^c\mathbf{x}) + \sum_{i=1}^{n} \int_{-\infty}^{\infty} \int_{c_{x_i}}^{x_i} \frac{\partial F(\xi)}{\partial \xi_i} d\xi_i p_i(x_i) dx_i$$
$$+ \sum_{1 \leq i<j \leq n} \int_{-\infty}^{\infty} \int_{-\infty}^{\infty} \int_{c_{x_i}}^{x_i} \int_{c_{x_j}}^{x_j} \frac{\partial F(\xi)}{\partial \xi_i, \xi_j} d\xi_i d\xi_j p_{ij}(x_i, x_j) dx_i dx_j + \dots$$
$$+ \int_{-\infty}^{\infty} \dots \int_{-\infty}^{\infty} \int_{c_{x_1}}^{x_1} \dots \int_{c_{x_n}}^{x_n} \frac{\partial f(\xi)}{\partial \xi_1, \dots, \xi_n} d\xi_1 \dots d\xi_n p_{1\dots n}(x_1, \dots, x_n) dx_1 \dots dx_n$$
(16)

where $p_i(x_i)$ is the PDF for the given variable. It should be noted that the higher order moments cannot be computed as the superposition of the independent partial moments of the increment functions, i.e.

$$\sigma^2 \neq \sum_{k=1}^{2^n-1} \sigma_k^2 \qquad (17)$$

Kubicek et al. [7] describes in detail the High Dimensional Model Representation and the associated SA theory and derivation.

5 DSMC and AD-HDMR Coupling

Since the dsmcFoam code operates within the openFOAM C++ environment while the AD-HDMR has been developed on MATLAB, the programming of a coupling interface had been required. In addition, the simulations have been performed on the ARCHIE-WeSt High Performance Computer, therefore also a job management and a drag convergence algorithm had to be programmed. The operational flowchart can be split into three different macro blocks shown in Fig. 6.

6 Aerodynamics Sensitivity and Uncertainty Quantification Analyses

In this work two different case studies performed on GOCE's geometric model have been considered. The first one is a broad range SA, while the second one is an UQ based on GOCE's real flight data. The setups have been chosen according to previous tests and by considering the best compromise between computational cost and results accuracy. For this work the input distributions are all uniform.

The broad range SA characterizes a set of parameters which are already known to have an effect on the FMF aerodynamics of satellites. The chosen ranges are

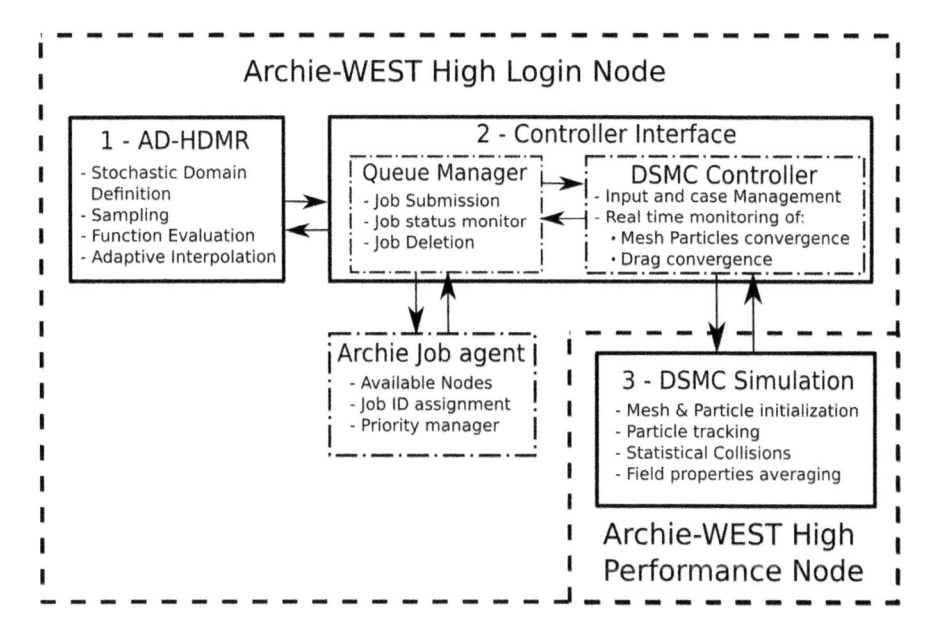

Fig. 6 DSMC and AD-HDMR coupling block diagram

Table 1 GOCE aerodynamics broad range sensitivity analysis input distributions

Input ID	Parameter	Min	Central values	Max	
1	SR	8.0	10.0	12.0	–
2	α	−5.0	0	5.0	–
3	N_2	0.1E+15	5.0E+15	10.0E+15	$\frac{atoms}{m^3}$
4	$\frac{T_{Wall}}{T_\infty}$	0.01	0.75	1.5	–
5	σ_{diff}	0.50	0.755	1.00	–

reported in Table 1. Since the AD-HDMR software had been developed with particular attention to deterministic functions, without any stochastic variability, this study will also be used to assess the reliability of the AD-HDMR when operating with a non-deterministic function.

Five different variables have been chosen: Molecular Speed Ratio, Side-slip, N_2 atoms, T_{ratio}, and σ_{diff}. By investigating broad input ranges, it has been possible to reduce the stochastic influence of the DSMC method, allowing the AD-HDMR code to perform a correct and smooth adaptive interpolation.

In the study case based on GOCE's orbital data (Table 2) the number of variables was higher and their range was narrower. This setup represents a challenging setup for the AD-HDMR, mainly due to the stochastic scattering on the DSMC side. With this investigation, it has been studied the influence of each variable and their interactions on the aerodynamics of GOCE at its flight conditions. The study has been performed with both approaches; MF and SF.

Table 2 On-orbit GOCE aerodynamics uncertainty quantification input distributions

Input ID	Altitude	Min 250	Mean 260	Max 270	[km]
1	SR	7.841	8.090	8.356	–
2	α	−5	0	5	–
3	O	8.16E+14	1.63E+15	2.45E+15	$\frac{atoms}{m^3}$
4	N_2	2.85E+14	5.40E+14	7.95E+14	$\frac{atoms}{m^3}$
5	O_2	9.01E+12	1.52E+13	2.15E+13	$\frac{atoms}{m^3}$
6	Ar	8.32E+10	1.85E+11	2.87E+11	$\frac{atoms}{m^3}$
7	Relax. coeff.	5	2500	5000	–
8	$\frac{T_{Wall}}{T_\infty}$	0.3209	0.3325	0.3440	–
9	σ_{diff}	0.7508	0.8754	1	–

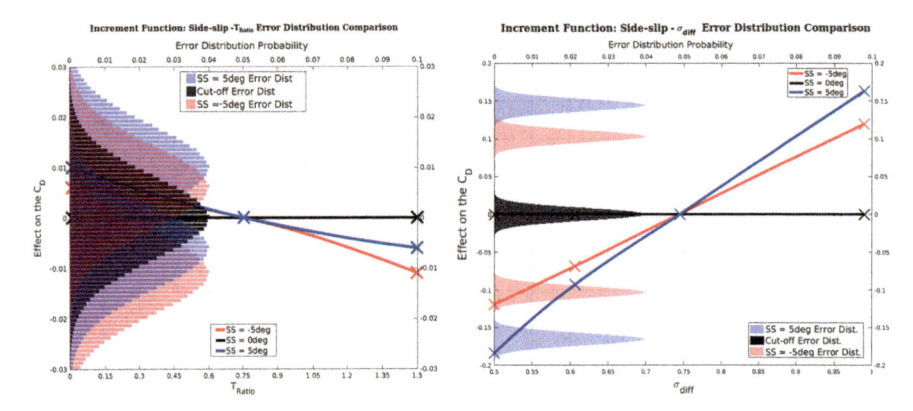

Fig. 7 Significance test applied to $Sideslip - T_{Ratio}$ (on the left) and $Sideslip - \sigma_{diff}$ interaction increment function

6.1 Broad-Range Sensitivity Analysis of GOCE Aerodynamics

A set of preliminary analyses has been required in order to verify that the investigation was carried out with an optimal setup. In this first analyses different tests have been carried out:

- a preliminary test on the speed ratio variation
- a preliminary test on the particle per cell optimal compromise
- a preliminary test on the convergence and control algorithm
- a post processing assessment for comparing the DSMC convergence error distribution to the parameters identified as significant by the AD-HDMR algorithm

For the sake of brevity, only the latest and most significant test has been reported in this work, i.e. the significance test on the identified surrogate models (Fig. 7). The significance test is applied to both a non-significant parameter (on the left), and a significant one (on the right). The example significance test has been applied on two different interactions: between side-slip-T_{Ratio} and side-slip-σ_{diff} parameters. In the figures, the black histogram represents the convergence error probability distribution determined by previous tests. The overlapping area (and probability) between the convergence error distributions and all the maximum and minimum sampled points distributions is computed and compared to an edge value. The input is considered negligible only when all the overlapping probability is below an input probability (equal to 95.5% for the broad range SA and 90.0% for the on-orbit UQ). The complete description of the significant outputs determination has been omitted for the sake of brevity.

6.1.1 Sensitivity Analysis Results

Given the input distributions shown in Table 1, the AD-HDMR coupling computed the SA results shown in Table 3. A brief description of the sensitivity analysis results is required: the coded characterized inputs are shown under the "Increment function" column. As an example, consider the code 1.2.5 is representative of a third order interaction between the variables 1–2–5 which are the SR, α and σ_{diff} respectively. The *Partial Mean* and the *Partial Variance*, are respectively the mean and variance of the increment function distributions of each different variable. The *Sensitivity Mean* is the sensitivity index on the mean C_D, according to each single variable surrogate model distribution weight on the overall mean C_D distribution; the *Sensitivity Variance* is the most important parameter, as it is representative of the variables increment function maximum absolute deviation, i.e.: the maximum contribution to the evaluated parameter. The algorithm evaluated the five single variables and adaptively detected six 2nd order interactions and one 3rd order interaction.

Table 3 GOCE drag coefficient preliminary sensitivity analysis results

Increment function	Partial mean	Partial variance	Sensitivity mean	Sensitivity variance	Input significance [%]
1	0.0168	0.0175	0.1021	0.4625	100
2	0.1387	0.0147	0.8429	0.3868	100
3	0	0	Neglected	Neglected	28.43
4	-0.0041	0.0011	0.0252	0.0283	100
5	0.0014	0.0015	0.0083	0.0406	99.97
4.5	0	0	Neglected	Neglected	49.53
2.5	-0.0005	0.002	0.0033	0.0535	100
2.4	0	0	Neglected	Neglected	49.52
1.5	-0.001	0.0005	0.0064	0.0129	100
1.4	0	0	Neglected	Neglected	55.43
1.2	-0.0019	0.0006	0.0118	0.0153	100
1.2.5	0	0	Neglected	Neglected	64.14

The results show that the significant variables, according to a 0.5% maximum convergence error and a 95.5% confidence level, are the following:

1. **[4/5] first order parameters:** SR, α, T_{Ratio}, σ_{diff}
2. **[3/6] second order interactions:** $\alpha - \sigma_{diff}$, SR-α, SR-σ_{diff}.
3. **[0/1] Third order interaction:** the only third order interaction that had been initially evaluated has been defined as negligible by the significance test.

In this analysis, it is possible to observe that the most influencing parameters according to the sensitivity on the variance are the *SR* (46.25%) and α(38.68%), followed by the $\alpha - \sigma_{diff}$ interaction (5.35%) and the σ_{diff}(4.06%). The remaining 5.65% is split among the T_{ratio}, and the 2nd order interactions between SR-σ_{diff} and SR-α. It must be reminded that the validity of these results is subjected to the analysis range (Table 1).

6.1.2 AD-HDMR Surrogate Models

Analyzing the resulting increment functions, two different SMs have been created and tested. It must be highlighted that The SMs are strictly valid within the investigated ranges. The first SM uses only four significant first order increment functions described by a set of second order polynomial functions (Eq. (18)). The second SM has been built with all the significant parameters, the complete model has not been reported.

$$\widetilde{C}_{D,1st} = 0.0123 SR^2 - 0.3604 SR + 0.0158\alpha^2 - 0.0093\alpha - 0.0383 T_{Ratio}^2$$
$$+ 0.1317 T_{Ratio} + 0.0965 \sigma_{diff}^2 - 0.4206 \sigma_{diff} + 5.8511 \qquad (18)$$

In order to verify the results, a simplified DSMC Latin Hypercube Sampling (LHS) has been run with the same precision and conditions as the AD-HDMR simulations. The LHS was performed with a relatively small number of initial inputs: 3000 samples. In Fig. 8 the comparison of the probability histograms obtained through the DMSC LHS, AD-HDMR (complete significant model) and the 1st order SM LHS is shown. The results show a good agreement among the different distributions. In Table 4 the major advantage of using the surrogate models, which is the huge computational times difference, is highlighted. In the same tables the mean C_D and standard deviation for each different case, are reported, showing a very good agreement between the methods.

6.2 GOCE On-Orbit Aerodynamic Uncertainty Quantification

In this paragraph the specific application of the AD-HDMR and dsmcFoam coupling to perform the UQ of GOCE aerodynamics during the drag-free orbiting phase, according to what has been described in Sect. 2, is described.

DSMC LHS - SM - AD-HDMR C_D Probability Histograms Comparison

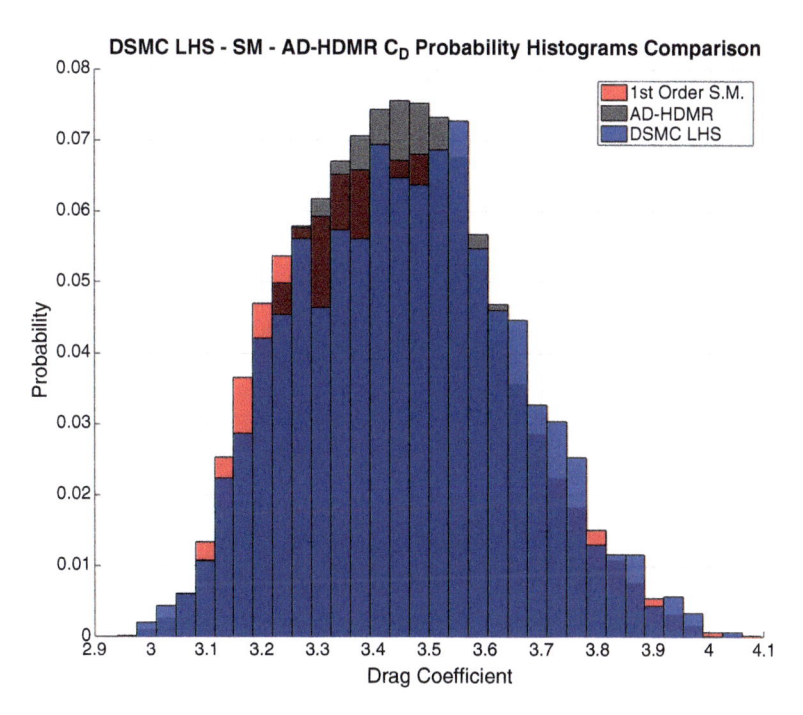

Fig. 8 Drag coefficient probability histograms comparison. DSMC LHS (3000 samples), 1st order SM LHS, and AD-HDMR

Table 4 AD-HDMR broad range sensitivity analysis results summary

	$\overline{C}_D$	σ_{C_D}	Run-time [core-hours]
DSMC LHS	3.4601	0.1701	3895
AD-HDMR	3.4374	0.1694	121
1st SM LHS	3.4393	0.1867	~ 0

6.2.1 Single-Fidelity Uncertainty Quantification and Sensitivity Analysis Results

The AD-HDMR SF approach has been run three different times, in this way it has been possible to evaluate the approach repeatability. The SA relative to the uniform distributions (Table 2) have led to the results reported in Table 5, for the sake of brevity only the significant inputs have been listed. The table shows also the $C_{D, central}$, which is the C_D computed for the input distributions central values. The results show that the most influencing parameter on the C_D uncertainty is the wind velocity (modeled with the side-slip angle), the residual mean sensitivity is split between the speed ratio (representative for the satellite velocity) and the interaction between the side-slip and the σ_{diff}.

Table 5 AD-HDMR SF sensitivity analysis of GOCE C_D using 90% significance limit of three different simulations

Increment function	Partial mean	Partial variance	Sensitivity mean	Sensitivity variance	Input significance [%]
1st SF simulation, $C_{D_{central}} = 3.477$					
1	−0.0009	0.0008	0.0055	0.0353	93.34
2	0.1628	0.0203	0.9891	0.9467	100
2.9	0.0009	0.0004	0.0054	0.0179	99.97
2nd SF simulation, $C_{D_{central}} = 3.470$					
1	0.0043	0.0006	0.0244	0.0262	92.96
2	0.1669	0.0218	0.9478	0.9547	100
2.9	−0.0049	0.0004	0.0278	0.0191	100
3rd SF simulation, $C_{D_{central}} = 3.488$					
1	−0.0028	0.0008	0.0168	0.039	96.22
2	0.1603	0.0197	0.9716	0.9451	100
2.9	0.0019	0.0003	0.0116	0.0159	99.99

By analyzing the single increment functions, it's been possible to observe that the effect of the Speed Ratio and the interaction between Side-slip and σ_{diff} is almost the same. This result shows that the Gas-Surface-Interaction model has a significant influence on the aerodynamic estimation, and within the considered interval it has the same effect of a ±237 m/s satellite velocity variation (computed with respect to the central point, V = 7710 m/s), therefore the epistemic uncertainty on the σ_{diff} cannot be neglected.

The three resulting AD-HDMR probability distributions have been compared in Fig. 9 (on the left). The distributions do not exactly match due to the fluctuations on the raw DSMC C_D samples. Indeed, the difference has been proven to fall within the interval of the convergence error. Contrarily to the broad range SA, in this case, the probability distribution does not resemble to have a normal distribution. Another interesting result is that the C_D uncertainty at the orbital altitude is dominated by the attitude or wind velocity, and the parameters that would have been considered in a complete model have lost their importance (molecular densities, wall and free flow temperature and vibrational relaxation coefficient). Therefore, as long as the flow is in a free molecular regime, the drag coefficient is independent from the density. From the AD-HDMR SF results the following surrogate models have been generated, leading to Eqs. (19) and (20):

$$\begin{cases} \Delta C_{D,SR} = -0.00861 * SR^2 - 0.04552 * SR + 4.4094 - C_{D_{Central}} \\ \Delta C_{D,\alpha} = 0.01905 * \alpha^2 + 0.00363 * \alpha + 3.4811 - C_{D_{Central}} \end{cases} \quad (19)$$

$$\begin{cases} \Delta C_{D,SR-\alpha} = 8.67E^{-5} \cdot \alpha^2 + 0.13956 \cdot \sigma_{diff}^2 - 0.02207 \cdot \alpha \cdot \sigma_{diff} \\ \qquad\qquad + 0.02009 \cdot \alpha + 0.08605 \cdot \sigma_{diff} - 0.18372 \end{cases} \quad (20)$$

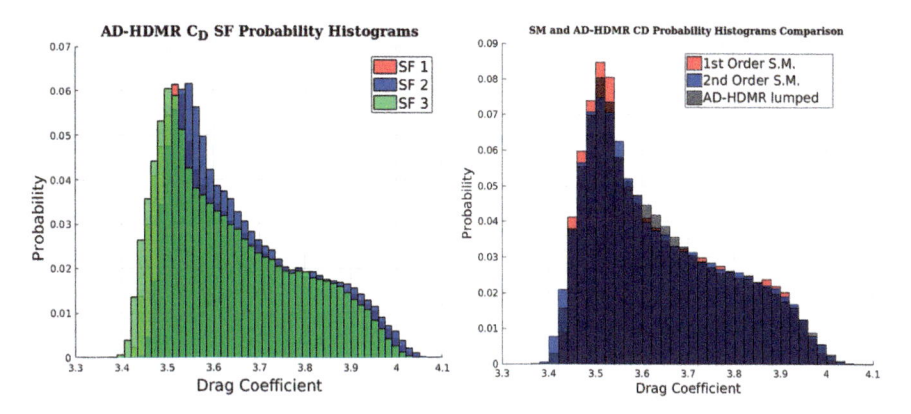

Fig. 9 *Left*: on-orbit drag coefficient UQ of GOCE performed with the AD-HDMR SF approach. *Right*: comparison of AD-HDMR and SMs probability histograms

where the $C_{D_{Central}}$ is the drag coefficient computed at the central interval sampling point ($C_{D_{Central}} = 3.478$), and the different $\Delta C_{D,x}$ are the increment functions relative to the shown parameters. In order to compute the actual C_D it must be computed the summation of all the different $\Delta C_{D,x}$ for a complete surrogate model, or just the $\Delta C_{D,x}$ of Eq. (19) for the SM of the single variables. The reader must remind that the validity of this surrogate model is strictly subjected to the investigated ranges (Table 2).

A comparison of the C_D probability distributions obtained with the three single fidelity test cases is shown in Fig. 9 (on the left). The distributions show a good level of consistency and their discrepancies are due to the minor convergence errors in the raw DSMC C_D samplings.

The created surrogate models have been compared to the lumped SF distributions provided by the AD-HDMR code. The comparisons (right of Fig. 9) have been performed running a LHS with the SMs (Eqs. (19) and (20)); the AD-HDMR distribution shown on the right (in black) is the result of lumping together the C_D distributions obtained with the 3 SF results shown in the left figure. It is possible to notice that even though the number of variables has been greatly reduced, there is an extraordinary match between SMs and the AD-HDMR complete model probability distribution. Meaning that using a very accurate convergence on the AD-HDMR setup may not provide any actual benefit, as it does also increase the number of samples required, therefore increasing the computational times. The three AD-HDMR single fidelity simulations required an average of 125 core-hours. The SF approach applied to this specific case required to evaluate ∼54 samples before reaching the convergence.

6.2.2 Multi-Fidelity Uncertainty Quantification and Sensitivity Analysis Results

The MF approach required some additional runs to define the influence of the low-fidelity (LF) DSMC case accuracy on the overall result. Even though the LF model accuracy should have a lower influence on the overall UQ, a lower accuracy on the raw DSMC C_D will have an impact on the single increment function distributions, which are built from the data on all the sampled points, whether they had been obtained with the low or HF model. This leads to a high impact on the sensitivity analyses, which are based on the increment function distributions. In addition, also the interpolated increment functions will be affected by the LF model accuracy, therefore, also the obtainable surrogate models will be affected by a lower accuracy.

The C_D probability distribution of four AD-HDMR MF simulation results are presented in Fig. 10. If compared to the single fidelity cases, the MF probability histograms show a larger variation range, and it can be assumed that it is mainly due to the higher convergence error and lower PPC employed on the LF model. The SA results reported in Table 6 show some similarities to the single fidelity cases: the speed ratio, α, and $\alpha - \sigma_{diff}$ interaction have been considered significant for each simulation, the same that had been obtained with the SF approach. Although, the 4th simulation has captured also one additional input; the interaction between the side-slip and Argon atomic density (ID 2–6). The interaction 2–6 has been captured only by the 4th simulation, even though its significance (which is computed for the

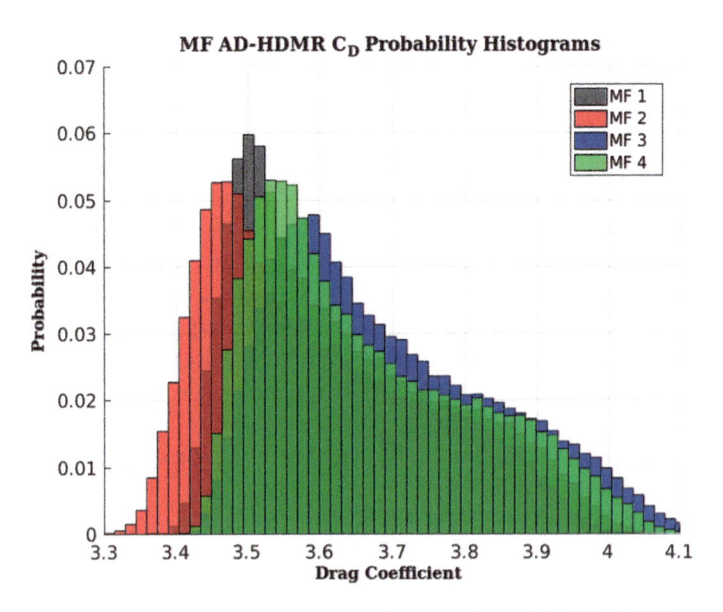

Fig. 10 On-orbit drag coefficient UQ of GOCE performed with the AD-HDMR MF approach

Table 6 AD-HDRM MF sensitivity analysis of GOCE C_D using 90% significance limit, significant inputs of four different simulations

Increment function	Partial mean	Partial variance	Sensitivity mean	Sensitivity variance	Input significance [%]
1st MF simulation, $C_{D_{central}} = 3.482$					
1	−0.0021	0.0007	0.0128	0.0316	94.07
2	0.1606	0.0196	0.9784	0.9498	100
2.9	−0.0014	0.0004	0.0087	0.0186	99.99
2nd MF simulation, $C_{D_{central}} = 3.488$					
1	−0.0054	0.0008	0.0312	0.0379	97.41
2	0.1652	0.0202	0.9632	0.9374	100
2.9	0.001	0.0005	0.0056	0.0247	100
3rd MF simulation, $C_{D_{central}} = 3.472$					
1	0.0075	0.0013	0.0419	0.0543	97.44
2	0.1701	0.0216	0.9556	0.9221	100
2.9	−0.0004	0.0006	0.0025	0.0237	99.99
4th MF simulation, $C_{D_{central}} = 3.468$					
1	0.0079	0.0009	0.0436	0.0379	99.11
2	0.1687	0.022	0.9339	0.9406	100
2.9	0.0023	0.0004	0.0129	0.0182	100
2.6	−0.0017	0.0001	0.0096	0.0033	96.43

maximum or minimum value of the increment function) is high, analyzing the raw samples C_D it has been found that this had been caused by the C_D fluctuations on the LF model.

The sensitivity indexes of the mean and variance computed by the MF simulations have all the same order of magnitude. Although, The fluctuations of the sensitivity on the mean are higher while the sensitivity of the variance has a more robust behavior. Since the sensitivity index on the variance is the most important parameter, this is deemed to be a good result.

All the simulations were performed on a single node of Archie-WEST HPC, the average MF simulation runtime was 164.4 core-hours. The MF approach evaluated an average of 36 HF and 46 LF DSMC simulations.

6.2.3 Single-Fidelity and Multi-Fidelity Results Comparison

In this paragraph it is presented a comparative analysis between the single-fidelity and multi-fidelity simulations that have been discussed in the previous sections. In Fig. 11 on the left, it is shown the comparison of SF and MF C_D probability histograms, which show a good agreement. Although, the MF C_D distributions appear to have a higher uncertainty. This is more likely due to the lower accuracy on the employed LF case. In the right figure, it is presented a comparison with a simplified Monte Carlo sampling evaluated on 2000 points (which required ∼2700

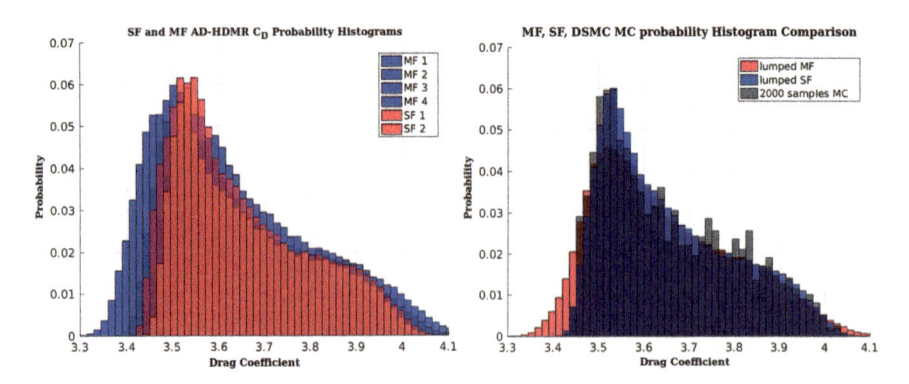

Fig. 11 *Left*: AD-HDMR C_D probability distributions, SF and MF comparison. *Right*: comparison between the SF and MF lumped distributions and a simplified MC sampling with 2000 samples

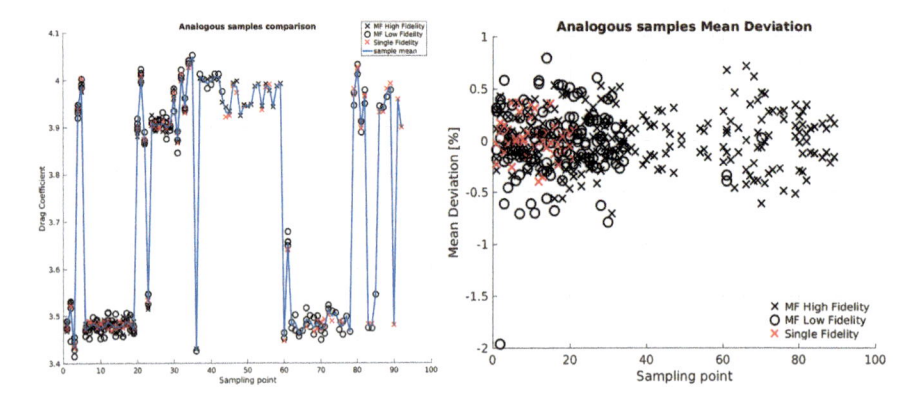

Fig. 12 *Left*: SF and MF analogous samples C_D distributions and their average value. *Right*: analogous samples C_D deviation from the mean

core-hours, more than 20 times more of the AD-HDMR UQ). For these histograms, 2 SF simulations and 4 MF simulations have been lumped together in order to have a more representative distribution. The SF and the MC probability histograms match is very good, and even though the MC samples are not enough to be considered as a proper validation case, they provide a good representation of the uncertainty within the considered inputs.

For this specific DSMC setup, the MF approach has not been proven advantageous in comparison to the SF. The reason for this must be researched on the number of samples required by the MF approach. In fact, the fluctuations on the LF C_D caused the AD-HDMR to evaluate a higher number of interaction between the inputs, making the MF more expensive than the SF approach. To better understand the correlation between the DSMC cases accuracy and the fluctuations on the results some additional analysis on the analogous sampled C_D points have been performed. In Fig. 12, on the left, are shown the C_D of analogous points having the same inputs

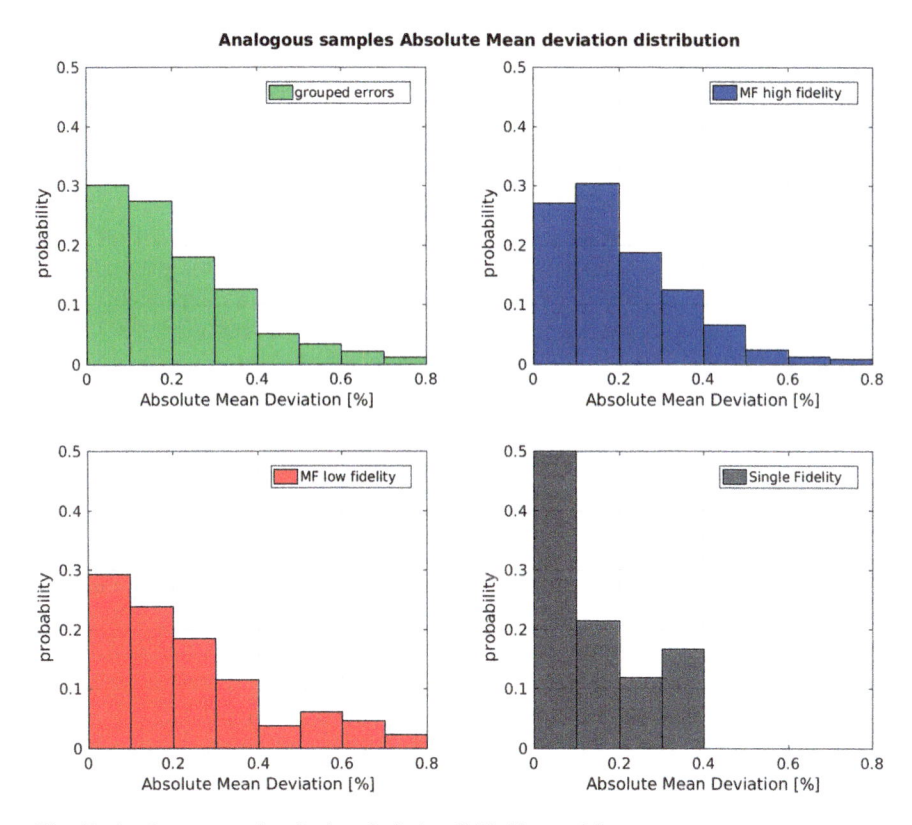

Fig. 13 Analogous samples absolute deviation divided by model

combination for all the SF and MF simulations; the blue line represents the average of those samples. In the figure on the right, it is reported the deviation with respect to the average of the analogous samples. As it was expected, the LF simulations have the highest deviation, but fall within the $\pm 1.0\%$, as it was set in the convergence algorithm, except for an out-layer which reaches the 2%. The HF model shows a lower average deviation instead, even though it does not always respect the $\pm 0.5\%$ convergence error. A better representation of the absolute deviation is shown in Fig. 13, where it is possible to see clearly how the deviation from the mean changes with the model. The green histogram represents the overall absolute deviation distribution for every simulated case.

7 Conclusions

The work focused on the DSMC aerodynamics analysis of GOCE satellite within the Free Molecular Flow regime which occurs while orbiting at the drag-free altitude, and two different analyses have been presented: a broad-range drag coefficient (C_D)

sensitivity analysis (SA) performed on five different environmental and modeling parameters, and the uncertainty quantification (UQ) of the satellite aerodynamics using a set of nine variables in a narrow range. The UQ and the SA have been performed with the newly developed adaptive derivative algorithm based on the cut-HDMR approach and the dsmcFoam software developed at the University of Strathclyde based on the OpenFOAM platform.

The AD-HDMR algorithm which was already extensively validated on deterministic functions, has been here applied on the DSMC stochastic methodology. The tool is able to operate by using a single-fidelity (SF) or a multi-fidelity (MF) approach, and the two have been evaluated, tested, and compared.

The HDMR-based tool has proven to be a useful tool for performing the UQ with a far lower number of function evaluations than a standard Monte Carlo sampling. In a single run, in addition to the UQ, the AD-HDMR computes also the SA indexes and the generation of surrogate models (SM) for the investigated variables and their interactions which are valid within the input distributions ranges.

In the specific application of the AD-HDMR to the DSMC, studying the results from a computational point of view has led to the conclusion that the SF approach represent the easiest implementation of the tool, indeed, only a single case and setup must be created, while, the use of the MF approach will increase difficulty of the study. In fact, building and evaluating a low-fidelity (LF) DSMC case will inevitably increase the statistical scattering. Additionally, the MF approach becomes advantageous with respect to the SF approach, when the LF case is at least 3~4 times computationally faster than the HF. Unfortunately, to decrease the computational time of a DSMC simulation, the setup conditions cannot respect the good DSMC practice. Therefore, a lower accuracy on the DSMC side causes higher fluctuations on the evaluated parameter, which may be detected by the AD-HDMR code as "false-interaction" among the variables. The detection of possible interactions will always be resolved with a higher number of function evaluation by the employment of both, HF and LF models. Eventually, the MF code will be more expensive than the simpler SF approach. Many different setup conditions have been tested for the LF case, but it has not yet been possible to find a compromise to sensibly reduce the MF approach computational time. Therefore, in light of all the previous considerations, it is suggested the use of the SF approach when applying the method to a non-deterministic method, tool, or function.

The proposed approach and results highlight the necessity of performing UQ of satellites and spacecraft aerodynamics. The obtained results show that even a small uncertainty in the attitudes could lead to a drastic change in the C_D. The uncertainty in the C_D is representative of how the aerodynamic coefficients could be affected by environmental, geometric, attitude and epistemic variables. In addition, the employed methodology represents an innovative way for treating the UQ by adopting surrogate models.

In the future, the coupling between the two codes will be further developed, introducing a Knudsen-based adaptive mesh generation process, making it possible to study aerodynamics and aero-thermodynamics within the transitional regime. The application of the AD-HDMR allows the generation of SMs which would provide a high-fidelity-based UQ for re-entry analyses. The SMs can be easily implemented into other low-fidelity aerodynamics simulators (e.g.: the ones based on the local panel theory) and re-entry analysis software, which as now, do not include any UQ methodology.

Acknowledgements The authors would like to state that the results were obtained using the EPSRC funded ARCHIE-WeSt High Performance Computer (www.archie-west.ac.uk). EPSRC grant no. EP/K000586/1.

References

1. Bailey, A., Hiatt, J.: Sphere drag coefficients for a broad range of Mach and Reynolds numbers. AIAA J. **10**(11), 1436–1440 (1972)
2. Bird, G.: Definition of mean free path for real gases. Phys. Fluids (1958–1988) **26**(11), 3222–3223 (1983)
3. Bird, G.A.: Direct simulation and the Boltzmann equation. Phys. Fluids (1958–1988) **13**(11), 2676–2681 (1970)
4. Cook, G.: Satellite drag coefficients. Planet. Space Sci. **13**(10), 929–946 (1965)
5. Gini, F., Otten, M., Springer, T., Enderle, W., Lemmens, S., Flohrer, T.: Precise orbit determination for the GOCE re-entry phase. In: Proceedings of the Fifth International GOCE User Workshop
6. Hirschfelder, J.O., Curtiss, C.F., Bird, R.B., Mayer, M.G.: Molecular Theory of Gases and Liquids, vol. 26. Wiley, New York (1954)
7. Kubicek, M., Minisci, E., Cisternino, M.: High dimensional sensitivity analysis using surrogate modeling and high dimensional model representation. Int. J. Uncertain. Quantif. **5**(5), 393–414 (2015)
8. Moe, K., Moe, M.M.: Gas–surface interactions and satellite drag coefficients. Planet. Space Sci. **53**(8), 793–801 (2005)
9. Moe, K., Moe, M.M., Wallace, S.D.: Improved satellite drag coefficient calculations from orbital measurements of energy accommodation. J. Spacecr. Rocket. **35**(3), 266–272 (1998)
10. OpenCFD Ltd: openFOAM Software Package, Ver. 2.3, Bracknell. http://www.openfoam.com/ (2016)
11. Opiela, J.N., Hillary, E., Whitlock, D.O., M., H.: Debris assessment software, user guide. Technical report, NASA Lyndon B. Johnson Space Center (2012)
12. Picone, J.M., Heding, A.E., Drob, D.P., Aikin, A.C.: NRLMSISE-00 empirical model of the atmosphere: statistical comparisons and scientific issues. J. Geophys. Res. Space Phys. **107**(A12) (2002)
13. Rochelle, W.C., Kinsey, R.E., Reid, E.A., Reynolds, R.C., Johnson, N.L.: Spacecraft orbital debris reentry: aerothermal analysis (1997)
14. Scanlon, T.J., Roohi, E., White, C., Darbandi, M., Reese, J.M.: An open source, parallel DSMC code for rarefied gas flows in arbitrary geometries. Comput. Fluids **39**(10), 2078–2089 (2010)
15. Scanlon, T.J., White, C., Borg, M.K., Palharini, R.C., Farbar, E., Boyd, I.D., Reese, J.M., Brown, R.E.: Open-source direct simulation monte carlo chemistry modeling for hypersonic flows. AIAA J. **53**(6), 1670–1680 (2015)
16. Steiger, C., Romanazzo, M., Floberghagen, R., Fehringer, M., Emanuelli, P.: The deorbiting of GOCE-a spacecraft operations perspective. In: ESA Special Publication, vol. 728, p. 19 (2015)
17. Whitfield, D.L., Stephenson, W.B.: Sphere drag in the free-molecular and transitional flow regimes. Technical report, DTIC Document (1970)